FATTY ACIDS AND OXIDATIVE STRESS IN NEUROPSYCHIATRIC DISORDERS

Fatty Acids and Oxidative Stress in Neuropsychiatric Disorders

Ravinder Reddy
and
Jeffrey K. Yao
Editors

Nova Biomedical Books
New York

For permission to use material from this book please contact us:
Telephone 631-231-7269; Fax 631-231-8175
Web Site: http://www.novapublishers.com

Library of Congress Cataloging-in-Publication Data
Fatty acids and oxidative stress in neuropsychiatric disorders / Ravinder Reddy and Jeffrey K. Yao (editors).
p. ; cm.
Includes bibliographical references and index.
ISBN-13: 978-1-60021-677-0 (hardcover)
ISBN-10: 1-60021-677-3 (hardcover)
1. Neurobehavioral disorders. 2. Fatty acids--Pathophysiology. 3. Oxidative stress--Pathophysiology. I. Reddy, Ravinder. II. Yao, Jeffrey K.
[DNLM: 1. Mental Disorders--therapy. 2. Neurodegenerative Diseases--therapy. 3. Fatty Acids, Essential--therapeutic use. 4. Mental Disorders--etiology. 5. Neurodegenerative Diseases--etiology. 6. Oxidative Stress--physiology. WM 400 F254 2007]
RC386.F38 2007
616.8--dc22

200701057

Published by Nova Science Publishers, Inc. ✢New York

Contents

Contributors

Gregor E Berger, MD
Unit for Neuroprotection in Young People (UNYP)
ORYGEN Research Centre
University of Melbourne, Victoria, Australia

Jaspreet Singh Brar, MBBS, MPH
Western Psychiatric Institute and Clinic
Department of Psychiatry
University of Pittsburgh Medical Center
Pittsburgh, Pennsylvania, USA

Peter F Buckley, MBBS
Department of Psychiatry and Health Behavior &
Departments of Radiology & Pharmacology and Toxicology
Medical College of Georgia &
Medical Research Service Line
VA Medical Center
Augusta, Georgia, USA

Undurti N Das, MD, FAMS
UND Life Sciences
Walpole, Massachusetts, USA

Billy W Day, PhD
Department of Pharmaceutical Sciences
School of Pharmacy & Department of Chemistry
University of Pittsburgh
Pittsburgh, Pennsylvania, USA

April A Dukes, BS, BA
Center for Neuroscience
University of Pittsburgh
Pittsburgh, Pennsylvania, USA

Ashraf H Elamin, PhD
Department of Chemistry and Biochemistry
California State University Long Beach
Long Beach, California, USA

Robin Emsley, MD
Departments of Psychiatry
Faculty of Health Sciences
University of Stellenbosch
Cape Town, South Africa

Denise Evans, MD
Department of Psychiatry and Health Behavior
Medical College of Georgia &
VA Medical Center
Augusta, Georgia, USA

Rohan Ganguli, MD
Departments of Psychiatry, Pathology & Health Services Administration
University of Pittsburgh Medical Center
Pittsburgh, Pennsylvania, USA

Antoni Y Hardan, MD
Western Psychiatric Institute and Clinic
Department of Psychiatry
University of Pittsburgh Medical Center
Pittsburgh, Pennsylvania, USA

Teresa G Hastings, PhD
Center for Neuroscience & Department of Neurology
University of Pittsburgh
Pittsburgh, Pennsylvania, USA

Sahebarao Mahadik, PhD
Department of Psychiatry and Health Behavior
Medical College of Georgia &
VA Medical Center
Augusta, Georgia, USA

Mark P Mattson, PhD
Laboratory of Neurosciences
National Institute on Aging Gerontology Research Center &
Department of Neuroscience
Johns Hopkins University School of Medicine
Baltimore, Maryland, USA

Mirabel McConchie, PhD
Unit for Neuroprotection in Young People (UNYP)
ORYGEN Research Centre
University of Melbourne, Victoria, Australia

Patrick D McGorry
Unit for Neuroprotection in Young People (UNYP)
ORYGEN Research Centre
University of Melbourne, Victoria, Australia

Meera Narasimhan, MD
Department of Psychiatry
University of South Carolina
Columbia, South Carolina, USA

Piet Oosthuizen, MB
Department of Psychiatry
Faculty of Health Sciences
University of Stellenbosch
Cape Town, South Africa

Ward A Pedersen, PhD
Center for Aging, Alzheimer's Disease
and Neurodegenerative Disorders
Departments of Neurology and Biomedical Sciences
Creighton University Medical Center,
Omaha, Nebraska, USA

Malcolm Peet, MBChB, MRCPsych
Section of Mental Health
School of Health and Related Research
University of Sheffield &
Swallownest Hospital
Sheffield, UK

Anilkumar Pillai, PhD
Department of Psychiatry and Health Behavior
Medical College of Georgia &
VA Medical Center
Augusta, Georgia, USA

Tina Proffitt, DPsy
Unit for Neuroprotection in Young People (UNYP)
ORYGEN Research Centre
University of Melbourne, Victoria, Australia

Ravinder Reddy, MD
Department of Psychiatry, School of Medicine
University of Pittsburgh
Western Psychiatric Institute & Clinic &
VA Pittsburgh Healthcare System
Pittsburgh, Pennsylvania, USA

Alexandra J. Richardson, DPhil
University Laboratory of Physiology
University of Oxford
Oxford, UK

Marion Ross
University Laboratory of Physiology
University of Oxford
Oxford, UK

Nitin Shendarkar, MD
Department of Psychiatry and Health Behavior
Medical College of Georgia &
VA Medical Center
Augusta, Georgia, USA

Ali F Sonel , MD
Cardiac Catheterization Laboratories
VA Pittsburgh Healthcare System &
Department of Medicine
University of Pittsburgh
Pittsburgh, Pennsylvania, USA

Jeff A Stanley, PhD
Department of Psychiatry and Behavioral Neurosciences
Wayne State University School of Medicine
Detroit, Michigan, USA

Martin Strassnig, MD
Western Psychiatric Institute and Clinic
Department of Psychiatry
University of Pittsburgh Medical Center
Pittsburgh, Pennsylvania, USA

Elizabeth A. Thomas, PhD
Department of Molecular Biology
The Scripps Research Institute
La Jolla, California, USA.

Victor S Van Laar, PhD
Center for Neuroscience
University of Pittsburgh
Pittsburgh, Pennsylvania, USA

Susan J van Rensburg, PhD
Department of Chemical Pathology
Faculty of Health Sciences
University of Stellenbosch
Cape Town, South Africa

Stephen J Wood, PhD
Melbourne Neuropsychiatry Centre
National Neuroscience Facility
University of Melbourne &
Brain Research Institute, Heidelberg, Victoria, Australia

Jeffrey K Yao, PhD
VA Pittsburgh Healthcare System &
Department of Psychiatry, School of Medicine
Western Psychiatric Institute & Clinic &
Department of Pharmaceutical Sciences, School of Pharmacy,
University of Pittsburgh
Pittsburgh, Pennsylvania, USA

Preface

Research into the role of oxidative stress and fatty acids in neuropsychiatric disorders has been ongoing for decades and has been written about before. Why, then, another volume on these topics? This book aims to examine how these seemingly disparate bodies of evidence from the fields of oxidative stress and essential fatty acid metabolism are, in fact, intimately connected to each other. Investigator from either side of the "fence" are just beginning to appreciate that making connections between the metabolism of essential fatty acids and the damaging oxidative processes provide fuller explanatory models, and ultimately offer plausible therapeutic approaches.

Therefore, for this book we invited researchers of oxidative stress (OS) and essential fatty acid (EFA) in neuropsychiatric disorders, exemplified by schizophrenia, mood disorders, Parkinson's disease and Alzheimer's disease. Together the chapters bring a fresh perspective on the relationship between the presence of oxidative stress in the afore-mentioned disorders, and the fatty acid deficits that have been identified in these disorders. Some authors report on the pathophysiological aspects of OS-EFA equation and others on the therapeutic successes and opportunities of essential fatty acid and antioxidant supplementation. A third group of investigators offer novel theoretical and methodological perspectives on the OS-FA story. In sum, we hope the readers will be illumined about the need to view oxidative stress and essential fatty acid research as inter-related fields of inquiry.

Ravinder Reddy
Jeffrey J. Yao

In: Fatty Acids and Oxidative Stress...
Editors: R. Reddy and J.K. Yao, pp. 1-4
ISBN: 978-1-60021-677-0

Chapter I

Oxidative Stress and Fatty Acids in Neuropsychiatric Disorders: How Are They Connected?

Ravinder Reddy and Jeffrey K. Yao

Introduction

The notion that free radicals, in the form of adrenochromes, may have a pathophysiological role in schizophrenia extends back over 50 years (Hoffer, 1954). Systematic research of oxidative stress in schizophrenia is however more recent (Mahadik and Mukherjee, 1996; Reddy and Yao, 1996), reviewed by Pillai and colleagues in chapter 3 of this volume.

Likewise, the possibility that fatty acids might have pathophysiological relevance for schizophrenia dates back almost three decades (Horrobin, 1977), and yet evidence supporting such a contention has been accumulating only in the last 15 years (Horrobin et al, 1999; 2003), reviewed by Peet in chapter 7.

However, research into oxidative stress or fatty acids has tended to evolve in relative isolation from each other. The fact that essential fatty acids (EFA) are highly vulnerable to oxidative damage has promoted the idea that these two biological systems need to be examined together, and that there are biological consequences of relevance for schizophrenia, as an example (Mahadik et al, 2003; Reddy and Yao, 2003) (Fig. 1).

The complexity of these systems, of course, precludes the likelihood of simple answers. It is only with the advent of recent advances in integrated biological methods (genomics, proteomics, metabolomics), comprehensively reviewed by Elamin and colleagues (chapter 16), that now permits practical examination of oxidative and fatty acid domains as interacting systems, as well as to identify potentially interesting biochemical *pathways*. Prabhakaran et al (2004), demonstrated the utility of these approaches and intriguing findings in schizophrenia implicating systems that are involved with generating oxidative stress and defending against it.

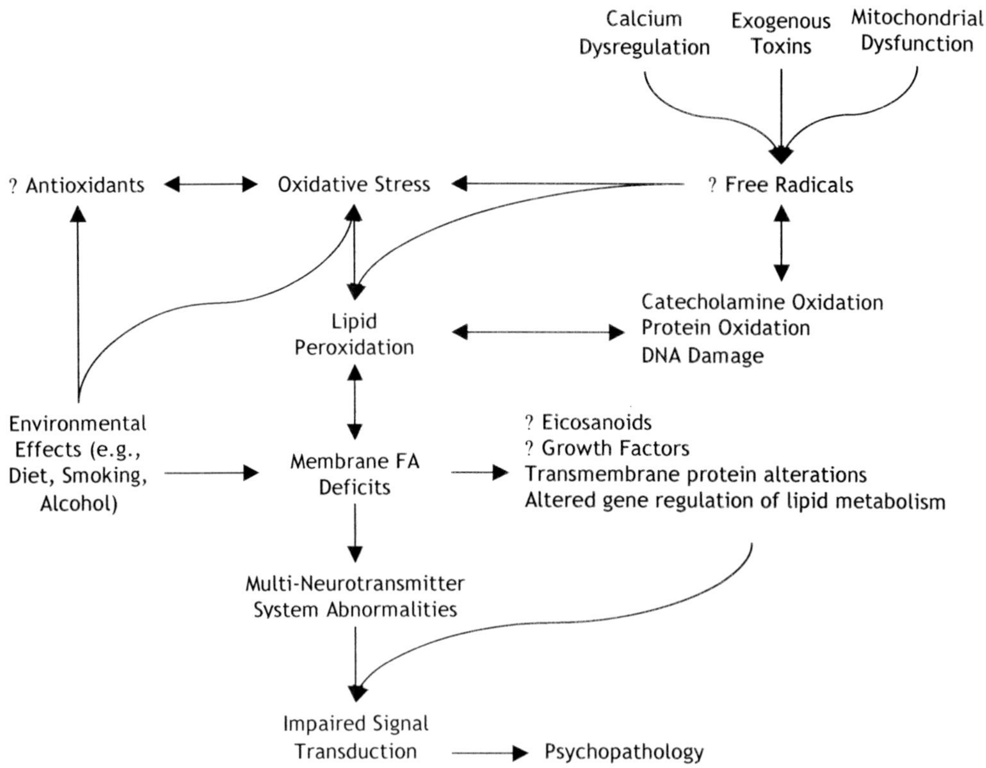

Figure 1. Simplified model of links between oxidative stress and fatty acid (FA) metabolism.

In recent years there has a broadening of inquiry to the larger group of neuropsychiatric disorders, such as neurodevelopmental disorders, Alzheimer's disease and Parkinson's disease, among others (Peet et al, 2003). Ward Pederson and Mark Mattson comprehensively discuss oxidative stress in Alzheimer's disease (Chapter 5), and specifically address the role of sphingolipid metabolism. Van Laar and colleagues provide an exhaustive review of oxidative stress in Parkinson's disease (Chapter 6). The neurodevelopmental disorders, such as autism, may well be the next frontier in applying what has been learned about EFA in adults to young patients. Early intervention may be key to modifying the course of illness, or mitigating the full expression of the illness, or perhaps even preventing the emergence of these disorders. Antonio Hardan and Jeff Stanley provide an overview of the role of EFA in neurodevelopmental disorders (Chapter 8), while Richardson and Ward provide a comprehensive review of the EFA treatment in neurodevelopmental disorders (Chapter 9).

The whole point of the research enterprise is to discover and test treatment options that utilize omega-3 fatty acid supplementation, antioxidants strategies, or a combination. The idea behind this approach, shown in a simplified model in Fig. 2, is to supplement omega-3 fatty acid intake (or utilize pharmaceutical approaches that would increase *in vivo* EFA availability).

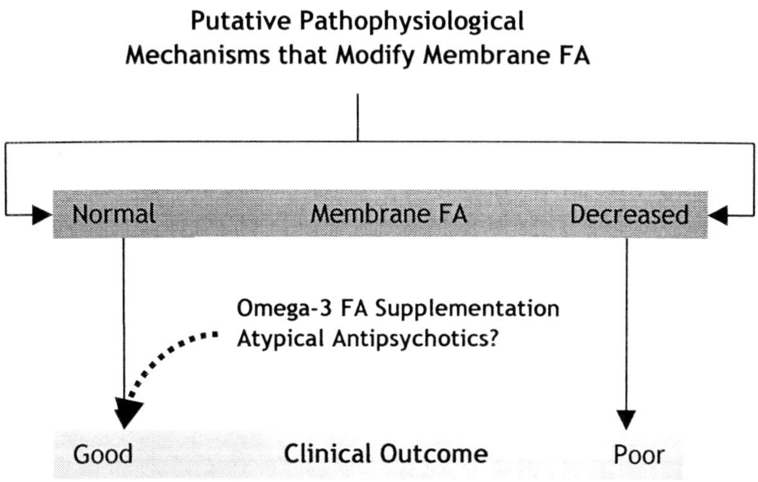

Figure 2. Working hypothesis: A primary pathophysiological mechanism, active to varying degrees, is associated with membrane fatty acid (FA) deficits, which in turn mediate a range of clinical outcomes. Therefore, substantially decreased membrane FA will be associated with poor outcome. Thus, rectifying membrane deficits may favorably modify clinical outcome.

To date, there have been some successes in omega-3 fatty acid supplementation trials. The treatment evidence in neurodevelopmental disorders is reviewed by Richardson and Ross (Chapter 9), and the evidence from schizophrenia and mood disorders is presented by Emsley and colleagues (Chapter 10). Berger and colleagues (Chapter 11) present thought-provoking data on the effects omega-3 supplementation on cognition in early schizophrenia. These are promising beginnings, and it is quite clear that much systematic work needs to be done, including controlled trials with large enough samples and longitudinal studies that examine the durability of therapeutic gains.

References

Hoffer A, Osmand H, Smythies J. Schizophrenia: new approach. *J Ment Sci* 1954; 100: 29-45

Horrobin DF. Schizophrenia as a prostaglandin deficiency disease. *Lancet*. 1977; 1(8018):936-7.

Peet M, Glen I, Horrobin D (Eds). *Phospholipid Spectrum Disorders in Psychiatry*. Marius Press; 1999

Peet M, Glen I and Horrobin DF (Eds). *Phospholipid Spectrum Disorders in Psychiatry and Neurology*. 2nd Edition; Marius Press; 2003

Mahadik SP, Mukherjee S. Free radical pathology and the antioxidant defense in schizophrenia. *Schizophr. Res*. 1996; 19:1-18.

Prabakaran S., Swatton J., Ryan M., *et al.* (2004). Mitochondrial dysfunction in schizophrenia: evidence for compromised brain metabolism and oxidative stress. *Mol Psychiatry*. 9: 687-697.

Reddy R, Yao J. Free radical pathology in schizophrenia: a review. *Prostaglandin Leukotri Essen Fatty Acids*. 1996; 55:33-43.

Reddy R, Yao JK (2003): Membrane protection in schizophrenia: Concepts and therapeutic approaches. In: Peet M, Glen AIM, Horrobin DF (Eds): *Phospholipid Spectrum Disorder in Psychiatry and Neurology, 2nd Ed.* Lancashire, UK, Marius Press, pp 299-314

In: Fatty Acids and Oxidative Stress...
Editors: R. Reddy and J.K. Yao, pp. 5-25
ISBN: 978-1-60021-677-0

Chapter II

Diet in Schizophrenia: an Overview

Martin Strassnig, Jaspreet Singh Brar and Rohan Ganguli

Background

As early as 1921, Kraepelin observed a correlation between improvement of psychopathology and body weight gain of his psychiatric patients (Heimann, 1980).

Over the years there have been numerous attempts to link dietary intake to the pathogenesis of mental disorders such as schizophrenia (Berlet et al, 1966). In fact, one of the major proponents of this theory, Nobel Prize laureate Linus Pauling, hypothesized that mental disease was for the most part caused by abnormal chemical reaction rates, as determined by genetic constitution and diet, and by abnormal molecular concentrations of essential substances. He also suggested that some individuals might require higher micronutrient intake than others to function normally and that the "normal" concentration may not be the optimum concentration (Pauling, 1968). His observations, although not undisputed, inspired further research into dietary habits and course and outcome of mental disorders.

Historically, a variety of dietary factors have been implicated in the pathology of Schizophrenia. For example, Dohan (1966) proposed cereal intake to be linked to schizophrenic symptomatology. A few years later, Dohan and Grasberger (1973) reported significantly earlier discharge of Schizophrenia patients after acute episodes when maintained on a cereal-free, milk free diet during their inpatient stay; wheat gluten content was implicated (Singh and Kay, 1976). Others hypothesized that abnormal tryptophan metabolism (Gilka, 1975), or abnormal Vitamin C pathways (Grant et al, 1973) play major roles in the pathogenesis of schizophrenia.

Concurrently, derived from Pauling's original thoughts, "orthomolecular" treatment options were explored (Autry, 1975) including addition of high-dose Niacin or Vitamin C during treatment of patients suffering from schizophrenic episodes. Results, however, remained inconclusive (Kety, 1976).

In 1988, Christensen and Christensen made the observation that Schizophrenia has a more benign outcome in countries with higher consumption of polyunsaturated fats. In a large-scale epidemiologic study, they found a highly significant correlation between fat intake and schizophrenic symptoms (Christensen and Christensen 1998) and concluded that the dietary ratio of saturated fat to polyunsaturated fatty acids (PUFAs) correlates with the long-term outcome of schizophrenia. For example, they found that the saturated fat to PUFA ratio in the United states was 1.89 : 1, whereas the ratio in India was 1: 0.29, and 1 : 0.1 in Nigeria. Essentially, industrialized countries with a higher saturated fat intake from land animal and bird sources had worse outcome of schizophrenia than countries with relatively higher PUFA intake from sources such as fish, seafood and vegetables. Similar observations of better outcome of schizophrenia in "developing" countries had been made before (WHO, 1979; Jablenski et al, 1992) and were recently re-confirmed (Hopper and Wanderling, 2000), yet have never been fully accounted for. In a similar epidemiologic investigation incorporating large population-based nutrition intake databases, Peet (2004) deduced that a high refined sugar intake may be involved in schizophrenia outcome.

Interest in the implications of dietary intake of patients with schizophrenia has recently grown again, accounted for mainly by two only peripherally related topics: First - tying into the previously mentioned correlations of fatty acid intake and Schizophrenia outcome - there is increasing evidence for involvement of essential fatty acids, especially polyunsaturated fatty acids in the pathology of schizophrenia (Horrobin et al, 1994). PUFAs constitute parts of all cell membrane phospholipids, and are especially abundant in neural cell membranes. Together with cholesterol, they form most of the cell membrane bi-layer, and are closely associated with receptors, ion channels and other cell signalling entities. Thus, PUFA imbalances may affect a variety of neuronal functions (Haag, 2003).

Second, patients suffering from Schizophrenia are especially prone to accumulate excess body weight which exposes them to serious health problems (Kurzthaler and Fleischhacker, 2001). The high body weight observed in patient may be in part medication-induced (Schwartz et al, 2004) or be effected by poor exercise habits (Callaghan, 2004), but has been most frequently linked to poor dietary choices (Strassnig et al, 2003). Clearly, a poor diet may also predispose Schizophrenia patients to insufficient intake of certain nutrients; among them PUFAs. Critically, whereas most fatty acids can be produced within the body, several polyunsaturated fatty acids (PUFA), such as α-linolenic acid (18:3n-3), docosahexaenoic acid (22:6n-3) and eicosapentaenoic acid (20:5n-3), are essential and need to be supplied through food intake.

Nutrition and Health

Obesity in mentally healthy Americans is increasing at an alarming rate (Hedley et al, 2004) and is now considered a major medical hazard (National Task Force on the Prevention and Treatment of Obesity, 2000). Obesity-related co-morbidities are expected to overtake tobacco-smoking as the number one cause for morbidity and mortality in the US (Marshall, 2004). The National Institute of Health, National Heart, Lung, and Blood Institute (1999) has established three distinct Body Mass Indeces (BMI, kg/m2) cut-off points (ie ≤25, 25-30, ≥30) to delineate between healthy and unhealthy body weight. According to these criteria, approximately 66 % of Americans are overweight (defined as a BMI above 25), and 30.6 % are obese (BMI ≥30); which leaves only about one-third of the entire population at healthy weight as defined by a BMI of less than 25 (Mokdad et al. 1999).

Physical health risks of excess body weight are numerous and include insulin resistance, diabetes mellitus, hypertriglycidemia, decreased levels of high-density lipoprotein cholesterol, and increased levels of low-density lipoprotein cholesterol. Obesity is associated with gallbladder disease as well as sleep apnea, chronic hypoxia and hypercapnia, degenerative joint disease and certain cancers. Obesity is an independent risk factor for death from coronary heart disease (Pi-Sunyer 1993). Apart from higher morbidity and mortality rates, overweight subjects are also more likely to experience impaired quality of life (Han et al, 1998).

While dietary variations are known to predict the prevalence of physical morbidity, and most certainly, dietary modifications and weight loss improve physical health (Moore, 2003), a possible influence of diet on mental health has been neglected. There is evidence of indirect contributions of obesity to psychopathology (Eremis et al, 2004, Kawachi, 1999; Foster et al, 1997). Schizophrenia patients experience overweight as distress (Strassnig et al, 2003). The already poor quality of life in schizophrenia patients is significantly influenced by overweight (Allison et al, 2003).

However, a more direct link between food intake and mental disease also seems to exist. Large clinical databases and national food intake data seem to indicate such trends (Christensen and Christensen, 1986; Jablenski, 1992).

Recently, Peet (2004) raised the intriguing hypothesis that the outcome of schizophrenia and the prevalence of depression may be influenced by dietary factors. His research essentially confirms the previous ecological observations (Jablenski et al, 2000) that a diet rich in saturated animal fat from meat and dairy products compared with a with low fat diet derived from fish and vegetable sources is associated with a worse outcome of schizophrenia. Peet (2004) also found an association between meat, egg, dairy product, sugar and alcohol intake and depressive and schizophrenic symptomatology. A similar conclusion was proposed by McIntosh and Lawrie (2004) in their appropriately termed paper "you are (associated with) what you eat". Such observations have since even caught the eye of the mainstream media (Underwood, 2001).

Dietary Habits and Schizophrenia

Similar to the pattern observed for the general population, patients with schizophrenia in North America tend to be overweight (Allison et al. 1999). Factually, the group of schizophrenia patients is significantly more overweight than the population; and often to a considerable degree (Homel et al, 2002, Aronne, 2001; Gopalaswamy and Morgan, 1985).

Schizophrenia patients have a higher than average mortality rate per se (McGlashan, 1988). Obesity and related co-morbidities only contribute to mortality and morbidity (Goldman, 1999). While the exact mechanisms of weight gain in schizophrenia are not known, it is generally accepted that the above-average obesity rates in patients is a result of poor dietary choices (Brown et al. 1999). Thus, the interesting question arises, whether the disproportionately high rates of obesity in patients stems from a diet different to the diet observed in the general population; and if so, what constitutes these differences.

Surprisingly, only a few studies have systematically examined the diet in patients with schizophrenia. In a nutritional assessment among people living in "mental health residential houses" in Sydney, Australia it was observed that the respondents had a significantly higher prevalence of obesity (including abdominal obesity) than the general population (Wallace and Tennant 1997). No subject met the recommended dietary intake for certain standardized "healthy" food groups.

Brown et al (1999) assessed the food intake of 102 subjects with schizophrenia living semi-independently in long term mental health housing in the UK. Dietary habits were ascertained through administration of a food frequency questionnaire, which usually focuses on food groups consumed within a specified time frame prior to examination; a week in the prevailing study. Food frequency data is generally averaged over the assessment period, resulting in mean daily intakes. Diets of schizophrenia patients were found to be higher in total fat and lower in fibre than a reference population matched for age, gender and social class. Intake of low, moderately and highly unsaturated fat was not different between patients and population. Notably, no patient met the official UK recommendations for a healthy diet including the intake of five portions of fruits or vegetables per day.

McCreadie et al (1998) examined 30 patients suffering from schizophrenia living in assisted-living facilities in Scotland for their dietary intake and compared the results to a control group matched for sex, age, smoking status and employment status. The average weekly food intake was obtained through a modified established food frequency questionnaire. Most patients in their sample were overweight or obese. Saturated fat intake was higher than recommended. Conversely, patients consumed significantly less energy, total fibre, retinol, carotene, vitamin C, vitamin E and alcohol than the reference population. They also consumed fewer fruit portions and vegetable portions.

In 2003, McCreadie et al. conducted a similar trial incorporating a larger sample of community-dwelling people with schizophrenia. One-hundred and two patients were assessed for dietary habits (with special emphasize on fruit and vegetables) and smoking behavior; resulting cardiovascular risk factors were estimated as well. Diet was retrospectively elicited via food frequency questionnaire. Again, a significant number of patients (73 %) were overweight. When compared to Scottish reference intakes, male patients consumed a diet with less cooked green and root vegetables, fresh fruit, skimmed or semi-skimmed milk, raw vegetables or salad, potatoes, pasta and rice, and pulses. Female patients ate less potatoes, pasta and rice, raw vegetables or salad, and also less skimmed milk products. Regarding biochemical markers, more than half of the patients (53 %) were outside the recommend Cholesterol to HDL ratio (indicating increased saturated fat intake). 42 % of subjects were outside the beta-carotene reference range, whereas serum Vitamin C content was not outside reference range. Fish consumption (rich in PUFAs) was not different between males and females and the respective reference populations. Smokers (70 % of the sample) had even lower intake of fruit and vegetables than non-smokers. The authors concluded that patients made poor dietary choices; with males especially susceptible to poor dietary patterns.

Gothelf et al (2002) prospectively monitored food intake and energy expenditure of adolescent male inpatients with Schizophrenia over four weeks. A dietician closely monitored food intake for 2 consecutive days. All foods, beverages and snacks were weighted before and after meals. Patients were allowed to choose freely what they wanted to eat. Computerized "food tables" were used to calculate caloric intake, protein-, carbohydrate-, and fat content. These authors found significantly increased caloric intake in a subgroup treated with the atypical antipsychotic Olanzapine compared to a subgroup receiving Haloperidol. The relative percentages of the macronutrients protein, carbohydrates, and fat did not differ.

Table 1. Macronutrient Intake in Patients with Schizophrenia compared to Representative US Population Standards

	All		Male		Female	
	Group N=146	Population	Patients N=78	Population	Patients N = 68	Population
	Mean ± SD	*Mean ± SD*	*Mean ± SD*	*Mean ± SD*	*Mean ± SD*	*Mean ± SD*
BMI (height/m²)	32.7± 7.9 ***		30.8± 7.2***		34.8± 8.1**	
Total Calories	3057 ±1132 **	2206 ±33.4	3201±1084	2642 ±57.4	2891± 1171***	1792 ±38.5
Protein Intake (g)	95±40	83±4.1	100±40	100±6.3	85±40*	67±4.1
Calories from Protein (%)	12.8±3.8	15.5±0.19	12.7±4	15.4±0.21	12.8±3.7	15.5±0.27
Carbohydrate Intake (g)	394±166***	263±2.7	413±53	309±3.8	372±178***	220±2.7
Calories from Carbohydrates (%)	51.8±10.6)	51.8±0.46	52.5±11.7	47.4±0.59	51.1±9.1	49.9±0.6
Total Fat Intake (g)	121±61*	86±1.3	126±66.8	103±1.9	116± 53**	70±0.9
Calories from Fat (%)	35.1±9.2	34.1±0.46	34.2±0.1	34.3±0.56	36±8.2	34±0.54

* P = < 0.05   ** P = < 0.005   *** P = < 0.001

Compared to cata for the general population from NHANES 1999 – 2000 (Wright et al, 2003) Wright JD, Wang CY, Kennedy-Stephenson J, Ervin RB. Dietary Intake of Ten Key Nutrients for Public Health, United States: 1999-2000. Advance Data from Vital and Health Statistics; no. 334. Hyattsville, Maryland: National Center for Health Statistics. 2003. Table adapted from Strassnig et al, 2003.

Rather, the subgroup treated with the atypical antipsychotic simply ate more of the same foods and ended up with significantly higher weight gain. As a side note, physical activity was also measured (via acceleromety monitor and a heart rate monitor); 90 % of subjects spent less than 10 continuous minutes a day in moderate activity, despite intensive staff efforts to promote activity.

Our group (Strassnig et al, 2003) examined a total of 146 adult community-dwelling outpatients suffering from schizophrenia for their dietary intake via a 24-hour diet recall. The methodology is different from a food frequency questionnaire and focuses on food intake in the 24 hour period prior to assessment. A semi-structured interview is conducted to elicit nutritional information. Food portions are estimated with standardized food models; specialized nutrition software is used to convert the intake data into individual nutritional values such as calories, total fat, protein; or PUFA and antioxidant intake. The method has longstanding proven validity and reliability (Young et al, 1952) and has been widely used in large epidemiologic studies (Kohlmeier 1992); such as the NHANES (National Health and Nutrition Survey). The approach allowed us to directly compare our patient's dietary intake to US population standards.

We used intake data from NHANES III (National Health and Nutrition Examination Survey, Cycle 3) representing the average nutrient intakes of the US population and subgroups (McDowell et al, 1994; Alaimo et al, 1994) for comparisons with age and gender matched Schizophrenia patients.

Table 1 shows major macronutrient intakes, such as of protein, fat and sugar of patient with Schizophrenia compared to age- and gender matched US population estimates.

Most patients in our sample were overweight or obese. Mean body mass index was 32.8 (±7.8). Mean BMI of male subjects was 30.8 (±7.3), and of female subjects 35.1 (±8). As a group, patients ate more calories; and also more protein, carbohydrates and saturated fat than the age and gender matched general population. Quantitatively, caloric intakes derived from these nutrients were higher. However, the relative percentages of protein, carbohydrates and fat were not different from the pattern observed in the general population. Thus, our patients did not appear to make dietary choices much different to choices subjects in the population would make. They rather ate more of the same food.

Conclusion

Patients with Schizophrenia tend to eat more food than corresponding subjects in matched populations. Their diet seems to especially contain more fat. Further, data appears to indicate that patients have a lower fruit and vegetable intake than matched controls from the population.

High fat intake per se has been linked to a variety of medical problems such as Coronary Heat Disease, Hypertension, and Cancer (Prentice, 2001). It may also predispose Schizophrenia patients to die prematurely from complications of these disorders (Weisburger, 2000). They already have a decreased life expectancy (Simpson and Tsuang, 1996) and the overtly high total fat intake together with a pattern of low fruit and vegetable consumption may only accelerate this trend.

The prevalence of obesity per se has been increasing in patients with Schizophrenia (Coodin, 2001). Several of the newer antipsychotics such as Olanzapine, Seroquel or

Clozapine seem especially implicated in the pathogenesis of excess weight gain (Casey et al, 2004). While certainly a step forward in the treatment of schizophrenic symptoms, these agents may represent poor choices in the treatment of already overweight individuals. Psychopathology may be well controlled, but the overall increased nutrient intake and especially increased fat intake induced by these agents may put patients at overt risks for serious medical co-morbidities and thus significantly contribute to morbidity and mortality (Ryan and Thakore, 2002).

Also, the patients' subjective assessment of quality of life is considered to be a critical outcome variable in the care of individuals with schizophrenia (Steinwachs et al. 1996). Patients already suffer from low quality of life due to the chronic nature of their illness (Ritsner et al, 2003). Overweight further impairs quality of life in the Schizophrenic population (Strassnig et al, 2003). Distress related to high body weight is a modifiable factor, and the Quality of Life of schizophrenic patients can be improved substantially by proper weight management, apart from obvious advantages for physical health (Kawachi, 1999).

Dietary Essentiality of PUFA

Saturated and monounsaturated fatty acids can be synthesized within the body. However, several polyunsaturated fatty acids (PUFAs), such as the ω-3 fatty acids α-linoleic acid (18:3) and the ω-6 fatty acid eicosapentaenoic acid (20:5), are essential and need to be supplied through dietary intake. PUFAs contribute about 7 % to the total energy intake and 19 – 22 % of energy from fat in the adult diet, with linoleic acid (18:2n-6) being the major PUFA, comprising 84 – 89 % of total PUFA energy content. α-linolenic acid (ALA; 18:3n-3) contributes about 9 – 11 %. Eicosapentaenoic acid (EPA; 20:5n-3) and docosahexaeonic acid (DHA; 22:6n-3) provide less than 2% The n-6 fatty acids 18:4(n-6) and arachidonic acid 20:4(n-6), contribute about 0.1 – 0.2 % of total energy intake.

A considerable decrease in dietary PUFA availability of Western countries has been noted, largely due to changes in dietary habits and consumption of more refined food products (Simoupulos, 2000; Holman, 1997). Most importantly, with the beginning of this century, industrialized countries discovered that conversion of liquid oils into solid fats was possible (Emken, 1984). Oils heated in the presence of metal catalysts are exposed to hydrogen gas. During this process known as "hardening", close to 50 % of double bonds of exposed fatty acids are shifted to a trans-configuration while others become saturated. The results are fats used in products such as margarine, vegetable shortenings or oils for deep frying. Partially hydrogenated vegetable oils are cheap, and are more resistant to oxidation, which considerably prolongs their shelf-life. Comparatively low cost and easy handling also seem main reasons for their abundant use nowadays. Unfortunately, many PUFAs are destroyed in the process, largely contributing to decline in PUFA supply in industrialized countries. Practically, the problem is of considerable clinical importance, and causes major public health concerns: Coronary Artery Disease, for example, has been associated with intake of hydrogenated fatty acids, and more than 300000 fatalities from CAD per year in the US alone may be attributed to hydrogenated fats (Willet and Ascherio, 1994).

Table 2. Major Dietary Sources for N-3 and N-6 Fatty Acids in %

	N-6 PUFA			N-3 PUFA			
Major Food Group	**18:2**	**20:4**	**Total N6**	**18:3**	**20:5**	**22:6**	**Total N3**
Meat, Poultry, Game	14.94	70.15	15.27	15.26	20.58	11.77	15.86
Cereal-based products *	14.73	1.00	14.64	23.8	0.15	0.4	20.85
Milk products	3.18	0.12	3.16	11.4	0	0	8.54
Vegetables products	11.64	0	11.54	9.81	2.22	0.66	8.54
Fish and Seafood	3.24	27.21	3.36	3.54	75.87	73.12	12.86
Eggs and Products	1.58	0.16	1.78	0.55	1.12	10.02	1.3
Cereals	14.15	0.24	14.03	12.86	0.05	0.33	11.1
Fats and oils	21.85	0	21.66	16.31	0	0	14.5
Legumes and Pulses	0.69	0	0.69	1.01	0	0.03	0.87

* ie cakes, pies, croissants, biscuits, lasagne

\+ ie bread, muffins, rice, pasta, breakfast cereals)

Adapted from Meyer BJ, Mann NJ, Lewis JL, Milligan GC, Sinclair AJ, Howe PRC. Dietary Intakes and Food Sources of Omega-6 and Omega-3 Polyunsaturated Fatty Acids. Lipids 2003, 38, 391-398.

Micronutrient Intake of Schizophrenia Patients

Dietary Intake of Polyunsaturated Fatty Acids

Evidence has accumulated, that PUFAs such as omega-3 fatty acids have beneficial effects for a wide variety of diseases when adequately supplied (Harris, 2004). n-3 omega fatty acids have been used in the secondary prevention of diverse medical problems such as coronary artery disease, hypertension, type 2 diabetes; and to a lesser extent in renal disease, rheumatoid arthritis, ulcerative colitis, Crohn's disease and COPD (Simopoulos, 1999). Cardiovascular health benefits of n-3 omega fatty acids have recently been acknowledged by the American Heart Association (Kris-Etherton et al, 2002). n-6 omega fatty acids do seem to have various health benefits as well, but less convincing evidence is available (Berry, 2001).

Various abnormalities in PUFA cell membrane phospholipid metabolism exist in schizophrenia patients (Emsley et al, 2003). A variety of reasons may be responsible for these observations such as altered phospholipid metabolism (Horrobin, 1998), or environmental factors such as smoking and inadequate dietary intake (Reddy and Yao, 2003). Intriguingly, the clinical outcome of schizophrenia patients with improvements in PUFA levels of RBC membranes during treatment may be better (Evans et al, 2003).

In fact, cell membrane PUFA composition reflects dietary intake in healthy persons (Abedin et al, 1999); thus, dietary fatty acid composition may influence membrane phospholipid composition. Consequently, dietary essential fatty acid supplementation may even restore membrane PUFA deficits (Simopuoulos, 1999). Yet, despite the described PUFA deficiency state observed in Schizophrenics, and a direct link between their intake and cell membrane composition, only a few studies have examined the actual dietary PUFA intake of these patients.

Mellor et al (1995, 1996) attempted to quantify essential fatty acid intake in inpatients with Schizophrenia via a "seven day weighted intake method" that consisted of weighing of meals and diet history diaries to record any snacks or food taken between meal times in addition to regular meals. Specialized software was used to analyze dietary fatty acid content. The authors found that patients ate well above the recommended dietary intakes for the parent EFAs. They conclude that the low EFA cell membrane content of could not be accounted for by simple dietary deficiency, but rather was a result of other factors.

Hibbeln et al (2003) measured erythrocyte PUFA levels of outpatients with schizophrenia prior and after supplementation with 3 g of EPA per day and accounted for possible environmental confounding factors such as smoking and dietary intake through assessment of both parameters. The diet was examined using a Food Frequency Questionnaire. Smoking status and quantitative smoking information was obtained. In total, data of 87 adult outpatients with schizophrenia was collected.

Total fat intake in male smokers was 83 (±42) g, in male non-smokers 84 (±44) g; female smokers ate 73 (±29) g fat and female non-smokers 81 (±81) g. PUFA intake, expressed as percentage of total fat intake was 18.3 (±4.2) % in male smokers, 18.6 (±15.9) % in female smokers; 17.5 (±3.6) % in male non-smokers, and 21.6 (±4.0) % in female non-smokers. Dietary intakes expressed in grams of LNA, EPA, DHA, LA and AA in smokers and non-smokers did not differ significantly. When comparing the percentages of the respective fatty acids to total fat intake, smokers consumed significantly less LNA than non-smokers. No difference for intake of EPA, DHA, or AA was observed. Dietary eicosapentaeonic acid (EPA) intake was correlated to erythrocyte EPA content in all subjects. Docosahexaeonic acid (DHA) intake predicted erythrocyte DHA content in both male smokers and non-smokers. It was concluded that smoking status contributed to PUFA intake via possible lifestyle influences associated with smoking, such as poorer self-care including worse diet (Koo, 1997).

In on of our own studies (Strassnig et al, in submission), we also examined fat and fatty acid intake of stable outpatients with schizophrenia. The 24-hour diet recall method was used to estimate dietary intake of various fats. Similar to the prior study, our data was analyzed separately for smokers and non-smokers in order to account for possible contributions of smoking status on dietary quality (James et al, 1997).

Consistent with observations of other groups (Brown, 1998; McCreadie 2003), Schizophrenia patients in our study had higher fat intake than the age- and gender matched population. Our patients ate more saturated fat (42±24.1 g) and slightly more monounsaturated fat (37.3 ± 22.2 g) than persons in the general population (29.9 ±0.68 g and 33.8 ±0.72 g) ($p \leq 0.001$ and ≤ 0.059, respectively). Trans-fatty acid intake was lower in patients (2.6±3.7 g) than in controls (5.4 ±0.16 g; $p \leq 0.001$). Intake of various essential fatty acids (EFAs) is described in **Table 3.** We did not find a significant differences in dietary intake of these micronutrients when corrected for smoking status and number of cigarettes smoked.

Table 3. Polyunsaturated Fatty Acid Intake in Patients with Schizophrenia

	Group N = 146	**Non-Smokers** N = 59	**Smokers** N = 87
	g *Mean ± SD*	g *Mean ± SD*	g *Mean ± SD*
PUFA	20.26±14.3	20.89±14.1	19.5±14.4
18:2(n-6)	16.6±12.5	15.6±10	17.18±13.9
18:03(n-3)	1.33±1.07	1.28±0.93	1.37±1.14
20:4(n-6)	0.13±0.15	0.11±0.14	0.14±0.16
20:05(n-3)	0.024±0.08	0.039±0.11	0.014±0.05
22:05)	0.02±0.06	0.03±0.08	0.01±0.03
22:06(n-3)	0.06±0.21	0.097±0.32	0.037±0.09

Table 4. Omega Fatty acid Intake in Patients

		Omega-3 (g)	**+/- SD**	**Omega-6 (g)**	**+/- SD**	**Ratio**
All	N = 146	1.15	1.1	14.11	12.4	1 : 12.27
Smokers	N = 87	1.14	1.09	14.8	14	1 : 12.98
Non-Smokers	N = 59	1.15	1.1	13.4	9.6	1 : 11.65

Limited availability of published NHANES III US population data on EFA consumption precluded extensive further analysis. However, population data for ALA, DHA, and EPA intakes was available and thus was used for comparison with our study sample. Briefly, consumption of α-linoleic acid (1.33 ±1.07 g), docosahexaenoic acid (0.06 ±0.21 g) and eicosapentaenoic acid (0.024 ±0.08 g) was not quantitatively different between patients and healthy controls (Table 4).

Again, there was no significant difference noted between intake of both omega-3 and –6 fatty acids when corrected for smoking status. There was a trend towards decreased intake of omega-3 fatty acids in male patients who also were smokers ($p = 0.65$). The ratio of omega-3 to omega-6 in intake of our patients was 1 : 12.27, compared to a US population ratio of about 1 : 9.8 (Kris-Etherton et al, 2000)

Dietary Antioxidant Intake

Polyunsaturated fatty acids are exquisitely sensitive to oxidation. Importantly, PUFAs are abundant in brain tissue. The brain depends on aerobic metabolism, but has relatively poor intrinsic antioxidant defence mechanisms. Oxidative stress in the brain has been implicated in the pathogenesis of a variety of psychiatric and neurologic disorders, including schizophrenia (Lohr, 1991). There is evidence of increased peroxidative stress during acute schizophrenic episodes (Fenton et al, 2001; Reddy, 2003). Further, some antipsychotics used in routine treatment including Haloperidol or Chlorpromazine, have oxidative capabilities per se, and extrapyramidal side effects from long term treatment with these compounds may in part be explained by neuronal degeneration through oxidative cell membrane damage (Jeding et al, 1995; Sachdev et al, 1999; Gunne and Arden, 1993).

Most patients with schizophrenia smoke at alarmingly high rates. Estimates run from 58 % (Kelly and McCreadie, 1999) up to 88% to 90 % (Hughes et al, 1986; Lohr and Flynn, 1992). Cigarette smoke contains countless highly reactive substances, such as NO- and Peroxyl radicals, among others (Church and Pryor, 1985.) Considering that each puff of cigarette smoke contains up to 10^{15} of such free radicals (Pryor 1997), smoking may be regarded a major contributor to oxidative stress in Schizophrenia patients.

The body (and the brain) usually take care of free radicals via several specialized enzymatic pathways, including the superoxide-dismutase and glutathion-peroxidase pathways. Several dietary compounds such as the Vitamins E, C, A, and various trace elements including Folate and Zinc have antioxidant properties and make important contributions to physiologic antioxidant defences. Increased oxidative stress contributes to PUFA depletion and cell membrane damage in Schizophrenia (Mahadik et al, 2001) and meets a priori poor antioxidant defences in patients suffering from Schizophrenia (Mahadik and Mukherjee, 1996, Phillips et al, 1993).

Thus, it seems important to estimate intakes of various antioxidants. To our knowledge, no study is available that estimates dietary supply of these compounds in patients with Schizophrenia. We therefore attempted to quantify antioxidant intake in our study sample (Strassnig et al, in submission). Again, intakes were compared to US population standards as per NHANES III. **Table 5** shows dietary antioxidant supply.

The actual intake of most antioxidants was not quantitatively different from the intake observed in the US population. Patients in showed a trend towards ingestion of more Vitamin C than the corresponding population. Zinc intake was lower in the patient sample. Smoking status did not predict antioxidant intake.

Table 5. Antioxidant Intake in Patients with Schizophrenia compared to Population Standards

	Population	Patients N = 146	Smokers N = 87	Nonsmokers N = 59
	Mean ± SD	*Mean ± SD*	*Mean ± SD*	*Mean ± SD*
Vitamin A IU	6063±231	5141±4509	5199±4474	5054±4600
Vitamin C (mg)	106.38±3.59	133.7±162.8 +	140±183	124±130
Vitamin E α-Tocopherol EQ	9.73±0.34	9.75±6.98	9.5±7	10.2±7.1
Folate (mg)	288.79±6.43	301.63±190.3	294±190	323±190
Zinc (mg)	12.13±0.27	10.75±87 **	11.4±7.6	9.9±6

+ Trend towards significance
** 0.005

Conclusion

Patients suffering from Schizophrenia do not appear to have insufficient dietary supply of polyunsaturated fatty acids. Concurrently (apart from Zinc), patients also do not fall short of adequate antioxidant supply. It rather seems that patients get adequate supply of these micro-nutrients through overall increased dietary intake observed previously (Strassnig et al, 2003). Positive smoking status and the associated lifestyle factors such as poorer diet, seemed to play only a minor role predicting dietary intake, and to predominantly pertain male patients.

Accelerated PUFA loss (Horrobin, 1996) and increased oxidative cell membrane damage in schizophrenia patients have been described previously (Mahadik and Scheffer, 1996), leading to a relative PUFA paucity in various tissues of Schizophrenia patients. The increased oxidative stress – whether inherent to the disease process, introduced via antipsychotic medication, or caused by high free radical load from disproportionately high smoking rates – and the described intrinsic membrane PUFA deficiencies may still predispose patients to fall short of both dietary PUFAs and antioxidants. A dietary intake of these micro-nutrients similar to that of the healthy population may hypothetically not provide enough supply for an increased demand.

The epidemiological observation of a better outcome of Schizophrenia in "developing countries" with a less refined diet and higher PUFA supply as opposed to high animal fat intake and low-in-PUFA diet in industrialized countries (Christensen and Christensen, 1988) may partially be explained by meeting the increased demand of patients with Schizophrenia. Additionally, the diet in "developing countries" has a more equal ratio of ω-3 to ω-6 fatty acids than the diet in industrialised countries, which contains approximately tenfold less a ω-3 than ω-6 fatty acids (1 : 9.8 for the US; Kris-Etherton et al, 2000),

It is known that ω-6 to ω-3 fatty acids competitively inhibit cellular elongase and desaturase enzymes and thus, a high ω-6 intake may down-regulate ω-3 metabolism (Emken et al, 1994). Predominance of dietary ω-6 fatty acids with a relative paucity of ω-3 PUFA, therefore, may contribute to subnormal cellular membrane concentrations of ω-3 fatty acids. Indirect confirmation comes from clinical supplementation trials with PUFAs in patients with Schizophrenia. The ω-3 fatty acids seem to have better efficacy (Peet, 2003).

Oxidative processes have been implicated in Schizophrenia pathology as early as in the 1950s. (Hoffer et al, 1954). Pro-oxidative properties of several antipsychotic medications only magnify this problem (Jeding, 1995). Moderately successful supplementation with antioxidants such as Vitamin E to ameliorate medication induced side-effects has been reported (Rotrosen et al, 1996). Unfortunately, no clinical trials have been pursued to test the efficacy of antioxidant supplementation on Schizophrenic symptoms per se.

Discussion

Overall, Schizophrenia patients do not have decreased intake of either nutrient. Rather, their caloric intake is higher and evidently causes the consistently high rates of overweight and obesity. Patients also do not seem to prefer or "crave" a certain nutrient group such as fat or sugar (Gothelf, 2002), but eat more of all available food. Yet, higher quantitative intake does not imply higher quality of foods consumed. Several studies point to decreased intake of "healthful food" such as fruit and vegetables. On the brighter side, the increased intake seems to help patients get a sufficient supply of micronutrients such as PUFAs and antioxidants. Smoking may play a minor additional role in male patients, predisposing them to an even poorer diet than respective non-smokers.

However, from a physical health perspective, the emerging pattern of above-average dietary intake (including increased fat and sugar intake) with subsequent excess body weight warrants further consideration. Health burdens of such a constellation are apparent.

Fortunately, excess body weight is a modifiable risk factor. In the clinical care of Schizophrenia patients, however, body weight has not been a primary concern. Usually, overweight becomes focus of attention only after medical co-morbidities occur. Such situations can be easily avoided. In contrast to popular belief, patients seem to be well aware of their weight status and willing to lose weight, but need proper guidance to do so (Strassnig et al, in print). Several structured weight loss programs tailored to the needs of patients with Schizophrenia are being developed (Ganguli et al, 2002; Vreeland et al, 2003). Additionally, patients tend to be sedentary (Gothelf et al, 2002) and encouragement to engage in physical activity seems to be a neglected intervention. (Callaghan, 2004). Just as for mentally healthy individuals, a combination of dieting and exercise seems the most appropriate way to achieve long-term weight loss (National Task Force on the Prevention and Treatment of Obesity, 2000).

Aside from apparent direct benefits of proper weight management of Schizophrenia patients such as decreased risk for coronary artery disease, hypertension, stroke (Blair et al, 2004), new-onset type II diabetes triggered by treatment with antipsychotic medication has become increasingly problematic in clinical settings (Melkersson and Dahl, 2004). Evidence exists that dietary composition modulates insulin resistance. For example, relatively high saturated fat intake such as observed in patients with Schizophrenia increases the likelihood to develop type 2 diabetes mellitus (Rivellese et al, 2003) and decreases insulin sensitivity

(Hu et al, 2001). There is convincing evidence that cutting down saturated fat intake per se improves insulin sensitivity (Summers et al, 2002).

Interestingly, essential fatty acid intake may exert direct effects on insulin sensitivity. For example, dietary substitution of ω-6 polyunsaturated fatty acid for saturated fat has been shown to enhance insulin sensitivity (Vessby et al, 2001), and ω-3 fatty acid content in muscle cell membranes correlates to insulin sensitivity in animal models (Storlien et al, 1991).

At least a subset of Schizophrenia patients may benefit from PUFA supplementation also. Proper dietary balance of omega-3 and omega-6 fatty acids seems to be an important key factor, but for previously discussed reasons, is hard to achieve. Indeed, omega-3 fatty acids have been shown to ameliorate symptoms that go along with Schizophrenia better than omega-6 fatty acids, when supplied along with regular antipsychotic treatment (Peet, 2003, Emsley et al, 2003).

Overall, there seems to be a complex interrelationship between proper dietary intake, health, and schizophrenic symptoms. Dietary modification should incorporate the promotion of a low-fat diet and fruit and vegetable intake (Diniz et al, 2004). Improvement in patient's health status may also be achieved through interventions such as smoking cessation programs or increased physical activity. Low fat and sugar diets may enhance cogntition (Molteni et al, 2002). Peet (2004) has recently proposed that a diet low in saturated fat and refined sugar, whilst high in PUFAs, may improve symptoms and outcome of schizophrenia (Peet, 2004). Thus, in addition to a reduction of dietary fat and sugar, PUFA supplementation may be warranted. Antioxidant supplementation has not been investigated thoroughly; and Vitamin E supplementation to ameliorate medication side effects has yielded mixed results. However, given the increasing evidence for oxidative damage from antipsychotic medication, investigation into alternative antioxidant approaches to this condition may be indicated (Lohr et al, 2003).

Limitations

Dietary intake surveys are only estimates, whether assessed via food frequency questionnaire, or via dietary recall. How accurate these surveys are in patients with schizophrenia remains to be established. However, patients in all the reported studies of dietary intakes were outpatients and thus, not acutely ill. At this point, there is no reason to believe data may not be accurate, especially since there are clear correlations between reported intakes and measured body weight and BMI.

References

Abedin L, Lien EL, Vingrys AJ, Sinclair AJ. The effects of dietary alpha-linolenic acid compared with docosahexaenoic acid on brain, retina, liver, and heart in the guinea pig. *Lipids* 1999; 34(5), 475-482.

Alaimo K, Mcdowell MA, Briefel RR, Bischof AM, Caughman CR, Loria M, Johnson CL. Dietary intake of vitamins, minerals, and fiber of persons ages 2 months and over in the

United States: Third National Health and Nutrition Examination Survey, Phase 1, 1998-91. *Adv Data* 1994; 258, 1-28.

Allison DB, Fontaine KR, Heo M, Mentore JL, Cappelleri JC, Chandler LP, Weiden PJ, Cheskin LJ. The distribution of body mass index among individuals with and without schizophrenia. *J Clin Psychiatry* 1999 ; 60(4), 215-220.

Allison DB, Mackell JA, McDonnell DD. The impact of weight gain on quality of life among persons with schizophrenia. *Psychiatr Serv* 2003; 54(4), 565-567.

Aronne LJ. Epidemiology, morbidity, and treatment of overweight and obesity. *J Clin Psychiatry* 2001; 62(Suppl 23), 13-22.

Autry JH. Workshop on orthomolecular treatment of schizophrenia: a report. *Schizophr Bull* 1975; (12), 94-103.

Berlet HH, Bull C, Himwich HE, Kohl H, Matsumoto K, Pscheidt GR, Spaide J, Tourlentes TT, Valverde JM. Effect of diet on schizophrenic behavior. *Proc Annu Meet Am Psychopathol Assoc* 1966; 54, 425-452.

Berry EM. Who's afraid of n-6 polyunsaturated fatty acids? Methodological considerations for assessing whether they are harmful. *Nutr Metab Cardiovasc Dis* 2001; 11(3),181-188.

Blair SN, LaMonte MJ, Nichaman MZ. Modifiable behavioral factors as causes of death. *JAMA* 2004; 291(24),2942.

Brown AS, Birthwhistle J. Excess mortality of mental illness. *Br J Psychiatry* 1996;169(3), 383-384.

Brown S, Birtwistle J, Roe L, Thompson C. The unhealthy lifestyle of people with schizophrenia. *Psychol Med* 1999; 29(3);697-701.

Callaghan P. Exercise: a neglected intervention in mental health care? *J Psychiatr Ment Health Nurs* 2004; 11(4), 476-483.

Callaghan P. Exercise: a neglected intervention in mental health care? *J Psychiatr Ment Health Nurs* 2004; 11(4), 476-83.

Casey DE, Haupt DW, Newcomer JW, Henderson DC, Sernyak MJ, Davidson M, Lindenmayer JP, Manoukian SV, Banerji MA, Lebovitz HE, Hennekens CH. Antipsychotic-induced weight gain and metabolic abnormalities: implications for increased mortality in patients with schizophrenia. *J Clin Psychiatry* 2004; 65(Suppl 7), 4-18.

Christensen O, Christensen E. Fat consumption and schizophrenia. *Acta Psychiatr Scand* 1988; 78(5), 587-591.

Christensen O, Christensen E. Fat consumption and schizophrenia. *Acta Psychiatr Scand* 1988; 78(5), 587-591.

Christensen O, Christensen E. Fat consumption and schizophrenia. *Acta Psychiatr Scand* 1988; 78(5), 587-591.

Church DF, Pryor WA. Free-radical chemistry of cigarette smoke and its toxicological implications. *Environ Halth Perspect* 1985; 64, 111-126.

Coodin S. Body mass index in persons with schizophrenia. *Can J Psychiatry* 2001; 46(6), 549-55.

Diniz YS, Cicogna AC, Padovani CR, Santana LS, Faine LA, Novelli EL. Diets rich in saturated and polzunsaturated fatty acids: metabolic shifting and cardiac health. *Nutrition* 2004; 20(2), 230-234.

Dohan FC, Grasberger JC. Relapsed schizophrenics: earlier discharge from the hospital after cereal-free, milk-free diet. *Am J Psychiatry* 1973; 130(6), 685-688.

Dohan FC. Cereals and schizophrenia: data and hypothesis. *Acta Psychiatr Scand* 1966;42, 125-152.

Emken EA, Adlof RO, Gulley RM. Dietary linoleic acid influences desaturation and acylation of deuterium-labeled linoleic and ALAs in young adult males. *Biochem Biophys* 1994; Acta 1213, 277-288..

Emken EA. Nutrition and biochemistry of trans and positional fatty acid isomers in hydrgenated oils. *Annu Rev Nutr* 1984; 4, 339-376.

Emsley R. Oosthuiyen P, van Rensburg SJ. Clinical potential of omega-3 fatty acids in the treatment of schizophrenia. *CNS Drugs* 2003; 17(15), 1081-1089.

Emsley R. Oosthuiyen P, van Rensburg SJ. Clinical potential of omega-3 fatty acids in the treatment of schizophrenia. *CNS Drugs* 2003; 17(15), 1081-1089.

Eremis S, Cetin N, Tamar M, Bukusoglu N, Akdeniz F, Goksen D. Is obesity a risk factor for psychopathology among adolescents? *Pediatr Int* 2004; 46(3), 296-301.

Fenton WS, Dickerson F, Boronow J, Hibbeln JR, Knable M. A placebo-controlled tiral of omega-3 fatty acid (ethyl eicosapentaenoic acid) supplementation for residual symptoms and cognitive impairment in schizophrenia. *Am J Psychiatry* 2001; 158(12), 2071-2074.

Foster GD, Sarwer DB, Wadden TA. Psychological effects of weight cycling in obese persons: a review and research agenda. *Obes Res* 1997;5(5), 474-488.

Ganguli R, Pandina G, Turkoz I, Berry S, Mahmoud R, Brar JS. The effectiveness of behavioral treatment for weight loss in patients with schizophrenia. Presented at the *43rd Annual Meeting of the New Clinical Drug Evaluation Unit (NCDEU)*, 2003, Boca, Raton, FL.

Gilka L. Schizophrenia, a disorder of tryptophan metabolism. *Acta Psychiatr Scand* 1975; 258(suppl), 1-83.

Goldman LS. Medical illness in patients with schizophrenia. *J Clin Psychiatry* 1999; 60(suppl 21), 10-15.

Gopalaswamy AK, Morgan R. Too many chronic mentally disabled patients are too fat. *Acta Psychiatr Scand* 1985; 72(3), 254-258.

Gothelf D, Falk B, Singer P, Kairi M, Phillip M, Zigel L, Poraz I, Frishman S, Constantini N, Zalsman G, Weizman A, Apter A. Weight Gain Associated With Increased Food Intake and Low Habitual Activity Levels in Male Adolescent Schizophrenic Inpatients Treated With Olanzapine. *Am J Psychiatry* 2002; 159, 1055 – 1057.

Gothelf D, Falk B, Singer P, Kairi M, Phillip M, Zigel L, Poraz I, Frishman S, Constantini N, Zalsman G, Weizman A, Apter A. Weight Gain Associated With Increased Food Intake and Low Habitual Activity Levels in Male Adolescent Schizophrenic Inpatients Treated With Olanzapine. *Am J Psychiatry* 2002; 159, 1055 – 1057.

Gothelf D, Falk B, Singer P, Kairi M, Phillip M, Zigel L, Poraz I, Frishman S, Constantini N, Zalsman G, Weizman A, Apter A. Weight Gain Associated With Increased Food Intake and Low Habitual Activity Levels in Male Adolescent Schizophrenic Inpatients Treated With Olanzapine. *Am J Psychiatry* 2002; 159, 1055 – 1057.

Grant FW, Cowen MA, Ozerengin MF, Bigelow N. Nutritional requirements in mental illness. I. Ascorbic acid retention in schizophrenia. A reexamination. *Biol Psychiatry* 1973; 5(3), 289-294.

Gunne LM, Andren PE. An animal model for coexisting tardive dyskinesia and tardive parkinsonism: a glutamate hypothesis for tardive dyskinesia. *Clin Neuropharmacol* 1993; 16(1), 90-95.

Haag M. Essential fatty acids and the brain. *Can J Psychiatry* 2003;48(3), 195-203.

Han TS, Tijhuis MAR, Jean MEJ, Seidell JC. Quality of Life in Relation to Overweight and Body Fat Distribution. *Am J Public Health* 1998; 88, 1814-1820.

Harris WS. Fish oil supplementation: evidence for health benefits. *Cleve Clin J Med* 2004; 71(3), 208-210, 215-218 passim.

Hedley AA, Ogden CL, Honson CL, Carroll MD, Curtin LR, Flegal KM. Prevalence of overweight and obesity among US children, adolescents, and adults, 1999-2002. *JAMA* 2004; 291(23), 2847-2850.

Heimann H. Nosology and physiopathology in psychiatry—current aspects of Kraepelin's views on pathology. *Confin Psychiatr* 1980; 23(4), 262-74.

Hibbeln JR, Makino KK, Martin CE, Dickerson F, Boronow J, Fenton WS. Smoking, Gender, and Dietary Influences on Erythrocyte Essential Fatty Acid Composition among Patients with Schizophrenia or Schizoaffective Disorder. *Biol Psychiatry* 2003; 53, 431-441.

Hoffer A, Osmand H, Smythies J. Schizophrenia: new approach. *J Ment Sci* 1954; 100,29–45.

Holman RT. ω-3 and ω-6 essential fatty acid status in human health and disease. In: Handbook of Essential Fatty Acid Biology: Biochemistry, *Physiology and Behavioral Neurobiology* (Yehuda S & Mostovsky DI, eds.) pp. 138-182. Humana Press, 1997, Totawa, NJ.

Homel P, Casey D, Allison DB. Changes in body mass index for individuals with and without schizophrenia, 1987-1996. *Schizophr Res* 2002; 55(3), 277-284.

Hopper K, Wanderling J. Revisiting the developed versus developing country distinction in course and outcome in schizophrenia: result from ISOS, the WHO collaborative follow-up project. *Schizophr Bull* 2000; 26, 835-846.

Horrobin DF, Glen AI, Vaddadi K. The membrane hypothesis of schizophrenia. *Schizophr Res* 1994; 13(3), 195-207.

Horrobin DF. Schizophrenia as a membrane lipid disorder which is expressed throughout the body. *Prostaglandins Leukot Essent Fatty Acids* 1996; 55(1-2), 3-7.

Horrobin DF. The membrane phospholipid hypothesis as a biochemical basis for the neurodevelopmental concept of schizophrenia. *Schizoph Res* 1998; 30(3), 193-208.

Hu FB, van Dam RM, Liu S. Diet and risk of type II diabetes: the role of types of fat and carbohydrate. *Diabetologia* 2001; 44, 805-817.

Hughes JR, Hatsukami DK, Mitchell JE, Dahlgren LA. Prevalence of smoking among psychiatric outpatients. *Am J Psychiatry* 1986; 143(8), 993-998.

Jablenksi A. Epidemiology of schizophrenia: the global burden of disease and disability. *Eur Arch Psychiatry Clin Neurosci* 2000; 250(6), 274-285.

Jablenski A, Sartorius N, Emberg G. et al. Schizophrenia: manifestations, incidence and course in different cultures: a World Health Organization 10-country study. *Psychological Medicine* 1992; Monograph supplement (Suppl 20), 1-97.

Jablenski A, Sartorius N, Emberg g. et al. Schizophrenia: manifestations, incidence and course in different cultures: a World Health Organization 10-country study. *Psychological Medicine*. Monograph supplement 1992, Suppl 20, 1-97.

James WP, Nelson M, Ralph A, Leather S. Socioeconomic determinants of health. The contribution of nutrition to inequalities in health. *BMJ* 1997; 314(7093), 1545-1549.

Jeding I, Evans PJ, Akanmu D, Dexter D, Spencer JD, Aruoma OI, Jenner P, Halliwell B. Characerization of the potential antioxidant and pro-oxidant actions of some neuroleptic drugs. *Biochem Pharmacol* 1995; 49(3), 359-365.

Jeding I, Evans PJ, Akanmu D, Dexter D, Spencer JD, Aruoma OI, Jenner P, Halliwell B. Characerization of the potential antioxidant and pro-oxidant actions of some neuroleptic drugs. *Biochem Pharmacol* 1995; 49(3), 359-365.

Kawachi I. Physical and psychological consequences of weight gain. *J Clin Psychiatry* 1999; 60(Suppl 21), 5-9.

Kawachi I. Physical and psychological consequences of weight gain. *J Clin Psychiatry* 1999; 60(Suppl 21), 5-9.

Kelly C, McCreadie RG. Smoking habits, current symptoms, and premorbid characteristics of schizophrenic patients in Nithsdale, Scotland. *Am J Psychiatry* 1999; 156(11), 1751-1757.

Kety SS. Editorial: Dietary factors and schizophrenia. *Ann Intern Med* 1976; 84(6), 745.

Kohlmeier, L. Dietary methodology considerations for NHANES III. *Vital & Health Statistics – Series 4: Documents & Committee Reports*, 27:81–84, 1992.

Koo LC. Diet and lung cancer 20+ years later: more questions than answers? Int J Cancer 1997; 10, 22–29.

Kraepelin E. *Manic Depressive Illness and Paranoia.* 1921, S. Livingstone, Edinburgh.

Kris-Etherton PM, Harris WS, Appel LJ, for the Nutrition Committee. Fish Consumption, Fish Oil, Omega-3 Fatty Acids, and Cardiovascular Disease. *Circulation* 2002; 106, 2747–2757.

Kris-Etherton, Taylor Shaffer D, Yu-Poth Shaomei, Huth P, Moriarty K, Vishell V, Hargrove RL, Zhao G, Etherton TD. Polyunsaturated fatty acids in the food chain in the United States. *Am J Clin Nutr* 2000; 71(suppl), 179S-188S.

Kris-Etherton, Taylor Shaffer D, Yu-Poth Shaomei, Huth P, Moriarty K, Vishell V, Hargrove RL, Zhao G, Etherton TD. Polyunsaturated fatty acids in the food chain in the United States. *Am J Clin Nutr* 2000; 71(suppl), 179S-188S.

Kurzthaler I, Fleischhacker WW. The clinical implications of weight gain in schizophrenia. *J Clin Psychiatry* 2001; (62)Suppl 7, 32-37.

Lohr JB, Flynn K. Smoking and schizophrenia. *Schizophr Res* 1992; 8(2), 93-102.

Lohr JB, Kuczensli R, Niculescu AB. Oxidative mechanisms and tardive dyskinesia. *CNS Drugs* 2003; 17(1), 47-62.

Lohr JB. Oxygen radicals and neuropsychiatric illness. Some speculations. *Arch Gen Psychiatry* 1991;48, 1067-1106.

Mahadik S, Evans D, Lal H. Oxidative stress and role of antioxidant ω-3 essential fatty acid supplementation in schizophrenia. *Prog Neuropsychopharmacol Biol Psychiatr* 2001; 25, 463-493.

Mahadik SP, Mukherjee S. Free radical pathology and antioxidant defense in schizophrenia: a review. *Schizophr Res* 1996; 19(1), 1-17.

Mahadik SP, Scheffer RE. Oxidative injury and potential use of antioxidants in schizophrenia. *Prostaglandins Leukot Essent Fatty Acids* 1996; 55(1-2), 45-54.

Marshall E. Epidemiology. Public enemy number one: tobacco or obesity? *Science* 2004; 304(5672), 804.

McCreadie R, MacDonald E, Blackock C, Tilak-Singh D, Wiles D, Halliday J, Paterson J. Dietary intake of schizophrenic patients in Nithsdale, Scotland: case-control study. *BMJ* 1998; 317, 784-785.

McCreadie RG; Scottish Schizohphrenia Lifestyle Group. Diet, smoking and cardiovascualr risk in people with schizophrenia: descriptive study. *Br J Psychiatry* 2003; 183, 534-539.

McDowell MA; Briefel RR, Alaimo K, Bischof AM, Caughman CR, Carroll MD, Loria CM, Johnson CL. Energy and macronutrient intakes of persons ages 2 moths and over in the United States: Thrid National Health and Nutrition Examination Survey, Phase 1, 1988-91. *Adv Data* 1994; 255, 1-24.

McGlashan TH. A selective review of recent North American long-term follow-up studies of schizophrenia. *Schizophr Bull* 1988; 14(4), 515-42.

McIntosh A, Lawrie S. Cross-national differences in diet, the outcome of schizophrenia and the prevalence of depression: you are (associated with) what you eat. *Br J Psychiatry* 2004; 184(5):381-338.

Melkersson K, Dahl ML. Adverse metabolic effects associated with atypical antipsychotics: literature review and clinical implications. *Drugs* 2004;64(7), 701-723.

Mellor JE, Laugharne JDE, Peet M. Omega-3 fatty acid supplementation in schizophrenic patients. *Human Psychopharmacology* 1996; 11, 39–46.

Mellor JE, Laugharne JDE, Peet M. Schizophrenic symptoms and dietary intake of n-3 fatty acids. *Schizophr Res* 1995; 18, 85–86.

Meyer BJ, Mann NJ, Lewis JL, Milligan GC, Sinclair AJ, Howe PRC. Dietary Intakes and Food Sources of Omega-6 and Omega-3 Polyunsaturated Fatty Acids. *Lipids* 2003; 38, 391-398.

Mokdad AH, Serdula MK, Dietz WH, Bowman BA, Marks JS, Koplan JP. The Spread of the Obesity Epidemic in the United States 1991-1998. *JAMA* 1999; 282(16), 1519-1522.

Molteni R, Barnard RJ, Ying Z. A high-fat, refined sugar diet reduces hippocampal brain-derived neurotropic factor, neuronal pasticity and learning. *Neuroscience* 2002, 112; 803-814.

Moore BJ. Supersized America: help your patients regain control of their weight. *Cleve Clin J Med* 2003;70(3), 237-40.

National Institutes of Health, National Heart, Lung, and Blood Institute. *Clinical Guidelines on the Identification, Evaluation and Treatment of Overweight and Obesity in Adults: The Evidence Report*. Bethesda, Maryland, National Institutes of Health, 1998.

National Task Force on the Prevention and Treatment of Obesity, 2000. Overweight, Obesity and Health Risk. *Arch. Intern. Med.* 160(7), 898-904.

National Task Force on the Prevention and Treatment of Obesity, 2000. Overweight, Obesity and Health Risk. *Arch. Intern. Med.* 160(7), 898-904.

Pauling L. Orthomolecular Psychiatry. Science 1968; 160, 265-271.

Peet M. Diet, diabetes and schizophrenia: review and hypothesis. *Br J Psychiatry* 2004; 184(suppl 47), s 102-s105.

Peet M. Diet, diabetes and schizophrenia: review and hypothesis. *Br J Psychiatry* 2004; 184(suppl 47), s102-s105.

Peet M. Diet, diabetes and schizpophrenia: review and hypothesis. *Br J Psychiatry* 2004 184(suppl 47), s102-s105.

Peet M. Eicosapentaenoic acid in the treatment of schizophrenia and depression: rationale and preliminary double-blind clinical trial results. *Prostaglandins Leukot Essent fatty Acids* 2003; 69(6), 477-485.

Peet M. Eicosapentaenoic acid in the treatment of schizophrenia and depression: rationale and preliminary double-blind clinical trial results. *Prostaglandins Leukot Essent fatty Acids* 2003; 69(6), 477-485.

Phillips M, Sabas M, Greenberg J. Increased pentane and carbon disulfide in the breath of patients with schizophrenia. *J Clin Pathol* 1993; 46, 861-864.

Pi-Sunyer FX. Medical hazards of obesity. *Ann Intern Med* 1993; 119(7 pt 2), 655-660.

Prentice AM. Overeating: the health risks. *Obes Res* 2001; 9(Suppl 4), 234S-238S.

Pryor WA. Cigarette smoke radicals and the role of free radicals in chemical carcinogenicity. *Environ Health Perspect* 1997;105, 875-882.

Reddy RD, Yao JK. Environmental factors and membrane polyunsaturated fatty acids in schizophrenia. *Prostaglandins Leukot Essent Fatty Acids* 2003 ; 69(6), 385-391.

Reddy RD, Yao JK. Environmental factors and membrane polyunsaturated fatty acids in schizophrenia. *Prostaglandins Leukot Essent Fatty Acids* 2003; 69(6), 385-391.

Ritsner M, Kurs R, Gibel A, Hirschmann S, Shinkarenko E, Ratner Y. Predictors of quality of life in major psychoses: a naturalistic follow-up *study J Clin Psychiatry* 2003; 64(3), 308-315.

Rivellese AA, Lilli S. Quality of dietary fatty acids, insulin sensitivity and type 2 diabetes. *Biomed Pharmacother* 2003; 57(2), 84-87.

Rotrosen, Adler L, Lohr J, Edson R, Lavori P. Antioxidant treatment of tardive dyskinesia. *Prostaglandins Leukotr Essential fatty acids* 1996, 55(1&2), 77-81.

Ryan MC, Thakore JH. Physical consequences of schizophrenia and its treatment: the metabolic syndrome. *Live Sci* 2002; 71, 239-257.

Sachdev P, Saharov T, Cathart S. The Preventative Role of Antioxidants (Selegiline and Vitamin E) in a Rat Model of Tardive Dyskinesia. *Biol Psychiatry* 1999; 46, 1672-1681.

Schwartz TL, Nihalani N, Jindal S, Virk S, Jones N. Psychiatric medication-induced obesity: a review. *Obes Rev* 2004; 5(20), 115-121.

Simopolous AP. Essential fatty acids in health and chronic disease. *Am J Clin Nutr* 1999; 70(suppl), S560–S569.

Simopoulos AP, Leaf A, Salem N Jr. Workshop statement on the essentiality of and recommended dietary intakes for Omega-6 and Omega-3 fatty acids. *Prostaglandins Leukot Fatty Acids* 2000, 63(3), 119-121.

Simopoulos AP. Essential fatty acids in health and chronic disease. *Am J Clin Nutr* 1999; 70(3 suppl), 560s-569s.

Simpson JC, Tsuang MT. Mortality among patients with schizophrenia. *Schizophr Bull* 1996; 22(3), 485-499.

Singh MM, Kay SR. Wheat gluten as a pathogenic factor in schizophrenia. *Science* 1976; 191, 401-402.

Steinwachs DM, Fischer EP, Lehman AF. Outcomes assessment: information for improving mental health care. New *Dir Ment Health Serv* 1996; 71, 49-57.

Storlien LH, Jenkins AB, Cisholm DJ, Pascoe WS, Khouri S, Kraegen EW. Influence of dietary fat composition on development of insulin resistance in rats. Relationship to muscle triglyceride and omega-3 fatty acids in muscle phospolipid. *Diabetes* 1991; 40(2), 280-289.

Strassnig M, Brar JS, Ganguli R. Body mass index and quality of life in community-dwelling patients with schizophrenia. *Schizophr Res* 2003; 62(1-2), 73-76.

Strassnig M, Brar JS, Ganguli R. Body mass index and quality of life in community-dwelling patients with schizophrenia. *Schizophr Res* 2003; 62(1-2), 73-76.

Strassnig M, Brar JS, Ganguli R. Nutritional assessment of patients with schizophrenia: a preliminary study. *Schizophr Bull* 2003; 29(2), 393-397.

Strassnig M, Brar JS, Ganguli R. Nutritional assessment of patients with schizophrenia: a preliminary study. *Schizophr Bull* 2003; 29(2), 393-397.
Strassnig M, Brar JS, Ganguli R. Nutritional assessment of patients with schizophrenia: a preliminary study. *Schizophr Bull* 2003; 29(2), 393-397.
Strassnig M, Brar JS, Ganguli R. Self-reported Body Weight Status and Dieting Practices in Community-Dwelling Patients with Schizophrenia. *Schizophr Res*, in print.

Strassnig M. Brar JS, Ganguli R. Dietary Fatty Acid and Antioxidant Intake in Community-dwelling Patients suffering from Schizophrenia, in submission.
Strassnig M. Brar JS, Ganguli R. Dietary Fatty Acid and Antioxidant Intake in Community-dwelling Patients suffering from Schizophrenia, in submission.
Summers LK, Fielding BA, Bradshaw HA, Ilic V, Beysen C, Clark ML, Moore NR, Frayn KN. Substituting dietary saturated fat with polyunsaturated fat changes abdominal fat distribution and improves insulin sensitivity. *Diabetologia* 2002; 45(3), 369-377.

Underwood A. Nourishing your brain. *Newsweek* 2001; 137(17), 60-61.

Vessby B, Unsitupa M, Hermansen K, Riccardi G, Rivellese AA, Tapsell LC, Nalsen C, Berglund L, Louheranta A, Rasmussen BM, Calvert GD, Maffetone A, Pedersen E, Gustaffson IB, Storlien LH; KANWU Study. Substituting dietary saturated for monounsaturated fat impairs insulin sensitivity in healthy men and women: The KANWU Study. *Diabetologia* 2001; 44(3), 312-319.

Vreeland B, Minsky S, Menza M, Rigassio Radler D, Roemheld-Hamm B, Stern R. A program for managing weight gain associated with atypical antipsychotics. *Psychiatr Serv* 2003;54(8), 1155-1157.

Wallace B, Tennant C. Nutrition and obesity in the chronic mentally ill. Aust N Z *J Psychiatry* 1998; 32(1), 82-85.

Weisburger JH Eat to live, not live to eat. Nutrition 2000; 16(9), 767-73.

Willet WC. Ascherio A. Trans fatty acids: are the effects only marginal? Am J Public Health 1994; 84, 722-724.

World Health Organization. Schizophrenia: an international follow-up study. New York: Wiley & Sons, 1979.

Young M, Hagan GC, Tucker RE, Foster WE. A comparison of dietary study methods. 2. Dietary history vs. seven-day record vs. 24-hr. recall. *J Am Diet Assoc* 1952; 28, 218.

SECTION I: OXIDATIVE STRESS

In: Fatty Acids and Oxidative Stress...
Editors: R. Reddy and J.K. Yao, pp. 29-42
ISBN: 978-1-60021-677-0

Chapter III

Oxidative Stress in Schizophrenia

Anilkumar Pillai, Ravinder Reddy and Jeffrey K. Yao

Introduction

Schizophrenia is a debilitating illness that affects up to 1% population around the world. Various etiological factors have been reported in the pathophysiology of schizophrenia. These include genetic factors (Kendler, 2003), neuronal maldevelopment (Jakob et al., 1986; Thome et al., 1998), impaired neurotransmission (Carlsson et al., 1999; Kornhuber et al., 1994) and viral infections (Pearce, 2001). These factors either alone or in combination can cause abnormalities in neurodevelopment leading to both structural and functional changes in the brain. Evidence is increasing to indicate that mitochondrial pathology and oxidative stress may be the most critical component in the pathophysiology and outcome of schizophrenia (Goff et al., 1995, Whatley et al., 1998, Ben-Shachar and Laifenfeld, 2004 and Bubber et al., 2004). A large number of studies have indicated increased production of free radicals, alterations in antioxidant defense mechanism, increased lipid peroxidation and higher levels of apoptotic markers in patients with schizophrenia. It is also well known that oxidative stress can lead to the type of membrane abnormalities that have been observed in schizophrenia. This review will provide a brief summary of evidence for oxidative stress and oxidative cell injury in schizophrenia.

Oxidative Stress and Brain Pathophysiology

Oxidative stress is defined as higher cellular levels of reactive oxygen species than the cellular antioxidant defense. Oxidative stress in brain occurs when the generation of ROS overrides the ability of endogenous antioxidant systems to remove excess ROS subsequently leading to cellular damage. Brain consumes approximately twenty percent of the total amount of oxygen in the body. But the enhanced metabolic rate in the brain leads to the generation of excessive levels of ROS.

Free Radicals

Free radicals are reactive chemical species with an unpaired electron that are produced through a variety of physiological and pathological processes. The radicals generated from molecular oxygen are generally names ad ROS, These include: superoxide (superoxide (O2.–), hydroxyl (.OH), hydrogen peroxide (H_2O_2), singlet oxygen (1O2) and nitric oxide (NO•). The superoxide is an anionic radical formed by the reduction of molecular oxygen through the acceptance of a single electron. Superoxide radicals are generally formed by the electron transport chain in mitochondria and microsomes. They are also formed by the metal ion-dependent oxidation of epinephrine and norepinephrine, and by the action of enzymes such as tryptophane hydroxylase, indoleamine dioxygenase, and xanthine oxygenase. Hydrogen peroxide is formed by the dismutation of superoxide radicals by the metalloenzyme, superoxide dismutase and also by enzymes such as L-amino acid oxidase, glycolate oxidase, and xanthine oxidase (Sies, 1991). Hydrogen peroxide and superoxide may undergo further transformations in the presence of transition metals like iron and copper to give rise to the highly reactive hydroxyl radicals, by the Haber–Weiss or Fenton reactions. Singlet oxygen is very reactive and is generally formed by the enzymatic activation of oxygen (as in prostaglandin biosynthesis) or by various physicochemical reactions like thermal decomposition of endoperoxides and dioxetanes. NO^- formed by action of the enzyme called nitric oxide synthase can interact with superoxide forming under acidic condition forming unstable peroxynitrous acid that spontaneously generating ^-OH and NOO^-.

Antioxidant Defense System

The increased production of ROS is generally inactivated by antioxidant defense system, which includes enzymatic as well as non-enzymatic molecules. The enzymatic antioxidant molecules are superoxide dismutase (SOD), catalase and glutathione peroxidase, and the non-enzymatic molecules include primarily dietary antioxidants and uric acid. The enzymatic antioxidants act on specific ROS and degrade them to less harmful products. SODs convert superoxide radicals to H2O2. There are two major SODs, a copper-zinc-SOD present in the cytosol and a manganese-SOD present in the mitochondria. Mitochondrial SOD comprises almost 60% of the total SOD, because twice as much superoxide is produced in mitochondria compared to in the cytosol. The H2O2 is coveted into water either by catalase; an iron-containing enzyme found primarily in peroxisomes, lysosomes, and mitochondria or by glutathione peroxidase, a selenium-containing enzyme found predominantly in cytosol with low levels in mitochondria.

GSH is the most important non-enzymatic antioxidant. The other non-enzymatic antioxidants include both lipid soluble and water-soluble scavengers. The major lipid soluble scavengers are vitamin E, β carotene, and coenzyme Q. Ascorbic acid (Vitamin C) and uric acid are the major water-soluble antioxidants. Ascorbic acid is a water-soluble antioxidant and present in high concentrations in fruits and vegetables. Ascorbic acid is a biological antioxidant that acts as a chain-breaking scavenger for peroxy radicals and acts synergistically with vitamin E. In brain, the basal ganglia, which are rich in dopamine and glutamate, are also rich in ascorbic acid that rank among the highest in the body (Oke et al., 1987). Therefore, the antioxidant function of ascorbic acid may serve as an essential defense line against dopamine- or glutamate-induced neurodegenerative process (Rebec, 1987)

Vulnerability of the Brain to Oxidative Stress

It is well established that the brain is uniquely and preferentially vulnerable to oxidative stress (Halliwell, 1992; Mahadik and Mukherjee, 1996). This may be accounted by the fact that the brain, which is highly rich in lipids (O'Brien and Sampson, 1965), is under higher oxidative stress due to its very high aerobic metabolism, leading to high rates of oxyradical generation. Lipids are highly susceptible to peroxidation (Bielski et al., 1983). Likewise, DNA is also vulnerable to free-radial-induced damage. The timing and the degree of free radical insult can have lasting consequences. For example, during early brain developmental neuronal injury may cause abnormal neurodevelopment, and ongoing oxidative stress may lead to neurodegeneration.

Oxidative Stress in Schizophrenia

It is well known that oxidative stress can lead to the type of membrane abnormalities that have been observed in schizophrenia. Thus, free radical-mediated pathological processes may provide an explanatory model for the specific membrane alterations in schizophrenia, indicating that a role for oxidative stress might exist for some domains of schizophrenic pathophysiology as has been proposed by several investigators.

Increased Free Radical Generation

Moreover levels of nitric oxide and superoxides (NO^- and $^-O_2$) can be determined indirectly as nitrate (NO^-_2) and nitrite (NO^-_3). Compared to controls, higher levels of nitrates are reported in serum (Taneli *et al.*, 2004), RBC (Herken *et al.*, 2001a) and post-mortem brain (Yao *et al.*, 2004) but not in CSF (Ramirez *et al.*, 2004). Therefore, its increase may support some of the reports on increased neuronal-NOS (Bernstein *et al.*, 2001; Karson *et al.*, 1996).

Altered Antioxidant Defense System

Although individual antioxidants play a specific role in the antioxidant defense system (AODS), these antioxidants may act cooperatively *in vivo* to provide synergistic protection to the organs against oxidative damage. Therefore, it is more meaningful to evaluate AODS by measuring both the individual levels and the overall antioxidant status. Plasma total antioxidant status (TAS) was significantly lower in schizophrenic patients than in normal controls. Individual plasma antioxidants, albumin, bilirubin (Yao et al., 2000a) and uric acid (Yao et al., 1998a), also showed decrements in patients even more prominent than the plasma TAS.

Recently Suboticanec et. al., (1990) have demonstrated that both plasma and urinary vitamin C levels were decreased in chronic schizophrenic patients, relative to normal controls, even after controlling for diet. After vitamin C supplementation for one month, group differences were no longer significant. It is thus suggested that vitamin C requirements for schizophrenic patients may be higher than for normal subjects. It is well known that fat-soluble vitamin E involves in oxidative metabolism and associated with diseases linked to oxidative stress. McCreadie et al., (1995) found lower ratios of vitamin E to cholesterol in schizophrenic patients compared with normal control subjects. Later, Brown et al. (1998) also reported decreased lipid-corrected vitamin E levels in schizophrenic patients with tardive dyskinesia, relative to healthy controls, but not in patients without dyskinesia.

Role of antioxidant enzymes in schizophrenia has been thoroughly studied, but results are inconsistent. Increased SOD activities have been reported in red blood cells (RBC) of schizophrenic patients (Michelson et al., 1977; Abdalla et al., 1986; Golse et al., 1978a; 1978b; Reddy et al., 1991; Yao et al., 1998b) but not by others (Sinet et al., 1983). Neuroleptic-naïve first-episode schizophreniform and schizophrenic patients show both increased SOD activity (Khan and Das, 1997) and decreased SOD activity (Mukherjee et al., 1996). It is possible that with progression of the illness, the SOD levels rise as a compensatory response to oxidative stress (Mukherjee et al., 1996).

On the other hand, levels of GPx were also found inconsistent in patients with schizophrenia. GPx activity was found to be lower, relative to normal controls, in neuroleptic-treated chronic schizophrenic patients (Stoklasova et al., 1986) in drug-free female schizophrenic patients (Abdalla et al., 1986) and in neuroleptic-naive psychotic children (Golse et al., 1977). In plasma, Zhang et al. (1998) have recently reported increased GPx activities in long-term neuroleptic free as well as neuroleptic-naïve schizophrenic patients, while Yao et al. (1999) did not find any significant difference between chronic schizophrenic patients and normal subjects, but is correlated with psychosis severity. Studies on skin fibroblasts from schizophrenic patients did not show any change in GPx activity as compared to normal controls (Zhang et al., 1998) suggesting that plasma GPx elevation in patients may be state-dependent changes, and not a consequence of course of illness, treatment effects or neuroleptics. Increased GPx activity has been found in erythrocytes of schizophrenia patients (Herken et al., 2001; Kuloglu et al., 2002). Lower GPx activity was reported in erythrocytes (Altuntas et al., 2000) and in plasma (Akyol et al., 2001), whereas no change in GPx activity was found in polymorphonuclear leucocytes of patients with schizophrenic psychoses (Srivastava et al., 2001). In platelets, Buckman et al. (1990) found that GSH-Px activity was inversely correlated with computed tomography (CT) scan measures of brain atrophy in patients with chronic schizophrenia, specifically nonparanoid schizophrenia with a predominance of negative symptoms. Increased catalase activity was reported in erythrocytes of schizophrenia patients (Herken et al., 2001), whereas no alteration of the activity of this enzyme was observed in leucocytes (Srivastava et al., 2001).

Recently Yao and colleagues evaluated the activity of SOD, CAT and GPx in schizophrenic patients using a within-subject, repeated measures, on-off haloperidol (HD) treatment design (Yao et al., 1998b; 1999). Among the three enzymes in erythrocytes and plasma, only RBC-SOD was found significantly higher in drug-free schizophrenic patients than in age, sex-matched normal volunteers. This finding is consistent with most previous reports of increased SOD activity that have been reported in erythrocytes of drug-free and treated schizophrenic patients (Reddy and Yao, 1999). Moreover, the enzyme activities were not significantly correlated with subject age, age of onset of illness, or the duration of illness.

In addition, none of the major antioxidant enzymes showed significant differences between relapsed and clinically stable patients.

In addition to the changes in peripheral activity of antioxidant enzymes, post-mortem studies also indicate altered antioxidant defense system in the brain of patients with schizophrenia. Increase in Mn SOD with no change in Cu, Zn-SOD has been reported in the frontal and temporal cortex of patients with schizophrenia (Loven *et al.*, 1996). However, a recent report showed increased levels of Cu, Zn- and Mn-SOD in frontal cortex and substantia innominata but not in rest of the brain (Michel *et al.*, 2004).

In addition, plasma levels of vitamin E (McCreadie *et al.*, 1995), vitamin C (Suboticanec *et al.*, 1990), albumin (Yao *et al.*, 2000a), and uric acid (Yao *et al.*, 1998a) are reported lower in schizophrenic patients.

Factors Influencing Antioxidant Defense System

Age

The rate of production of ROS and the changes in antioxidant defense system roughly correlate with the life span and is though to represent a major determinant in defining the rate of aging (Finkel and Holbrook, 2000). It has been shown that the increased SOD activities are related significantly with aging in the brain (Vanella et al., 1982; Mavelli et al., 1978; Hiramastsu et al., 1992), reflecting a self-protection against an increased production of superoxide radicals in the brain. On the other hand, decreased glutathione is found in the brain with aging, suggesting an increased susceptibility to oxidative damage with accelerated aging (Mori and Packer, 1998). Moreover, strong correlations have been observed between the accumulation of oxidized proteins and life-span in cases, where is there is increased activity of antioxidant enzymes suggesting a causative role for ROS in the aging process (Sohal and Weindruch, 1996; Finkel and Holbrook, 2000). Antioxidant supplementation has been used to prolong life span in experimental animals (Clapp, 1979; Comfort, 1991; Heidrick, 1984). It is thus surmised that the aging process may be simply the sum of the deleterious radical reactions going on continuously throughout life in cells and tissue (Harman, 1981).

Yao et al., (2000a) examined the relationship between plasma antioxidant protein levels and clinically relevant demographic features, e.g., age, age of onset of illness, duration of illness, days on drug treatment, length of drug-free period, and body mass index. The results indicate that reduction of plasma antioxidant proteins in schizophrenia appears to be age-related. In healthy volunteers, plasma albumin levels were not correlated with age. By contrast, there was an inverse correlation with age in schizophrenic patients, although there was no significant difference in mean age between the two groups. Such an inverse correlation with age was in agreement with findings in plasma antioxidant uric acid (Yao et al., 1998a). However, plasma total bilirubin levels were positively correlated with age in healthy volunteers, whereas such an age-related increase was not observed in schizophrenia. Furthermore, these reductions in plasma albumin and bilirubin were not related to other clinically relevant demographic factors, particularly duration of illness. Taken together, age appears to be an important factor modifying antioxidant defense system in schizophrenia.

Neuroleptic Treatment

Though oxidative injury is a part of the primary illness (Mahadik *et al.*, 1998; McCreadie *et al.*, 1995) it can be exacerbated by certain neuroleptics. Studies done to compare the antioxidant enzyme status in haloperidol treated patients to that in the same individuals following haloperidol withdrawal indicated that SOD and GPx activities were shown to be higher in haloperidol-withdrawal patients than in haloperidol-treated patients (Yao et al., 1998b). Abdalla et al (1986) found that SOD and GSH-Px activities were not significantly different between schizophrenic patients on and off neuroleptic treatment. Buckman et al. (1990) also did not find any effect on platelet GSH-Px activity with neuroleptic treatment. On the other hand, neuroleptic discontinuation in chronic schizophrenic patients is often associated with worsening of symptoms, probably a consequence of alterations in the dopaminergic systems due to lifting the chronic D_2-receptor blockade. Worsening symptoms and accompanying stress that may be due to a hyperdopaminergic state may lead to increased superoxide radical production and consequently increased SOD activity as observed in this study. The observed increase in GSH-Px activity can be compensatory due to increased SOD-induced production of hydrogen peroxide. It is well recognized that non-specific stress can induce oxidative stress (Ohtsuka et al., 1994; Scarpellini et al., 1994). Thus, alterations in SOD activity in this study may be a consequence of changed behavioral state rather than direct effects of haloperidol, with secondary increases in GSH-Px activity.

Studies in rats with neuroleptics treatment indicate that neuroleptics have differential effects on various antioxidant enzymes in brain. Haloperidol and fluphenazine has been found to increase the oxidative stress by altering the levels of antioxidants, and cause oxidative injury that includes apoptosis through mitochondrial action in the brain (Cadet and Perumal, 1990; *Mahadik et al., 2003*) and in culture (Behl *et al.*, 1995; *Noh et al., 2000*). However chlorpromazine has been found to alter the antioxidant enzymes but reduces the lipid peroxidation (Roy *et al.*, 1984). Recently Mahadik and colleagues have studied the temporal and differential effects of first-generation antipsychotics (FGAs) such as haloperidol and chlorpromazine and second-generation antipsychotics (SGAs) such as risperidone, olanzapine and clozapine on various antioxidant enzymes and lipid peroxidation in rat brain (Parikh et al., 2003; Pillai et al., 2006). The results from these studies indicate that short-term (<14 days) treatment with both FGAs and SGAs do not alter the antioxidant defense enzymes as well as levels of lipid peroxides, whereas continued treatment with FGAs but not the SGAs until 90 days, altered the antioxidant enzymes and increased the lipid peroxides in rat brain. However continued treatment beyond 90 days up to 180 days altered the antioxidant defense even by SGAs. It was also interesting that switching animals after 90 days on FGAs to SGAs, the levels of antioxidant enzymes were significantly normalized with parallel reduction in levels of lipid peroxides. These data are very useful in planning long-term use as well as switch over of these antipsychotics for the treatment management of schizophrenia.

Cigarette Smoking

The prevalence of cigarette smoking in schizophrenic patients is between 70% - 90%, compared to that of 35 - 54% for all psychiatric patients and 30 - 35% for general population (Hughes et al., 1986; Lohr and Flynn, 1992; Olincy et al., 1987). Cigarette smoke is found to contain very high levels of both ROS and NO radicals (Pryor & Stone, 1993) and high levels of lipid peroxides are reported in smokers compared to non-smokers (Morrow *et al.*, 1995). Chronic cigarette smoking is associated with oxidative stress due to the increased free radical burden, reflected in alterations of various oxidative stress indices (Stegmayr et al., 1993), and may contribute directly to decreases in antioxidants. It has been shown that cigarette smoking does not appear to reduce the levels of plasma antioxidant enzyme levels and individual antioxidants, with exception of bilirubin (Yao et al., 1998a; 2000a).

Dietary Status

High caloric intake is one of the most important risk factors for increased oxidative stress and oxidative cell injury in individuals with schizophrenia. Plasma antioxidants of dietary origin (e.g., tocopherols, ascorbic acid, carotenoids, etc) are influenced directly by nutritional supplements as well as food and alcohol consumption (Becker et al., 1994; Parfitt et al., 1994; Lecomte et al., 1994). Studies have shown correlation between plasma uric acid and body mass index in patients with schizophrenia and not in control subjects (Yao et al, 1998a). Since individuals with schizophrenia have altered membrane pathology, it is advantageous to receive EPUFAs to restore the membrane levels of EPUFAs and thus to protect the cell membrane from oxidative damage.

Indices of Free Radical Pathology in Schizophrenia

Increased Lipid Peroxides

A number of oxygenated compounds like aldehydes including malondialdehyde are produced during the attack of free radicals to membrane lipoproteins and polyunsaturated fatty acids. Because of the high lipid content, the central nervous system is more susceptible to free radical attack. Increased levels of TBARS have been reported in plasma (Peet *et al.*, 1993) and CSF (Lohr *et al.*, 1990; Pall *et al.*, 1987; Tsai *et al.*, 1998) in some chronic schizophrenic patients treated with neuroleptics. Increased levels of TBARS were found in plasma/serum in both patients with chronic and never medicated first episode schizophrenia (Arvindakshan *et al.*, 2003b; Khan *et al.*, 2002; Mahadik *et al.*, 1998) and in patients with tardive dyskinesia (Peet et al., 1993). Moreover, Mahadik et al.1998 reported an elevated plasma lipid peroxides at the onset of psychosis in never-medicated, first-episode schizophrenic patients. Moreover, the changes in the levels of TBARS were found to correlate with the severity of abnormal involuntary movements (Peet *et al.*, 1993), with negative symptoms, and with RBC glutathione peroxidase (Mahadik *et al.*, 1998).

DNA Damage

Increased reactive oxygen species cause peroxidative damage to DNA and cause apoptosis (Wood & Youle, 1994). Recent reports indicate the presence of apoptotic markers like increase in Bax / Bcl-2 ratio in the temporal cortex (area 21, middle temporal gyrus) of patients with schizophrenia compared to matched controls (reviewed by Jarskog et al., 2005).

Development of Tardive Dyskinesia

Free radical-mediated pathology has been implicated in the development of tardive dyskinesia (TD) (Reddy and Yao, 1999; Cadet et al., 1986; Cadet and Lohr, 1989). Lohr et al. (1990) found a significantly higher level of lipid peroxidation products (diene conjugates) in the cerebrospinal fluid of schizophrenic patients with TD than those without TD. They also demonstrated that the severity of TD was positively correlated with levels of diene conjugates, and more critically, high levels of diene conjugates were associated with subsequent development of TD. Recently, Peet et al. (1993) found a highly significant correlation between *plasma* levels of lipid peroxidation products and dyskinesia severity. Zubenko and Cohen (1986) have shown that platelet membrane fluidity is altered in patients with TD, but not in similarly neuroleptic-treated patients without TD. Moreover, increased oxidative stress has been reported in the CSF of schizophrenia patients receiving neuroleptics (Pall et al., 1987). In addition, several studies have shown the protective effect of vitamin E on TD symptoms (Egan et al., 1992; Dabiri et al., 1994).

Reduced Membrane Phospholipid EPUFAs

Phospholipid abnormalities have also been found in medication-free schizophrenic patients (Keshavan et al., 1993a), and decreases in the levels of key membrane phospholipids, phosphatidylcholine, phosphatidylserine, phosphatidylinositol and phosphatidylethanolamine were found in fibroblasts from neuroleptic-naive schizophrenic patients (Mahadik et al., 1994). A number of studies have reported low levels of phospholipids (Keshavan *et al.*, 1993; Rotrosen & Wolkin, 1987;), esterified phospholipid PUFAs, particularly arachidonic and docosahexaenoic acids (Glen *et al.*, 1994; Mukherjee *et al.*, 1994; Yao *et al.*, 1994) in erythrocyte membranes and in brain homogenates (Horrobin *et al.*, 1991) from patients with schizophrenia. Reduced levels of phosphatidylcholine and phosphatidylethanolamine were also found in postmortem brain tissue from schizophrenic patients (Yao *et al.*, 2000b). More recently, reduced levels of particularly AA and DHA with increased levels of lipid peroxides have been reported in drug-naïve first episode patients at early stages of psychosis (Evans *et al.*, 2003; Khan *et al.*, 2002). Arvindakshan *et al.*, 2003 reported decreased levels of RBC DHA but not AA, and also no increase in plasma lipid peroxides in schizophrenia patients. Reddy et al. (1999) found a significant reduction in RBC arachidonic acid (AA) levels at neuroleptic-naïve baseline (Reddy et al., 1999). Moreover, a significant correlation was demonstrated between RBC phospholipid PUFAs and ^{31}P magnetic resonance spectroscopic measures of phospholipid metabolites in the combined right and left frontal lobe, but not other brain regions including caudate, occipital, parietal and temporal areas (Yao et al., 2000b). Specifically, both total and individual PUFA (20:4, 22:5, and 22:6) were significantly ($p<0.02$) and positively correlated with phosphomonoesters levels supporting the notion that decreased membrane fatty acids in peripheral tissue may be associated with similar changes in the brain. Studies on the effect of neuroleptics on membrane phospholipids indicate that membrane AA and DHA levels were higher in patients treated with SGAs as compared to FGAs (Arvindakshan *et al.*, 2003b; Evans *et al.*, 2003; *Horrobin, 1999*; Khan *et al.*, 2002).

Conclusions

In summary, oxidative stress plays an important role in the pathophysiology of schizophrenia and the effects are mediated through the well-recognized effects of lipid peroxidation on neuronal and mitochondrial membrane components (Floyd, 1996; Jesberger and Richardson, 1991). The life style and socioeconomic status of the schizophrenic patients act as prooxidative stress. The membrane alterations and function by the disease itself or by the treatment can be normalized in part by adjunctive treatment with supplements like antioxidants and/or fatty acids. The efficacy of a combined antioxidant and EFA supplementation strategy may be greater that either approach alone. EFA supplementation alone may be sufficient to repair the loss of membrane fatty acids, but not in reducing or preventing ongoing oxidation of fatty acids by free radicals (Aravindakshan et al., 2003). A major question remains about the timing and duration of such interventions: whether the greatest impact will be after the diagnosis of schizophrenia has been established (early versus late in the illness) or whether early intervention (e.g., in utero) could prevent the disorder. The time to develop treatment trials with adjunctive antioxidants would be extremely short because there already exist many antioxidant compounds with relatively low known toxicity, and essential fatty acid supplementations (Reddy and Yao, 1999).

References

Abdalla DS, Manteiro HP, Olivera JA, Bechara EJ. Activities of superoxide dismutase and glutathione peroxidase in schizophrenic and manic depressive patients. *Clin Chem* 1986; 32: 805-7

Adler LA, Peselow E, Rotrosen J, Duncan E, Lee M, Rosenthal M, Angrist B. Vitamin E treatment of tardive dyskinesia. *Am J Psychiatry* 1993; 150: 1405-7

Arvindakshan M, Sitasawad S, Debsikdar V, Ghate M, Evans D, Horrobin DF, Bennett C, Ranjekar PK, Mahadik SP. Essential polyunsaturated fatty acid and lipid peroxide levels in never-medicated and medicated schizophrenia patients. *Biol Psychiatry* 2003; 53, 56-64.

Becker K, Boetticher D, Leichsenring, M. Antioxidant vitamins in malnourished Nigerian children. *Int J Vit Nutr Res* 1994; 64: 306-10

Behl C, Rupprecht R, Skutella T. Holsboer F. Haloperidol-induced cell death--mechanism and protection with vitamin E in vitro. *Neuroreport* 1995; 7, 360-364.

Ben-Shachar D, Laifenfeld D. Mitochondria, synaptic plasticity, and schizophrenia. *Int Rev Neurobiol.* 2004; 59:273-96.

Bubber P, Tang J, Haroutunian V, Xu H, Davis KL, Blass JP, Gibson GE. Mitochondrial enzymes in schizophrenia. *J Mol Neurosci.* 2004;24(2):315-21.

Buckman TD, Kling A, Sutphin MS, Steinberg A, Eiduson S. Platelet glutathione peroxidase and monoamine oxidase activity in schizophrenics with CT scan abnormalities: Relation to psychosocial variables. *Psychiatr Res* 1990; 31: 1-14

Cadet JL, Lohr JB, Jeste DV. Free radicals and tardive dyskinesia. *TINS* 1986; 9: 107-8

Cadet JL, Perumal AS. Chronic treatment with prolixine causes oxidative stress in rat brain. *Biol Psychiatry* 1990; 28, 738-740.

Carlsson A, Waters N, Carlsson ML. Neurotransmitter interactions in schizophrenia--therapeutic implications. *Biol Psychiatry.* 1999; 46(10):1388-95.

Clapp NK, Satterfield LC, Bowles ND. Effects of the antioxidant butylated hydroxytoluene (BHT) on mortality in BALB/c mice. *J Gerontol.* 1979; 4(4):497-501.

Comfort A. Effect of ethoxyquin on longevity of c3h mice. *Nature* 1991; 229:254-5

Corrigan FM, Van Rhijn AG, Mackay AV, Skinner ER, Horrobin DF. Vitamin E treatment of tardive dyskinesia (letter). *Am J Psychiatry* 1993; 150: 991-2

Dabiri LM, Pasta D, Darby JK, Mosbacher D. Effectiveness of vitamin E for treatment of long-term tardive dyskinesia. *Am J Psychiatry.* 1994; 151(6):925-6.

Dakhale GN, Khanzode SD, Khanzode SS, Saoji A. Supplementation of vitamin C with atypical antipsychotics reduces oxidative stress and improves the outcome of schizophrenia. *Psychopharmacology (Berl).* 2005; 182(4):494-8.

Egan MF, Hyde TM, Albers GW, Elkashef A, Alexander RC, Reeve A, Blum A, Saenz RE, Wyatt RJ. Treatment of tardive dyskinesia with vitamin E. *Am J Psychiatry* 1992; 149: 773-77

Elkashef AM, Wyatt RJ. Tardive dyskinesia: possible involvement of free radicals and treatment with vitamin E. *Schizophr Bull* 1999; 25: 731-40

Evans DR, Parikh VV, Khan MM, Coussons C, Buckley PF, Mahadik SP. Red blood cell membrane essential fatty acid metabolism in early psychotic patients following antipsychotic drug treatment. *Prostaglandins Leukot Essent Fatty Acids*, 2003; 69, 393-399.

Finkel T, Holbrook NJ. Oxidants, oxidative stress and the biology of ageing. *Nature.* 2000 ; 408(6809):239-47.

Floyd RA. Mitochondrial damage in neurodegenerative disease. In: Packer L, Hiramatsu M, Yoshikawa T. editors. *Free radicals in brain physiology and disorders*. San Diego: Academic Press, 1996: 313-29

Glen AIM, Glen EMT, Horrobin DF, Vaddadi KS, Spellman M, Morse-Fisher N., Skinner EK. A red cell membrane abnormality in a sub-group of schizophrenic patients: Evidence for two diseases. *Schizophr Res*, 1984; 12, 53-61.

Goff DC, Tsai G, Beal MF, Coyle JT. Tardive dyskinesia and substrates of energy metabolism in CSF. *Am J Psychiatry.* 1995; 152(12):1730-6.

Harman, D. The Aging. *Proc Natl Acad Science USA* 1981; 78: 7124-8

Heidrick ML, Heidrick LC, Cook DE. Effect of dietary 2-mercaptoethanol on the life span, immune system, tumor incidence and lipid peroxidation damage in spleen lymphocytes of aging bc3f mice. *Mech Aging* 1984; 27: 341-58

Herken H, Uz E, Ozyurt H, Akyol O. Red blood cell nitric oxide levels in patients with schizophrenia. *Schizophrenia Research*, 2001; 52, 289-290.

Hiramatsu M, Kohno M, Edamatsu R, Mitsuta K, Mori A. Increased superoxide dismutase activity in aged human cerebrospinal fluid and rat brain determined by electron spin resonance spectrometry using spin trap method. *J Neurochem* 1992; 58: 116-1164

Horrobin DF, Manku MS, Hillman S, Glen IAM. Fatty acid levels in brains of schizophrenics and normal controls. *Biol Psychiatry*, 1991; 30, 795-805.

Hughes R, Hatsukami K, Mitchell E, et al. Prevalence of smoking among psychiatric outpatients. *Am J Psychiatry* 1986; 143

Jakob H, Beckmann H. Prenatal developmental disturbances in the limbic allocortex in schizophrenics. *J Neural Transm*. 1986; 65(3-4):303-26.

Jarskog LF, Glantz LA, Gilmore JH, Lieberman JA. Apoptotic mechanisms in the pathophysiology of schizophrenia. *Prog Neuropsychopharmacol Biol Psychiatry*, 2005, 29, 846-858.

Jesberger J A, Richardson J S. Oxygen free radicals and brain dysfunction. *Intl J Neurosci* 1991; 57: 1-17

Kendler KS. The genetics of schizophrenia: chromosomal deletions, attentional disturbances, and spectrum boundaries. *Am J Psychiatry*. 2003; 160(9):1549-53.

Keshavan MS, Mallinger AG, Pettegrew JW, Dippold C. Erythrocyte membrane phospholipids in psychotic patients. *Psychiatry Res*, 1993; 49, 89-95.

Khan MM, Evans DR, Gunna V, Scheffer RE, Parikh VV, Mahadik SP. Reduced erythrocyte membrane essential fatty acids and increased lipid peroxides in schizophrenia at the never-medicated first-episode of psychosis and after years of treatment with antipsychotics. *Schizophr Res,* 2002; 58, 1-10.

Kornhuber J, Weller M. Current status of biochemical hypotheses in the pathogenesis of schizophrenia. *Nervenarzt*. 1994; 65(11):741-54.

Lecomte E, Herbeth B, Pirollet P, Chancerelle Y, Arnaud J, Musse N, Paille F, Siest G, Artur Y. Effect of alcohol consumption on blood antioxidant nutrients and oxidative stress indicators. *Am J Clin Nutr* 1994; 60: 255-61

Lohr JB, Flynn K. Smoking and schizophrenia. Schizophrenia Research 1992; 8: 93-102

Lohr JB, Underhill S, Moir S, Jeste DV. Increased indices of free radical activity in the cerebrospinal fluid of patients with tardive dyskinesia. *Biol Psychiatry*, 1990; 28, 535-539.

Loven DP, James JF, Biggs L, Little KY. Increased manganese-superoxide dismutase superoxide dismutase activity in postmortem brain from neuroleptic-treated psychotic patients. *Biol Psychiatry*, 1996; 40, 230–232.

Mahadik SP, Mukherjee S, Correnti EE, Kelkar HS, Wakade CG, Costa RM, Scheffer R.

Mahadik SP, Mukherjee S, Correnti EE, Scheffer R, Mahadik J. Elevated plasma lipid peroxides at the onset of non-affective psychosis. *Biol Psychiatry*, 1998; 43, 674-679.

Mahadik SP, Evans D, Lal H. Oxidative stress and role of antioxidant and omega-3 essential fatty acid supplementation in schizophrenia. *Prog Neuropsychopharmacol Biol Psychiatry*, 2001; 25, 463-493.

Mavelli I, Mondovi B, Federico R, Rotilio G. Superoxide dismutase activity in developing rat brain. *J Neurochem*. 1978; 31(1):363-4.

McCreadie RG, MacDonald E, Wiles D, Campbell G, Paterson JR. The Nithsdale Schizophrenia Surveys. XIV: Plasma lipid peroxide and serum vitamin E levels in patients with and without tardive dyskinesia, and in normal subjects. *Brit J Psychiatry,* 1995; 167, 610-617.

Mellor JE, Laugharne JDE, Peet M. Omega-3 fatty acid supplementation in schizophrenia patients. *Hum Psychopharmacol* 1996; 11: 39-46

Michel TM, Thome J, Martin D, Nara K, Zwerina S, Tatschner T, Weijers HG, Koutsilieri E. Cu, Zn- and Mn-superoxide dismutase levels in brains of patients with schizophrenic psychosis. *J Neural Trans*, 2004; 111, 1191-1201.

Mori A, Packer L. Disorders of the oxidative stress defense system in senescence-accelerated mice (SAM0, and prolongation of life span by supplementation with antioxidants and free radical spin traps. In: Packer L, Ong A, editors. Biological oxidants and antioixidants: molecular mechanisms and health effects. *Champaign Illinois*. 1998; 37: 327-34

Morrow JD, Frei B, Longmire AW, Gaziano JM, Lynch SN, Shyr Y, Strauss WE, Oates JA, Roberts LJ. Increase in circulating products of lipid peroxidation (F2-isoprostanes) in smokers: smoking as a cause of oxidative damage. *New Engl J Med*, 1995; 332, 1198-1203.

Mukherjee S, Mahadik SP, Horrobin DF, Jenkins K, Correnti E, Scheffer R. Membrane fatty acid composition of fibroblasts from schizophrenic patients. *Biol Psychiatry*, 1994; 35, 700.

Noh JS, Kang HJ, Kim EY, Sohn S, Chung YK, Kim SU, Gwag BJ. Haloperidol-induced neuronal apoptosis: role of p38 and c-Jun-NH(2)-terminal protein kinase. *J Neurochem*, 2000; 75, 2327-2334.

Ohtsuka Y, Yabunaka N, Fujisawa H, Watanabe I, Agishi Y. Effect of thermal stress on glutathione metabolism in human erythrocytes. *Euro J Applied Physiol Occupational Physiol* 1994; 68: 87-91

Olincy, Young A, Freedman. Increased levels of the nicotine metabolite continine in schizophrenic smokers compared to other smokers. Biol Psychiatry. 1997; 42: 1-5

Pall HS, Williams AC, Blake DR, Lunec J. Evidence of enhanced lipid peroxidation in the cerebrospinal fluid of patients taking phenothiazines. *Lancet*, 1987; 2, 596–599.

Parfitt VJ, Rubba P, Bolton C, Marotta G, Hartog M, Mancini M. A comparison of antioxidant status and free radical peroxidation on plasma lipoproteins in healthy young persons from Naples and Bristol. *Eur Heart J* 1994; 15: 871-6

Parikh V, Khan MM, Mahadik SP. Differential effects of antipsychotics on expression of antioxidant enzymes and membrane lipid peroxidation in rat brain. *J Psychiat Res*, 2003; 37, 43-51.

Pillai A, Parikh V, Terry AV Jr, Mahadik SP. Long-term antipsychotic treatments and crossover studies in rats: Differential effects of typical and atypical agents on the expression of antioxidant enzymes and membrane lipid peroxidation in rat brain. *J Psychiatr Res*. 2006 Mar 23.

Srivastava N, Barthwal MK, Dalal PK, Agarwal AK, Nag D, Srimal RC, Seth PK,Dikshit M. Nitrite content and antioxidant enzyme levels in the blood of schizophrenia patients. *Psychopharmacology (Berl)*. 2001;158(2):140-5.

Pearce BD. Schizophrenia and viral infection during neurodevelopment: a focus on mechanisms. *Mol Psychiatry*. 2001; 6(6):634-46.

Peet M. New Strategies for the treatment of schizophrenia: ω-3 polyunsaturated fatty acids. In: Peet M, Glen I, Horrobin D, editors. Phospholipid spectrum disorder in psychiatry. Lanchashire UK: Marius Press. 1999; 18: 189-92

Peet M, Laugharne J, Rangarajan N, Reynolds GP. Tardive dyskinesia, lipid peroxidation, and sustained amelioration with vitamin E treatment. *Int Clin Psychopharmacol*, 1993; 8, 151-153.

Mahadik SP, Mukherjee S, Correnti EE, Kelkar HS, Wakade CG, Costa RM, Scheffer R. Plasma membrane phospholipid and cholesterol distribution of skin fibroblasts from drug-naive patients at the onset of psychosis. *Schizophr Res*. 1994; 13(3): 239-47.

Puri BK, Richardson AJ, Horrobin DF, Easton T, Saeed N, Oatridge A, Hajnal JV, Bydder GM. Eicosapentaenoic acid treatment in schizophrenia associated with symptom remission, normalisation of blood fatty acids, reduced neuronal membrane phospholipid turnover and structural brain changes. *Intl J Clin Pract.* 2000; 54: 57-63

Rebec GV. Ascorbate. An antioxidant neuroprotectant and extracellular neuromodulator. In: Connor JR. editor. *Metals and Oxidative Damage in Neurological Disorders.* New York: Plenum Press, 1997: 149-73

Reddy R, Yao JK. Schizophrenia: role of oxidative stress and essential fatty acids. In: Basu, Tk, Temple NJ, Garg ML, editors. Antioxidants in Human Health. *Cab International* 1999; 27: 351-66

Reddy RD, Yao JK, Zeigler M, et al. Essential fatty acids and outcome in first-episode schizophrenia. *Schizophr Res.* 1999; 36: 108

Rotrosen J, Wolkin A. Phospholipid And Prostaglandin Hypotheses Of schizophrenia. In H.Y. Meltzer (Ed.), Psychopharmacology: *The Third Generation of Progress* (pp. 759-764). 1987; New York: Raven Press.

Roy D, Pathak DN, Singh R. Effects of chlorpromazine on the activities of antioxidant enzymes and lipid peroxidation in the various regions of aging rat brain. *J Neurochem,* 1984; 42, 628-633.

Scarpellini F, Sbracia M, Scarpellini L. Psychological stress and lipoperoxidation in miscarriage. *Ann NY Acad Sci* 1994; 709: 210-3

Shah S, Vankar GK, Telang SD, et al. Eicosapentaenoic acid (EPA) as an adjunct in the treatment of schizophrenia. *Schizophr Res* 1998; 20: 158

Shriqui CL, Bradwejn J, Annable L, Jones BD. Vitamin E in the treatment of tardive dyskinesia: a double-blind placebo-controlled study. *Am J Psychiatry* 1992; 149: 391-93

Sies HE. Oxidative Stress: Oxidants and Antioxidants. 1991; Academic Press, London.

Sohal RS, Weindruch R. Oxidative stress, caloric restriction, and aging. *Science.* 1996; 273(5271):59-63.

Srivastava N, Barthwal MK, Dalal PK, Agarwal AK, Nag D, Srimal RC, Seth PK, Dikshit M. *Nitrite content and antioxidant enzyme levels in the blood of schizophrenia*

Stegmayr B, Johansson I, Huhtasaari F, Moser U, Asplund K. Use of smokeless tobacco and cigarettes – Effects on plasma levels of antioxidant vitamins. *Int Vit Nutr Res* 1993; 63: 195-200

Suboticanec K, Folnegovic V, Korbar M, Mestrovic B, Buzina R. Vitamin C status in chronic schizophrenia. *Biol Psychiatry,* 1990; 28, 959-966.

Taneli F, Pirildar S, Akdeniz F, Uyanik BS, Ari Z. Serum nitric oxide metabolite levels and the effect of antipsychotic therapy in schizophrenia. *Arch Med Res,* 2004; 35, 401-405.

Thome J, Foley P, Riederer P. Neurotrophic factors and the maldevelopmental hypothesis of schizophrenic psychoses. Review article. *J Neural Transm.* 1998;105(1):85-100.

Tsai G, Goff DC, Chang RW, Flood J, Baer L, Coyle JT. Markers of glutaminergic neurotransmission and oxidative stress associated with Tardive dyskinesia. *Am J Psychiatry,* 1998; 155, 1207-1213.

Vanella A, Geremia E, D'Urso G, Tiriolo P, Di Silvestro I, Grimaldi R, Pinturo R. Superoxide dismutase activity in aging rat brain. *Gerontology* 1982; 228: 108-13

Whatley SA, Curti D, Das Gupta F, Ferrier IN, Jones S, Taylor C, Marchbanks RM. Superoxide, neuroleptics and the ubiquinone and cytochrome b5 reductases in brain and lymphocytes from normals and schizophrenic patients. *Mol Psychiatry*. 1998; 3(3): 227-37.

Wood KA, Youle RJ. Apoptosis and free radicals. Ann NY Acad Sci, 1994; 738, 400-407.

Yao JK, Reddy R, McElhinny LG, van Kammen DP. Effect of haloperidol on antioxidant defense system enzymes in schizophrenia. *J Psychiat Res* 1998b; 32: 385-391.

Yao JK, Reddy R, van Kammen DP. Abnormal age-related changes of plasma antioxidant proteins in schizophrenia. *Psychiatry Res*. 2000a; 97(2-3):137-51.

Yao JK, Reddy R, van Kammen DP. Reduced level of plasma antioxidant uric acid in schizophrenia. *Psychiatry Res*. 1998a; 80(1):29-39.

Yao JK, Reddy R, van Kammen DP. Human plasma glutathione peroxidase and symptom severity in schizophrenia. *Biol Psychiatry* 1999; 45: 1512-15

Yao JK, van Kammen DP, Welker JA. Red cell membrane dynamics in schizophrenia. II. Fatty acid composition. *Schizophr Res*, 1994; 13, 217-226.

Yao JK, Leonard S, Reddy RD. Membrane phospholipid abnormalities in postmortem brains from schizophrenic patients. *Schizophr Res*, 2000b; 42, 7-17.

Yao JK, Leonard S, Reddy RD. Increased nitric oxide radicals in postmortem brain from patients with schizophrenia. *Schizophr Bull*, 2004; 30, 923-934.

Yao JK, Reddy RD, van Kammen DP. Oxidative damage and schizophrenia: an overview of the evidence and its therapeutic implications. *CNS Drugs*, 2001; 15, 287-310.

Zubenko GS, Cohen BM. A cell membrane correlate of tardive dyskinesia in patients treated with phenothiazines. *Psychopharmacology* 1986; 88: 230-36

In: Fatty Acids and Oxidative Stress...
Editors: R. Reddy and J.K. Yao, pp. 43-54
ISBN: 978-1-60021-677-0

Chapter IV

Antipsychotic Agents and Oxidative Stress

Sahebarao Mahadik, Anilkumar Pillai and Peter Buckley

Introduction

First generation and second generation antipsychotic medications (FGAs and SGAs respectively) form the backbone of disease management for schizophrenia. The decision to choose which type of antipsychotic medication - and which drug to use within either class - is complex. At the present time, the evidence for selecting one SGA during treatment over another is incomplete and there is still a dearth of comparative studies between these agents. Additionally, we are early on the learning curve of understanding the adverse effect profile of SGAs with SGAs (Casey et al, 2004). There is growing concern about weight gain (Allison et al, 1999) and metabolic disturbances. Therefore, it is likely that practice patterns and the information accrued to determine such changes will evolve further and at a rapid pace over the next 5 years.

The use of SGAs has now become the standard of care (Lehman et al, 2003; Miller et al, 2004) due to occurrence of fewer extrapyramidal side effects. The therapeutic effect of haloperidol has been considered due to its primary action on dopamine receptors. However SGAs have very complex receptor profiles of their action (Figure 1). These drugs may exert their antipsychotic effects by a variety of mechanisms including, preferential high affinity binding to 5-hydroxytryptamine $(5\text{-HT})_{2A}$-serotonin and D_4-dopamine receptors and low affinity binding and fast dissociation from D_2-dopamine receptors (Kapur et al, 2003). Partial agonist effects on dopamine and serotonin receptors is another new mechanism of action (Davies et al, 2004) and this adds to the complexity. This has been the basis for the drug development, since such strategy is targeted to a primary defect in receptor function. However, there is general agreement that the pathophysiology of schizophrenia is complex (well beyond a single neurotransmitter system) and likely involves abnormal neurodevelopment - which in turn may alter multi-neurotransmitter function/response by more complex mechanisms such as altered neural circuitry and number of neuronal connections. Moreover, a poorly developed brain may also be more susceptible to environmental stresses and, if untreated these effects may trigger progressive deterioration (Lieberman et al, 2001).

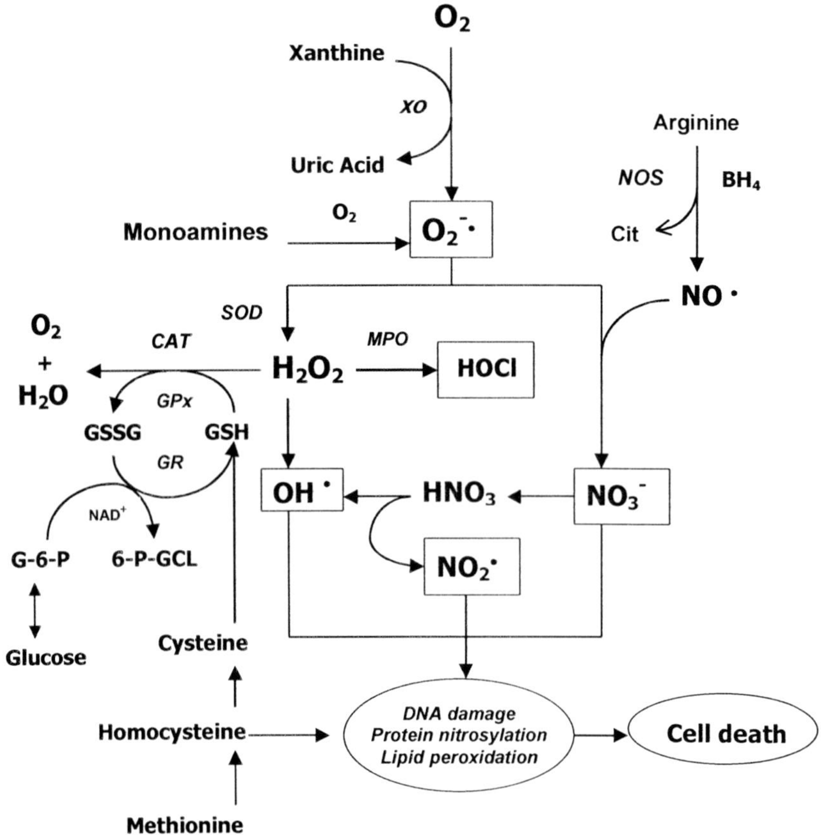

Figure 1.

As a consequence, pathophysiological and the search for effective treatment strategies has broadened and become more complex. The role of phospholipid membrane pathology represents one change in the paradigm for evaluating the pathophysiology of schizophrenia and its treatment (Mukherjee and Mahadik, 1994). There is mounting evidence to support the membrane pathology (structural and functional disorganization), primarily due to altered membrane phospholipid metabolism pre-dates the onset of schizophrenia (Horrobin et al., 1994; Horrobin, 1998; Mahadik and Evans, 2003). Furthermore, altered phospholipid metabolism seems to be a result of reduced synthesis as well as increased peroxidative breakdown of essential polyunsaturated fatty acids (EPUFAs), primarily arachidonic acid (AA) and docosahexaenoic acid (DHA). EPUFAs are excessively used for signal transduction and are selectively susceptible for peroxidation under oxidative stress, which is increased in schizophrenia (Mahadik et al., 1996a; Reddy and Yao, 1996; Mahadik et al., 1999a). Tone key interpretation of these findings is that antipsychotic treatments would be likely then not to impair the phospholipid metabolism but rather these medications should trigger biological processes toward correction of the pre-existing phospholipid changes in the brain. This is a provocative hypothesis. The purpose of this chapter is to review the effects of antipsychotics on the phospholipid metabolism in animals and human, and present an experimental strategy aimed at correcting phospholipid metabolism and consequently improving outcome in schizophrenia.

It is proposed that antipsychotics have differential effects on the membrane phospholipid metabolism, as a consequence of their differences neurotransmitter receptor profile. Antipsychotics may change the phospholipid breakdown by peroxidation due to oxidative stress and/or by phospholipases, or may change their synthesis by inhibiting synthetic enzymes or fatty acid transport proteins such as apolipoprotein D (ApoD), an arachidonic acid binding protein. Moreover, antipsychotics may also change the levels of growth factors that may then change the expression of genes involved in phospholipid breakdown and synthesis. The evidence for these claims is reviewed below.

Differential Effects of Antipsychotics on Phospolipid Metabolism in Brain

Antipsychotics, Oxidative Stress and Phospholipid Peroxidation

Cellular oxidative stress is defined as the higher levels of free radicals (also called as reactive oxygen species, ROS) than the antioxidant capacity of the cell. Oxidative stress generally leads to ROS-mediated peroxidation of membrane lipids, particularly EPUFAs and DNA, and inactivation of a large number of vital membrane transport proteins, receptors and enzymes leading to decreased potential of cellular growth, function and survival (Mahadik et al., 1999a). An elaborate discussion of the oxidative stress and peroxidative cell membrane injury in schizophrenia is provided in another Chapter in this book. Antipsychotics have been implicated in causing an increase in oxidative stress and thereby the development of tardive dyskinesia (TD) after chronic use of these medications (Cadet and Lohr, 1987). In addition, oxidative stress and oxidative injury are also considered part of the pathogenesis of schizophrenia (Mahadik and Mukherjee, 1996; Reddy and Yao, 1996; Mahadik et al., 1999a). However, since antipsychotics differ in their relative antioxidant or pro-oxidant effects (Jeding et al., 1995), the choice of antipsychotic may be very important.

Antipsychotics can increase the oxidative stress by several mechanisms. Directly, antipsychotics may be metabolized in the brain to form a free radical such as in case of haloperidol (Figure 2). Indirectly, these may either alter the expression of antioxidant enzymes (superoxide dismutase, SOD; glutathione peroxidase, GSHPOD; catalase, CAT) or increase the caloric intake and reduce the caloric expenditure, together can increase the oxidative stress and oxidative cell injury. Haloperidol administration to rats has been reported to decrease the levels of reduced glutathione and increase lipid peroxidation in the brain (Shivakumar and Ravindranath, 1993). Haloperidol and fluphenazine have also been found to alter the levels of antioxidant enzymes (Murthy et al., 1989; Cadet and Perumal, 1990). We have also found that chronic treatment with haloperidol for both 45 as well as 90 days reduced the expression of key antioxidant defense enzymes and increased the lipid peroxidation in the rat brain (Parikh et al., 2003a). Interestingly, chronic treatment up to 90 days with either SGA (risperidone, olanzapine, or clozapine) did not alter the levels of antioxidant enzymes and did not increase the lipid peroxides or affect the membrane EPUFAs. In cultured PC12 cells, olanzapine has been found to induce the expression of SOD

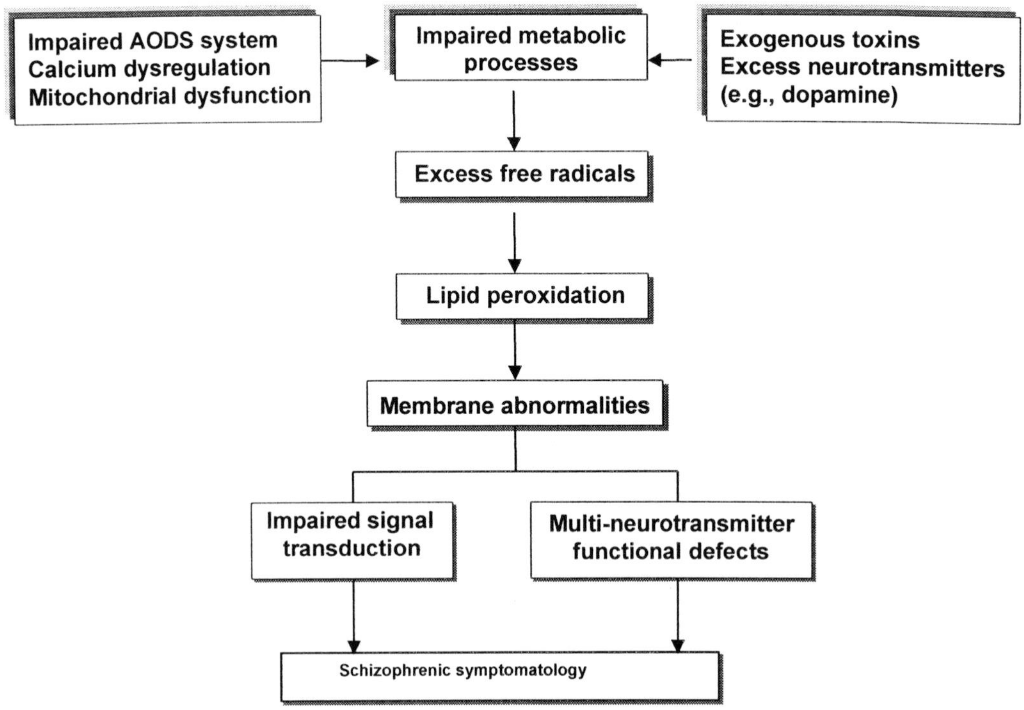

Figure 2.

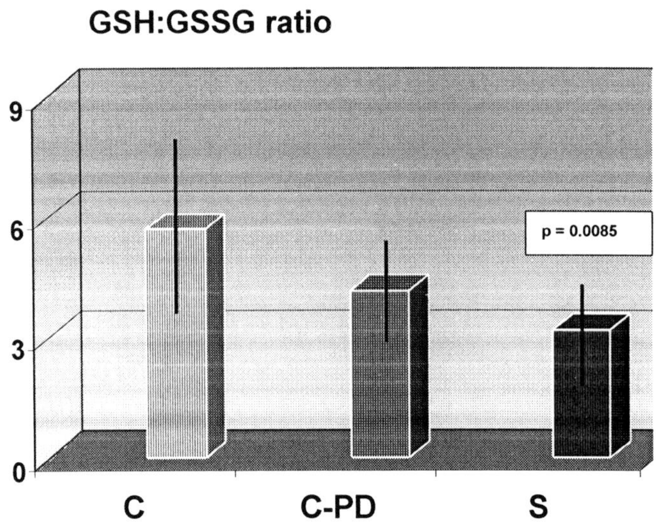

Figure 3.

(Li et al., 1999). These findings further suggest that altered levels of antioxidant enzymes reported for the last over 40 years in schizophrenic patients, may be at least partly a consequence of treatment with FGAs (Mukherjee et al., 1996). Thus, the differential effects of antipsychotics on lipid peroxidation may affect the membrane EPUFA status and thereby clinical outcome of schizophrenia (see later). SGAs are now the standard of care for treating schizophrenia (Lehman et al, 2004; Jones and Buckley, 2005). However, these antipsychotics have been less extensively studied with respect to their impact on membrane functions and oxidative cellular injury.

Antipsychotic Effects on Phospholipases

Phospholipids are degraded by several phospholipases, which are regulated by pathophysiological signals (Figure 4). Role of phospholipases, particularly PLA2, which specifically hydrolyzes the EPUFAs, has been implicated in the altered phospholipid metabolism in schizophrenia (Gattaz et al., 1990; Horrobin, 1998). Antipsychotics may also further differentially affect the levels of phospholipases. Treatment with haloperidol, chlorpromazine or trifluoperazine has been found to decrease and treatment with sulpiride, fluphenazine or thioridazine has been found to decrease the PLA2 in the rat brain (Trzeciak et al., 1995). The inhibition of PLA2 by haloperidol is reported to be associated with the decreased AA release in rat brain (Myers et al., 2001). Also, chlorpromazine but not the clozapine has been found to inhibit the phospholipase C in rat brain (Dwivedi and Pandey, 1999). Inhibition of phospholipases and thereby protection of membrane EPUFAs by some antipsychotics has been suggested as one their mechanisms of antipsychotic action for these agents (Gattaz et al., 1990). SGAs have also been found to increase the membrane EPUFA levels in schizophrenia patients. However, their effects on phospholipases have not yet been investigated.

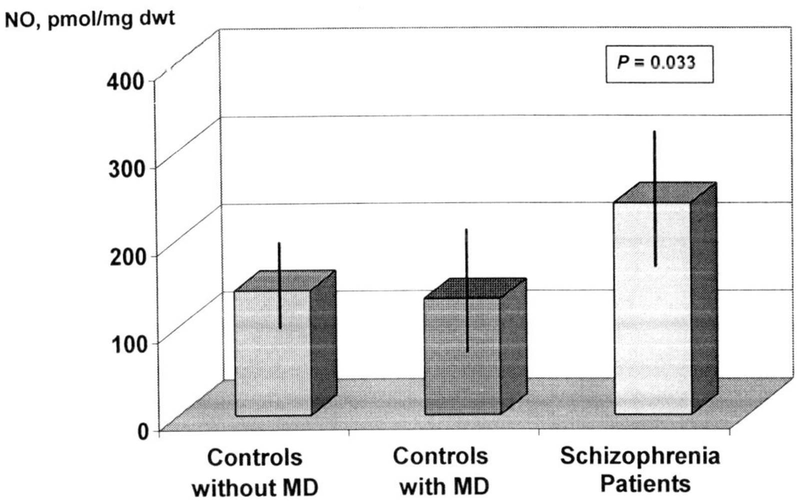

Figure 4.

Antipsychotic Effects on Neurotrophic Factors

A direct role of neurotrophic factors in phospholipid metabolism has not yet been thoroughly investigated. However, since these factors regulate growth, survival and plasticity of neural cells - which involves the synthesis of large quantities of membrane phospholipids - they may have a critical role. A total of six neurotrophic factors, which include nerve growth factors (NGF), brain-derived neurotrophic factors (BDNF), neurotropin-3 (NT-3), neurotropin-4 (NT-4) and neurotropin-5 (NT-5) have been found to play roles in specific neuronal (cholinergic, dopaminergic, GABAaminergic, etc) and glial growth and survival (Barde, 1989; Levi-Montalccini, 1987; Rylett and Williams, 1994). Among these, NGF and BDNF are the most extensively studied and best-characterized neurotrophic factors considered in the pathophysiology of schizophrenia. Our group has reported that altered levels of neurotrophic factors may be associated with their mechanisms of antipsychotic actions of antipsychotics (Parikh et al., 2004a; 2004b).

Loss of choline-acetyltransferase (ChAT) activity has been reported in rat (Mahadik et al., 1988; Mahadik and Mukherjee, 1995) as well as in post-mortem brains from schizophrenic patients (Miller and Chouinard, 1993). Moreover, loss of ChAT was associated with impaired cognitive performance in rats. This also may be relevant to the cognitive deficits in schizophrenia. Chronic haloperidol treatment has been found to reduce the NGF levels and ChAT activity in the rat brain (Angelucci et al., 2000). We have also investigated the effects of up to 90 days of treatment of rats with clozapine, olanzapine, risperidone, or haloperidol on the levels of NGF and BDNF in the brain (Parikh et al., 2004a; 2004b). Haloperidol treatment resulted in a significant reduction in the level of NGF as well as BDNF in the brain. Treatment with SGAs had no appreciable effect or caused a slight increase in these trophic factors. These data provide preliminary support for the notion that these SGAs, but not FGAs, are neuroprotective (Parikh et al., 2004a; 2004b).

Antipsychotics and Neuroplasticity: Sprouting, Synaptogenesis and Neurogenesis

Normal phospholipid metabolism is critical for neuroplasticity, which constitutes re-growth of neuropils, synaptogenesis, and even neurogenesis in adult brain. Excessive loss of neuropils and synapses, "excessive pruning" has been suggested in schizophrenia (Keshavan et al. 1994a) and now there is some evidence to support this (Goldman-Rakic and Selemon 1997; Shenton et al., 2001). In addition, the morphometric changes in brain with antipsychotics suggest that these medications may have differential effects on plasticity that may be related to their antipsychotic effects (Keshavan et al. 1994a, 1994b; Shenton et al. 2001). In this context, it is important to investigate the effects of antipsychotics on the neuroplasticity in adult brain. Following upon evidence that neurogenesis occurs in the adult brain under a variety of situations (Malberg et al. 2000), attention has focused in psychiatric research now to investigate the effects of antipsychotics on neurogenesis. We have found preliminary evidence for neurogenesis, particularly in the subventricular zone and hippocampus in adult rats treated with SGAs but not with haloperidol (Wakade et al., 2002). Some of the newly divided cells migrated in the direction of their historical path and did differentiate into both neurons and glia. Dr. Weinberger's group at the National Institute of

Mental health have also found a similar but not sustained effect with low dose (but not higher doses) of clozapine when they examined the effects of clozapine and haloperidol on the expression of new cells in the dentate gyrus region of the hippocampus. Kodama and colleagues (2004) at Yale have reported that both olanzapine and fluoxetine increase cell growth. However, the combination of fluoxetine-olanzapine did not confer any additional impact on neurogenesis. Wang and colleagues (2004) at Vanderbilt University found that olanzapine but not haloperidol enhanced the production of new cells in the prefrontal cortex but not in the nucleus accumbens. Again, the effect was modest and not sustained/enhanced by continued treatment. So while the findings to date are mixed, they still represent a replication of effect from 4 independent laboratories. This raises a whole host of questions - how does this occur? Are these cells able to become functional? Is this effect sustainable (most studies have examined only short-term exposure)? Can all SGAs do this? Does one SGA cause more neurogenesis than another? What, if any, are the clinical implications of all this? For example, we appreciate that SGAs can have subtle effects in enhancing cognition performance in patients with schizophrenia (Harvey et al, 2004) - is this the result of an antipsychotic-enhanced neurogenesis? It is suggested that SGAs may have neuroprotective effects through enhanced neuroplasticity (Parikh et al., 2004a; 2004b). The key processes for neuroprotection and to subsequently trigger neuroplasticity are; control of oxidative stress, induction of neurotrophic factors, and essential nutrients, such as EPUFAs (Mahadik, 1992). If antipsychotic can induce the first two changes, this may trigger a cascade of changes in neuroprotective processes.

Oxidative Stress, Antipsychotics and RBC EPUFAs

Lower plasma levels of lipid peroxides and higher levels of RBC membrane EPUFAs have been reported in patients treated with SGAs (Khan et al 2002; Arvindakshan et al., 2003a). Treatment with SGAs may reduce oxidative stress and oxidative injury and thereby contributed to increased RBC membrane EPUFAs (Mahadik and Evans, 2003). One relevant observation is that studies undertaken in patients from India did not show significant reductions in RBC membrane AA (Arvindakshan et al., 2003a). On the other hand, studies in patients from the USA did not show any reduction skin fibroblast AA levels (Mahadik et al., 1996a). Since AA is predominantly susceptible for peroxidation, this may be indirect evidence that changes in AA levels are more related to the oxidative stress. Clearly, such findings need to be replicated in patient populations of diverse ethnic background.

Phospholipases, Antipsychotics and RBC EPUFAs

Increased PLA2 has been considered as a mechanism of reduced EPUFAs in schizophrenia (Horrobin, 1998). However, inconsistent findings have been reported on the levels of PLA2 activities and its genetic polymorphism in schizophrenia (Gattaz et al., 1990; Albers et al., 1993; Katila et al., 1997; Ross et al., 1999; Choudhari et al., 2001). Also, as discussed earlier, the reduction of PLA2 after treatment with antipsychotics which was reported in rat brain has not been replicated in patients with schizophrenia (Volz et al., 1999). It is likely then that increased PLA2 may be a result of increased lipid peroxides. This is consistent with an earlier observation of a positive correlation between these measures and with the treatment PLA2 may act as a repair enzyme for membrane EPUFA peroxidation (Scheffer et al., 1999). Therefore, it will be helpful to study the differential effects of SGAs on PLA2 and to examine how these drugs influence on the changes with levels of plasma lipid peroxides and RBC membrane EPUFAs.

Neurotrophic Factors, Antipsychotics and RBC EPUFAs

There is increasing focus on the role for neurotrophic factors in schizophrenia (Weickert et al, 2003; Parikh et al., 2003b), reduced plasma NGF levels have been reported in patients, who were free of neuroleptics for short duration (Bersani et al., 1999) as well as treated with haloperidol (Aloe et al., 1997). We have also reported reduced plasma NGF levels in drug-naïve first episode psychotic patients (FEP) compared to control subjects, where as a medicated group of patients with chronic schizophrenia had significantly higher levels of NGF than the FEP group (Parikh et al., 2003b). Patients were treated mainly with CLZ for 6 months to three years. These effects of SGAs on NGF levels may be advantageous.

Therapeutic Implications

Efforts to improve understanding of the mechanism of action of antipsychotics should draw upon current concepts in the pathophysiology of schizophrenia. One such approach is to consider the impact of abnormal neurodevelopment upon phospholipid and altered membrane EPUFA metabolism in the brain. In such a schema, it is proposed that newer drug treatments for schizophrenia (in addition to their interactions with the neurotransmitter receptors) may be restorative of phospholipid metabolism. In order to achieve that, these drugs should be able to control the oxidative stress as well as to trigger the expression of growth factor and promote their actions through membrane phospholipid-mediated signal transduction. In addition, there is early information to suggest that dietary supplementation with agents that can independently control the oxidative stress, correct the EPUFA metabolism and induce neurotrophic factors may augment the effects of antipsychotic therapy. A few studies have already reported the augmentation of antipsychotic efficacy of antipsychotics with supplementation of EPUFAs such as EPA and DHA (Peet and Horrobin, 2002) as well as with a combination of antioxidants (Arvindakshan et al., 2003b). Supplementation may work synergistically on mechanisms to trigger neuroplasticity. It will be important to test such hypotheses in rigorous scientific studies.

References

Albers M, Meurer H, Marki F, Klotz J. Phospholipase A2 activity in serum of neuroleptic-naive psychiatric inpatients. *Pharmacopsychiatry.* 1993;26:94-98.

Allison DB, Mentore JL, Heo M, Chandler LP, Cappelleri JC, Infante MC, Weiden PJ. Antipsychotic-induced weight gain: a comprehensive research synthesis. *Am J Psychiatry.* 1999;156:1686-1696.

Aloe L, Iannitelli A, Bersani G, Alleva E, Angelucci F, Maselli P, Manni L. Haloperidol administration in humans lowers plasma nerve growth factor level: evidence that sedation induces opposite effects to arousal. *Neuropsychobiology.* 1997;36:65-68.

Angelucci F, Aloe L, Gruber SH, Fiore M, Mathe AA. Chronic antipsychotic treatment selectively alters nerve growth factor and neuropeptide Y immunoreactivity and the distribution of choline acetyl transferase in rat brain regions. *Int J Neuropsychopharmacol.* 2000;3:13-25.

Arvindakshan M, Sitasawad S, Debsikdar V, Ghate M, Horrobin D, Bennett C, Ranjekar PK, Mahadik SP. EPUFA and lipid peroxide levels in never-medicated and medicated schizophrenics. *Biol Psychiatry.* 2003a;53:56-64.

Arvindakshan M, Ghate M, Ranjekar PK, Evans DR, Mahadik SP. Supplementation with a combination of ω-3 fatty acids and antioxidants (vitamins E and C) improves the outcome of schizophrenia. *Schizophr Res.* 2003b; 62:195-204.

Barde YA. Trophic factors and neuronal survival. *Neuron.* 1989;2:1525-1534.

Bersani G, Iannitelli A, Maselli P, Pancheri P, Aloe L, Angelucci F, Alleva E. Low nerve growth factor plasma levels in schizophrenic patients: a preliminary study. *Schizophr Res.* 1999;37:201-203.

Cadet JL, Perumal AS. Chronic treatment with prolixine causes oxidative stress in rat brain. *Biol Psychiatry.* 1990;28:738-740.

Cadet JL, Lohr JB. Free radicals and the developmental pathophysiology of schizophrenic burnout. *Integr Psychiatry.* 1987;5:40-48.

Casey DE, Haupt DW, Newcomer JW, Henderson DC, Sernyak MJ, et al. Antipsychotic-Induced Weight Gain and Metabolic Abnormalities: Implications for Increased Mortality in Patients with Schizophrenia. *J Clin Psych.* 2004;65(suppl 7):4-18.

Choudhari KV, Brandstaetter B, Semwal P et al. Association studies of cytosolic phospholipase A2 polymorphisms and schizophrenia among two independent family-based samples. *Psychiatric Gen.* 2001;11:207-212.

Davies HM, Hopper DW, Hansen T, Liu Q, Childers SR. Synthesis of methylphenidate analogues and their binding affinities at dopamine and serotonin transport sites. *Bioorg Med Chem Lett.* 2004;14:1799-1802.

Dwivedi Y, Pandey GN. Effects of treatment with haloperidol, chlorpromazine, and clozapine on protein kinase C (PKC) and phosphoinositide-specific phospholipase C (PI-PLC) activity and on mRNA and protein expression of PKC and PLC isozymes in rat brain. *J Pharmacol Exp Ther.* 1999;291:688-704.

Gattaz WF, Hubner CV, Nevalainen TJ, Thuren T, Kinnunen PK. Increased serum phospholipase A2 activity in schizophrenia: a replication study. *Biol Psychiatry.* 1990;28:495-501.

Goldman-Rakic PS, Selemon LD. Functional and anatomical aspects of prefrontal pathology in schizophrenia. *Schizophr Bull.* 1997;23:437-458.

Horrobin DF, Glen A I M, Vaddadi KS. The membrane hypothesis of schizophrenia. *Schizophr Res.* 1994;13:195-208.

Horrobin DF. The membrane phospholipid hypothesis as a biochemical basis for the neurodevelopmental concept of schizophrenia. *Schizophr Res.* 1998;30:193-208.

Harvey PD, Meltzer H, Simpson GM, Potkin SG, Loebel A, Siu C, Romano SJ. Improvement in cognitive function following a switch to ziprasidone from conventional antipsychotics, olanzapine, or risperidone in outpatients with schizophrenia. *Schizophr Res.* 2004;66: 101-113.

Jeding I, Evans PJ, Akanmu D, Dexter D, Spencer JD, Aruoma OI, Jenner P, Halliwell B. Characterization of the potential antioxidant and pro-oxidant actions of some neuroleptic drugs. *Biochem Pharmacol.* 1995;49:359-365.

Jones P, Buckley P. In Clinical Practice - Schizophrenia. *Elsevier Publishers, London, England.* 2005 (in press).

Kapur S, VanderSpek SC, Brownlee BA, Nobrega JN. Antipsychotic dosing in preclinical models is often unrepresentative of the clinical condition: a suggested solution based on in vivo occupancy. *J Pharmacol Exp Ther.* 2003;305:625-631.

Katila H, Appelberg B, Rimon R. No differences in phospholipase-A2 activity between acute psychiatric patients and controls. *Schizophr Res.* 1997;26:103-105.

Keshavan MS, Anderson S, Pettegrew JW. Is schizophrenia due to excessive synaptic pruning in the prefrontal cortex? The Feinberg hypothesis revisited. *J Psychiatr Res.* 1994a;28:239-65.

Keshavan MS, Bagwell WW, Haas GL, Sweeney JA, Schooler NR, Pettegrew JW. Changes in caudate volume with neuroleptic treatment. *Lancet.* 1994b;344:1434.

Khan MM, Evans DR, Gunna V, Scheffer RE, Parikh VV, Mahadik SP. Reduced erythrocyte membrane essential fatty acids and increased lipid peroxides in schizophrenia at the never-mediated first-episode of psychosis and after years of treatment with antipsychotics. *Schizophr Res.* 2002;58:1-10.

Kodama M, Fujioka T, Duman RS. Chronic olanzapine or fluoxetine administration increases cell proliferation in hippocampus and prefrontal cortex of adult rat. *Biol Psychiatry.* 2004;56: 570-580.

Lehman AF, Fischer EP, Postrado L, Delahanty J, Johnstone BM, Russo PA, Crown WH. The Schizophrenia Care and Assessment Program Health Questionnaire (SCAP-HQ): an instrument to assess outcomes of schizophrenia care. *Schizophr Bull.* 2003;29:247-256.

Levi-Montalcini R. The nerve growth factor 35 years later. *Science.* 1987;237:1154-1162.

Li XM, Chlan-Fourney J, Juorio AV, Bennett VL, Shrikhande S, Keegan DL, Qi J, Boulton AA. Differential effects of olanzapine on the gene expression of superoxide dismutase and the low affinity nerve growth factor receptor. *J Neurosci Res.* 1999;56:72-75.

Lieberman JA, Perkins D, Belger A, Chakos M, Jarskog F, Boteva K, Gilmore J. The early stages of schizophrenia: speculations on pathogenesis, pathophysiology, and therapeutic approaches. *Biol Psychiatry.* 2001;50:884-897.

Mahadik SP. Gangliosides: New generation of neuroprotective agents. In: Marangos PJ, Lal H. *Emerging Strategies in neuroprotection.* Birkhauser, Boston, USA; 1992:187-223.

Mahadik SP, Evans DR. Is schizophrenia a metabolic brain disorder? Membrane phospholipid dysregulation and its therapeutic implications. *Psychiatr Clin North Am.* 2003;26:85-102.

Mahadik SP, Laev H, Korenovsky A, Karpiak SE. Haloperidol alters rat CNS cholinergic system: enzymatic and morphological analyses. *Biol Psychiatry.* 1988;24:199-217

Mahadik SP, Mukherjee S. Free Radical pathology and antioxidant defense in schizophrenia: a review. *Schizophr Res.* 1996;19:1-18.

Mahadik SP, Mukherjee S, Horrobin D, Jenkins K, Correnti EE, Scheffer RE. Plasma membrane phospholipid fatty acid composition of cultured skin fibroblasts from schizophrenic patients: comparison with bipolar and normal controls. *Psychiatry Res.* 1996a;63:133-142.

Mahadik SP, Mukherjee S. Monosialoganglioside cotreatment prevents haloperidol treatment-associated loss of cholinergic enzymes in rat brain. *Biol Psychiatry.* 1995;38:246-254.

Mahadik SP, Sitasawad V, Mulchandani M. Membrane peroxidation and the neuropathology of Schizophrenia. In: Peet M, Glen AL, Horrobin DF eds. *Phospholipid Spectrum Disorders in Psychiatry.* Carnforth: Marius Pres.; 1999a:99-111.

Malberg JE, Eisch AJ, Nestler EJ, Duman RS. Chronic antidepressant treatment increases neurogenesis in adult rat hippocampus. *J Neurosci.* 2000;20:9104-9110.

Miller AL, Crismon ML, Rush AJ, Chiles J, Kashner TM, Toprac M, Carmody T, Biggs M, Shores-Wilson K, Chiles J, Witte B, Bow-Thomas C, Velligan DI, Trivedi M, Suppes T, Shon S. The Texas medication algorithm project: clinical results for schizophrenia. *Schizophr Bull.* 2004;30:627-647.

Mukherjee S, Mahadik SP. A new paradigm for schizophrenia? *Schizophr Res.* 1994;13: 191-194.

Mukerjee S, Mahadik SP, Scheffer R, Correnti EE, Kelkar H. Impaired antioxidant defense at the onset of psychosis. *Schizophr Res.* 1996;19:19-26.

Miller R, Chouinard G. Loss of striatal cholinergic neurons as a basis for tardive and L-dopa-induced dyskinesias, neuroleptic-induced supersensitivity psychosis and refractory schizophrenia. *Biol Psychiatry.* 1993;34:713-738.

Murthy JN, Laev H, Karpiak S, Mahadik SP. Enzymes of oxyradical metabolism after haloperidol treatment of rat. *Soc Neurosci (Abstr).* 1989;15:139.

Myers CS, Contreras MA, Chang MC, Rapoport SI, Appel NM. Haloperidol downregulates phospholipase A2 signaling in rat basal ganglia circuits. *Brain Res.* 2001;896:96-101.

Parikh V, Khan MM, Mahadik SP. Differential effects of antipsychotics on expression of antioxidant enzymes and membrane lipid peroxidation in rat brain. *J Psychiatr Res.* 2003a;37:47-51.

Parikh V, Evans DR, Khan MM, Mahadik SP. Nerve growth factor in never-medicated first-episode psychotic and medicated chronic schizophrenic patients: possible implications for treatment outcome. *Schizophr Res.* 2003b;60:117-123.

Parikh V, Khan MM, Mahadik SP. Olanzapine counteracts reduction of brain-derived neurotrophic factor and TrkB receptors in rat hippocampus produced by haloperidol. *Neurosci Lett.* 2004a;356:135-139.

Parikh V, Terry AV, Khan MM, Mahadik SP. Modulation of nerve growth factor and choline acetyltransferase expression in rat hippocampus after chronic exposure to haloperidol, risperidone, and olanzapine. *Psychopharmacology (Berl).* 2004b;172:365-374.

Peet M, Horrobin DF in association with the E-E Multicenter Study Group. A dose-ranging exploratory study of the effects of ethyl-eicosapentaenoate in patients with persistent schizophrenic symptoms. *J Psychiatr Res.* 2002;36:7-18.

Reddy R, Yao JK. Free radical pathology in schizophrenia: a review. *Prostaglandins Leukotrienes and EFAs.* 1996;55:33-43.

Ross BM, Turenne S, Moszczynska A, Warsh JJ, Kish SJ. Differential alteration of phospholipase A2 activities in brain of patients with schizophrenia. *Brain Res.* 1999;821:407-413.

Rylett RJ, Williams LR. Role of neurotrophins in cholinergic-neurone function in the adult and aged CNS. *Trends Neurosci.* 1994;17:486-490.

Scheffer RD, Bradley J, Mahadik SP. Elevated lipid peroxidation and phospholipase A2: Possible mechanism of the lower cell membrane essential polyunsaturated fatty acid in schizophrenia. *Biol Psychiatry (Abstr).* 1999;43:101.

Shenton ME, Dickey CC, Frumin M, McCarley RW. A review of MRI findings in schizophrenia. *Schizophr Res.* 2001;49:1-52.

Shivakumar BR, Ravindranath B. Oxidative stress and thiol modification induced by chronic administration of haloperidol. *J Pharmacol Exp Ther.* 1993;265:1137-1141.

Trzeciak HI, Kalacinski W, Malecki A, Kokot D. Effect of neuroleptics on phospholipase A2 activity in the brain of rats. *Eur Arch Psychiatry Clin Neurosci.* 1995;245:179-182.

Volz HP, Rossger G, Riehemann S, Hubner G, Maurer I, Wenda B, Rzanny R, Kaiser WA, Sauer H. Increase of phosphodiesters during neuroleptic treatment of schizophrenics: a longitudinal 31P-magnetic resonance spectroscopic study. *Biol Psychiatry.* 1999;45:1221-1225.

Wakade CG, Mahadik SP, Waller JL, Chiu FC. Atypical neuroleptics stimulate neurogenesis in adult rat brain. *J Neurosci Res.* 2002;69:72-79.

Wang HD, Dunnavant FD, Jarman T, Deutch AY. Effects of antipsychotic drugs on neurogenesis in the forebrain cortex of the adult rat. *Neuropsychopharmacology.* 2004;29:1230-1238.

Weickert CS, Hyde TM, Lipska BK, Herman MM, Weinberger DR, Kleinman JE. Reduced brain-derived neurotrophic factor in prefrontal cortex of patients with schizophrenia. *Mol Psychiatry.* 2003;8:592-610.

In: Fatty Acids and Oxidative Stress...
Editors: R. Reddy and J.K. Yao, pp. 55-76
ISBN: 978-1-60021-677-0

Chapter V

Oxidative Stress in Parkinson's Disease

Victor S. Van Laar, April A. Dukes, and Teresa G. Hastings

Introduction

Parkinson's disease (PD) is a neurodegenerative disorder afflicting approximately one million Americans, with a prevalence of 1% in people age 65 and older [Lang and Lozano, 1998a, 1998b]. Symptoms of the disorder typically include a range of movement disorders such as postural deficits, bradykinesia, rigidity, and resting tremor. The pathology of PD is characterized by the loss of pigmented dopaminergic neurons in the substantia nigra pars compacta (SNpc) and their projections to the caudate and putamen, as well as the formation of proteinaceous cytoplasmic inclusions called Lewy bodies [reviewed in Samii et al., 2004].

Despite years of study, the cause of the degeneration in PD is still unknown. A number of genes have been identified in connection with rare familiar forms of PD, providing insights to the potential mechanisms of the disease. However, 85-90% of PD cases are considered sporadic [Nussbaum and Polymeropoulos, 1997; Bajaj et al., 1998; Gasser, 2001], with little understanding of the causes for the neurodegeneration. One potential contributor and possible cause of the neuronal depletion suggested by many lines of evidence is oxidative stress.

Oxidative stress is typically denoted by an overabundance of highly reactive oxygen-based molecules, known as reactive oxygen species (ROS), in the cellular environment [Halliwell, 1992]. Oxidative stress has been implicated in the neurodegenerative diseases of Alzheimer's disease [reviewed in Butterfield et al., 2001; reviewed in Honda et al., 2004] and amyotrophic lateral sclerosis [reviewed in Carri et al., 2003], and has been associated with Huntington's disease [reviewed in Butterfield and Kanski., 2001]. Parkinson's disease is no exception. Various studies have implicated oxidative stress and mitochondrial dysfunction as potential mechanisms in PD neurodegeneration [Betarbet et al., 2002; reviewed Dauer and Przedborski, 2003; reviewed in Jenner, 2003]. Evidence from PD patients as well as experimental models of PD and studies of genetic mutations associated with PD have lead to an increasing body of evidence supporting oxidative stress as a contributor to PD pathogenesis. In this chapter, we discuss the evidence that links oxidative stress to PD, as well as current thoughts on potential pathways of oxidative stress in PD [Hastings and Berman, 2000; Fiskum et al., 2003; Dawson and Dawson, 2003; Greenamyre and Hastings, 2004].

Evidence for Oxidative Stress in PD

Postmortem studies have found evidence for an increased oxidative environment and increased oxidative damage in PD brain [reviewed in Jenner, 2003]. In addition, many models, such as the rotenone model [Betarbet et al., 2000], the dopamine-induced toxicity model [Rabinovic et al., 2000], the MPTP model [Przedborski and Vila, 2003], and the 6-hydroxydopamine model [reviewed in Deumens et al., 2002], lead to selective dopaminergic degeneration through toxins that increase oxidative stress. Also, there is a known systemic deficiency in mitochondrial Complex I found in association with some PD patients, which could lead to increased oxidative stress (described further below) [Blandini et al., 1998; Shoffner et al., 1991]. Therefore, the role of oxidative stress in PD has become a major research focus in order to determine the cause of and find treatment for the devastating effects of substantia nigra (SN) degeneration in PD.

Antioxidant Enzymes

Several indicators of increased oxidative stress have been observed in PD brain, including changes in antioxidant enzymes and levels of antioxidants. There is some question of whether levels of catalase and glutathione peroxidase, enzymes that reduce hydrogen peroxide, are changed, with evidence for both deficits [Ambiani et al., 1975; Kish et al., 1985] and no change [Marttila et al., 1988; Sian et al., 1994b] in enzyme levels reported in PD. However, there is consistent evidence that the activity of superoxide dismutase (SOD), the enzyme responsible for conversion of the free radical superoxide into hydrogen peroxide (H_2O_2), is increased selectively in the SN of PD patients [Marttila et al., 1988; Saggu et al., 1989]. Elevation in the manganese-dependent mitochondrial form of SOD has been observed in the cerebrospinal fluid (CSF) of PD patients, however whether the copper/zinc-dependent cytosolic form of SOD is also elevated remains to be elucidated [Yoshida et al., 1994]. The increases in the mitochondrial form of SOD observed in PD may be compensatory for increased ROS production by impaired mitochondria.

Glutathione

In addition to changes in antioxidant enzymes, the major cellular antioxidant glutathione (GSH) is also altered in PD. Although levels of other antioxidants, ascorbic acid and alpha-tocopherol, are unchanged in PD, an estimated 40% decrease in GSH has been observed in PD SN, but not other brain regions [Perry et al., 1982; Perry and Yong, 1986; Riederer et al., 1989; Dexter et al., 1992; Sian et al., 1994a]. Since reduced GSH levels occur in incidental Lewy body disease, which is thought to be a pre-parkinsonian state, decreased GSH is thought to be an early event in PD [Jenner and Olanow, 1996]. Surprisingly, the decrease in total GSH is not associated with increases in oxidized GSH (GSSG) [Sofic et al., 1992]. However, levels of most enzymes involved in GSH synthesis and GSH-dependent reactions remain unchanged, such as gamma-glutamylcysteine synthetase (the rate-limiting enzyme for

GSH synthesis), GSH reductase (the enzyme that reduces GSSG), or GSH transferase (enzymes that catalyze the conjugation of GSH to various electrophilic compounds, including quinones [reviewed in van Bladeren, 2000]) in PD [Sian et al., 1994b]. An increase in the levels of gamma-glutamyl transpeptidase, an enzyme associated with the translocation and degradation of GSH, was observed in PD SN [Sian et al., 1994b], which may be a compensatory mechanism to import GSH precursor molecules or to breakdown excess GSSG [reviewed in Jenner and Olanow, 1996]. It has also been suggested that reduced GSH levels precede increased iron levels observed in PD, and thus may contribute to increased iron oxidation [Jenner and Olanow, 1996].

Iron

Reduced transition metals, iron (II) in particular, can catalyze reactions with H_2O_2 that lead to the production of iron (III) and ROS, including the highly reactive and toxic hydroxyl radical [Riederer et al., 1989; Youdim et al., 1989]. Increased amounts of iron have been found in PD SNpc [Dexter et al., 1989b; Dexter et al., 1991; Hirsch et al., 1991; Sofic et al., 1991; Hirsch and Faucheux, 1998], with increases in iron (III) [Riederer et al., 1989]. In addition, decreased ferritin, the protein that binds and stores iron, and decreased copper, a metal found in antioxidant enzymes such as SOD, were also found in PD SN [Dexter et al., 1989b; Dexter et al., 1991]. These findings suggest that an excess iron burden, combined with lower antioxidant capacity in PD, may promote ROS formation and contribute to the neurodegenerative process [Dexter et al., 1991].

DNA Oxidative Damage

The four nucleotide bases that comprise DNA are highly susceptible to damage by ROS in conditions of high oxidative stress [Halliwell and Gutteridge, 1999]. Oxidation of DNA may cause mutations and strand breakage, which can lead to cell death. Increased levels of oxidized DNA have been found in many neurodegenerative disorders, including Alzheimer's disease [Gabbita et al., 1998], Huntington's disease [Browne et al., 1997], amyotrophic lateral sclerosis [Ferrante et al., 1997], and PD [reviewed in Halliwell, 2001]. Guanine is the most sensitive to oxidative modification, and is the only base in PD that has conclusive evidence for increased oxidation. Increased levels of the oxidized form of deoxyguanosine, 8-hydroxy-2'-deoxyguanosine, have been found in the caudate, putamen, SN, and cerebral cortex of Parkinson's disease patients as compared to control patients [Sanchez-Ramos, et al., 1994]. In addition, decreases in 2,6-diamino-4-hydroxy-5-formamidopyrimidine (the reduced form of the intermediate 8-hydroxy-guanine radical) are paired with increases in 8-hydroxyguanine (the oxidized form of the intermediate 8-hydroxy-guanine radical) in PD SN [Alam et al., 1997b], suggesting an increased oxidative environment. Cytoplasmic RNA oxidation is also selectively increased in the SN of PD patients, with smaller increases in cytoplasmic 8-hydroxyguanine found in multiple system atrophy-Parkinsonian type and dementia with Lewy bodies patients [Zhang et al., 1999]. In PD patients, mitochondrial DNA oxidation was increased [Zhang et al., 1999], coupled with increased mitochondrial 8-oxo-dGTPase, the enzyme that removes oxidized deoxy-guanosine-triphosphate from the DNA precursor pool, suggesting that mitochondrial DNA was also damaged by oxidation [Shimura-Miura et al., 1999]. The increased amount of DNA oxidation, both nuclear and mitochondrial, may lead to cell death due to DNA damage, and lends evidence to the role of oxidative stress in PD pathology.

Lipid Peroxidation

Another indication of an increased oxidative environment is the oxidation of unsaturated lipids. Lipid peroxidation occurs when ROS react with the double bond of unsaturated fatty acids, generating highly reactive lipid peroxyradicals. The peroxyradicals may then react with other fatty acids in a chain reaction, leading to the formation lipid peroxidation byproducts such as 4-hydroxy-2,3-nonenal (HNE), acrolein, malondialdehyde, and F_2-isoprostanes [reviewed in Barnham et al., 2004]. Increased HNE levels have been observed in many neurodegenerative disorders, including Alzheimer's disease [Butterfield et al., 2002], amyotrophic lateral sclerosis [Pedersen et al., 1998], with high levels of HNE found in PD plasma and CSF [Selley, 1998]. HNE can react with cysteine, lysine, and histidine residues, forming HNE-protein conjugates. Increased HNE-protein conjugates have been observed in PD midbrain [Yoritaka et al., 1996]. Increases in malondialdehyde have also been observed in PD [Dexter et al., 1989a]. Increases in other markers of lipid peroxidation, such as fatty acid hydroperoxides and cholesterol lipid hydroperoxides were also observed in PD substantia nigra [Dexter et al., 1994]. Lipid peroxidation, in addition to producing the reactive peroxyradicals which can damage other cellular components, and specifically the HNEs have been shown to lead to DNA fragmentation and caspase-dependent apoptosis [Liu et al., 2000]. In addition, lipid peroxidation affects membrane fluidity and permeability, leading to dysfunction of membrane-associated ion channels, changes in the number and affinity of membrane receptors, and alterations in organelles such as breaks in lysosomal membranes and the release of their proteolytic enzymes [reviewed in Farooqui and Horrocks, 1998]. Therefore, oxidative damage to lipids may have serious effects that could contribute to the cell death observed in PD.

Protein Oxidative Damage

Oxidative damage to proteins has been well documented [reviewed in Beal, 2002]. The most common measure of protein oxidation is the formation of protein carbonyls, which have been found to be increased in PD brain, including the SN [Alam et al., 1997a; Floor and Wetzel, 1998], basal ganglia, cortex, globus palladum, and cerebellum [Alam et al., 1997a]. In addition to protein damage by ROS, there is also evidence for protein damage by reactive nitrogen species in PD. Increases in nitrates in CSF and increased brain nitrosyl adducts were found in PD [Giasson et al., 2000]. In addition, 3-nitrotyrosine modified proteins were detected in Lewy bodies and other protein aggregates in PD brain SN [Good et al., 1998], indicating that protein in the aggregates were oxidized by peroxynitrite. Increased amounts of nitrated MnSOD have been found in PD CSF [Aoyama et al., 2000], which, like other important proteins if inactivated, could lead to even more oxidative stress and cell death. Also, advanced glycation end-products, or damaged proteins that have been post-translationally modified by sugars and then oxidized in a process called glycoxidation, are increased in PD [Castellani et al., 1996]. Glycoxidated protein formation has been shown to be potentiated by oxidative stress [Baynes, 1991]. Protein oxidation and glycoxidation may lead to protein dysfunction and aggregation, which can be detrimental to normal cellular processes, especially in the case of antioxidant proteins.

The evidence for both an oxidative environment and for oxidative damage to DNA, lipids, and proteins strongly suggests a role for oxidative stress in PD.

Potential Pathways of PD Oxidative Stress: The Cause or the Effect of PD?

Observations from PD patients combined with studies of experimental PD models have hinted at pathogenic pathways directly related to oxidative stress, including mitochondrial dysfunction and oxidative protein modification. Whether these pathways are a cause for PD or the result of PD pathogenesis remains to be elucidated, but their involvement is apparent.

Mitochondrial Dysfunction

Mitochondria are involved in many cellular processes and are critical for energy production in the cell. Dysfunction of mitochondria has been associated with various neurodegenerative diseases [reviewed in Orth and Schapira, 2001] including Parkinson's disease [reviewed in Orth and Schapira, 2002 and reviewed in Fiskum et al., 2003]. Though neither the initial cause of PD-associated mitochondrial dysfunction nor the degenerative pathway that the mitochondrial dysfunction itself leads to are known, data suggest that oxidative stress is involved [Lenaz et al., 2002; Beal, 2003; reviewed in Fiskum et al., 2003].

Decreased NADH dehydrogenase (Complex I) activity in the mitochondrial electron transport chain (ETC) has been directly linked to PD. In the SNpc of PD patients, Complex I exhibited reduced activity in post mortem studies, with activity decreased up to 40% [Schapira et al., 1990; Janetzky et al., 1994]. Studies of platelets and, less consistently, muscle tissue from PD patients have also shown reduced Complex I activity, suggesting a systemic effect [Shoffner et al., 1991; Blandini et al., 1998]. The systemic effect of Complex I inhibition is supported by the rotenone toxicity model of PD [Sherer et al., 2003b]. Chronic administration of the high-affinity Complex I inhibitor rotenone to rats induced a systemic Complex I inhibition and resulted in PD effects; including dopaminergic cell degeneration in the SN, alpha-synuclein positive cytoplasmic inclusions, and motor deficits [Betarbet et al., 2000; Sherer et al., 2003b]. These results suggest a susceptibility of the dopaminergic SN neurons to impaired mitochondrial Complex I activity.

Deficiencies in mitochondrial complex function have been linked to oxidative stress. Both oxidative and nitrative species have been shown to inhibit complexes in the mitochondrial electron transport chain (ETC) [reviewed in Heales et al., 1999; reviewed in Halliwell, 2001; Nulton-Persson and Szweda, 2001]. In addition, the ETC is a known source of ROS generation and potential oxidative stress in cells [Lenaz et al., 2002; Fiskum et al., 2003; reviewed in Turrens, 2003]. Mitochondrial ETC Complexes I and III are associated with formation of the ROS free radical superoxide [reviewed in Orth and Shapira, 2002; reviewed in Fiskum et al., 2003]. Superoxide is typically converted by SOD to H_2O_2, which may lead to generation of the toxic hydroxyl radical; or likewise superoxide may react with nitric oxide forming reactive peroxynitrite [Halliwell, 1992; reviewed in Quijano et al., 1997]. In addition, inhibition of Complex I has been demonstrated to facilitate the generation of ROS *in vitro* and potentially contribute to oxidative stress [Pitkanen and Robinson, 1996; Votyakova and Reynolds, 2001].

ROS generation is also evident in models of PD using known Complex I inhibitors, specifically the pesticide rotenone and the PD-inducing toxin 1-methyl-4-phenyl-1,2,3,6 tetrahydropyridine (MPTP) [reviewed in Lenaz et al., 2002; Beal, 2003]. The pesticide rotenone is a high-affinity Complex I inhibitor [reviewed in Degli Esposti, 1998]. As previously mentioned, chronic administration of rotenone leads to systemic inhibition of Complex I, and results in degeneration and deficits characteristic of PD [Betarbet et al., 2000;

Sherer et al., 2002; Sherer et. al., 2003a,b]. MPTP is a toxin known to cause Parkinsonism in humans and primates [Langston et al., 1983; reviewed in Beal, 2001]. The MPTP metabolite, MPP+, is selectively taken up by dopaminergic neurons, where it induces toxicity as a mitochondrial Complex I inhibitor [Mizuno et al., 1987; Nicklas et al., 1985]. Both toxins *in vitro* result in production of ROS and cellular toxicity [Lotharius and O'Malley, 2000; reviewed in Lenaz et al., 2002; Beal, 2003]. In addition, the rotenone model demonstrates *in vivo* oxidative damage to cells in the form of protein carbonyl formation [Sherer et al., 2003b]. MPTP and MPP+ toxicity can be attenuated *in vivo* by antioxidants [Park et al., 2004] and by agents improving mitochondrial energy production, such as coenzyme-Q 10 and nicotinamide [Schulz et al., 1995]. The antioxidant alpha-tocopherol and coenzyme-Q 10 also protected against rotenone toxicity in rats [Sherer et al., 2003b]. Together, the MPTP and rotenone models of PD suggest the involvement of mitochondrial dysfunction and oxidative stress in PD development.

Several lines of evidence suggest a genetic link to Complex I dysfunction in PD. One supporting study utilized cybrid cell lines with normal nuclear genomes but mtDNA from PD patients [Swerdlow et al., 1996]. The PD mtDNA cybrids had a Complex I deficit and a higher sensitivity to MPP+, suggesting possible genetic defects in mtDNA encoding Complex I subunits associated with PD pathogenesis [Swerdlow et al., 1996; Sheehan et al., 1997]. In contrast, a study examining single-nucleotide polymorphisms in mtDNA comparing PD and control patients led to the discovery of a polymorphism associated with Complex I and a reduced risk of PD [van der Walt et al., 2003]. The study found a polymorphism associated with the NADH dehydrogenase 3 subunit of Complex I, resulting in an amino acid change from threonine to alanine. Based on genotyping, this polymorphism was associated with a significantly lower susceptibility to PD, suggesting a protective effect, though the mechanism is not known.

Oxidative stress can also lead to dysfunction of other critical mitochondrial proteins. The tricarboxylic acid cycle enzyme alpha-ketoglutarate has been shown to be sensitive to the oxidative state of the mitochondria [Nulton-Persson and Szweda, 2001; Kumar et al., 2003; Nulton-Persson et al., 2003]. Interestingly, decreased alpha-ketoglutarate immunostaining has been noted in postmortem PD brain [Mizuno et al., 1994]. Other studies have shown that creatine kinase, the enzyme responsible for maintaining ATP and creatine phosphate energy stores in the cell, is susceptible to oxidative inhibition [Koufen et al., 1999; Miura et al., 1999; Cornelio et al., 2004; da Silva et al., 2004]. Depletion of cellular ATP levels has been evidenced in mouse brain synaptosomes in the MPTP model of PD [Scotcher et al., 1990], suggesting that inhibition of mitochondrial respiratory systems could lead to depleted cellular ATP levels and bioenergetic deficits [reviewed in Olanow and Tatton, 1999; Halliwell, 2001; Tretter et al., 2004].

Mitochondrial-generated free radicals, as from Complex I deficiencies and inhibition, and increased oxidative stress may have detrimental effects on general cellular function, and on the mitochondria itself [Lenaz et al., 2002; Beal, 2003; Fiskum et al., 2003; reviewed in Turrens, 2003]. Both mtDNA and mitochondrial lipid membranes are susceptible to age- and disease-related oxidative damage, as discussed above [Zhang et al., 1999; reviewed in Barnham et al., 2004]. Damage to either has severe effects on mitochondrial function, altering protein expression and membrane fluidity. Also, as previously mentioned, various mitochondrial proteins are susceptible to oxidative modification and inhibition by ROS. In PD, oxidative stress and genetic factors may lead to mitochondrial dysfunction, causing increased ROS production. Mitochondria can be further damaged by increased ROS, leading to additional dysfunction and energy deficits, which may impart neuronal vulnerability in PD.

DA Oxidation and Protein Modification

Noting the selectivity of dopaminergic neuron degeneration in PD pathogenesis and the ability of DA to form reactive metabolites, DA has been implicated as a contributor to oxidative stress and PD pathogenesis [Hastings and Berman, 2000]. Both DA metabolism and DA oxidation are known to lead to ROS and free radical generation, as well as protein oxidation and modification, all thought to be contributors to DA-induced toxicity [reviewed in Hastings and Berman, 2000].

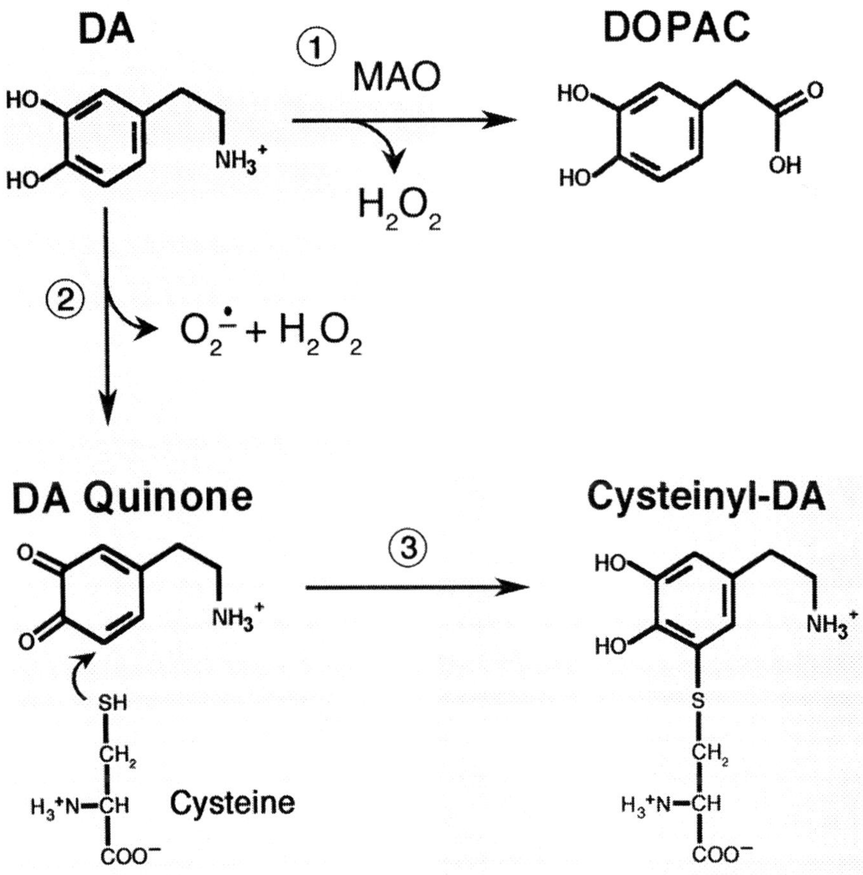

(1) Deamination of DA by MAO leads to the formation of DOPAC and H_2O_2. (2) DA can also readily oxidize to form DA quinone, producing superoxide($O_2^{\cdot}$) and H_2O_2. (3) The DA quinone will react with sulfhydryl groups, covalently bonding to free cysteines and protein cysteinyl residues in the cell to form a cysteinyl-DA conjugate.

Figure 1. Metabolism and Oxidation of DA.

DA oxidation can occur through two separate pathways. Through the metabolism of DA, deamination of DA by monoamine oxidase (MAO) leads to the formation of an aldehyde metabolite [Maker et al., 1981]. The aldehyde is quickly oxidized to form dihydroxyphenylacetic acid (DOPAC), and H_2O_2 is formed as a byproduct (figure 1) [Cooper et al., 1991]. Through a second pathway, DA, a catechol, has a chemical structure allowing it to readily oxidize to form the DA semiquinone and the DA quinone, producing superoxide and H_2O_2, respectively (Figure 1) [Bindoli et al., 1992; Monks et al., 1992; reviewed in Hastings and Berman, 2000].

The DA quinone is electron deficient, open to nucleophilic attack, and thus will react with sulfhydryl groups that have readily accessible electron pairs at physiological pH [Tse et al., 1976; Graham, 1978; Monks et al., 1992]. This reaction is very rapid and can modify thiols and protein cysteinyl residues in the cell, either oxidizing the sulfhydryl or covalently modifying the residue to form a cysteinyl-DA conjugate (Figure 1) [Tse et al., 1976; Fornstedt et al., 1990; Hastings and Zigmond, 1994], adding to the toxic potential of DA and DA quinone [Hastings and Berman, 2000]. DA modification of thiols forming increased levels of cysteinyl-DA has been shown *in vitro* and *in vivo* [Hastings and Zigmond, 1994; Hastings et al., 1996; Rabinovic et al., 1998; Rabinovic et al., 2000]. In addition, post-mortem studies found increased levels of cysteinyl-DA in SN brain tissue of PD patients [Fornstedt et al., 1989; Spencer et al., 1998].

Evidence for the oxidation of DA *in vivo* can be found in the cells of the SN in the form of the dark pigment neuromelanin. Neuromelanin is formed from the polymerized products of oxidized DA and DA-modified cysteine [Wakamatsu et al., 2003], demonstrating a regular occurrence of DA oxidation in pigmented SN cells. Interestingly, the neuromelanin-containing cells of the substantia nigra are those preferentially lost in PD [Hirsch et al., 1988]. In relevance to oxidative stress, melanins have been shown to interact with iron, leading to increased lipid peroxidation [Ben-Shachar et al., 1991]. However, neuromelanin may also have a protective role in removing excess catecholamines, oxidized DA products, and chelating harmful metals [reviewed in Zecca et al., 2003].

The ability of DA to exhibit toxic effects in various *in vitro* cell culture studies is well known [Mytilineou et al., 1993; Offen et al., 1995; Jones et al., 2000; Koshimura et al., 2000]. DA toxicity has also been demonstrated *in vivo* through intrastriatal injection of DA into rats, as exhibited by losses in striatal dopaminergic projections and increases in protein modification as cysteinyl-DA [Hastings et al., 1996; Rabinovic et al., 1998; Rabinovic et al., 2000]. While the exact mechanism of the DA-induced toxicity is not known, the toxicity of DA has been linked to oxidative stress. In several cell culture studies, antioxidants prevented DA toxicity [Hoyt et al., 1997; Passi et al., 1987]. Similarly, co-administration of antioxidants with DA intrastriatal injections prevented striatal dopaminergic terminal loss and significantly reduced cysteinyl-DA formation [Hastings et al., 1996].

Several proteins relevant to cell function, dopaminergic cells in particular, are known to be modified by exposure to DA and DA quinones. The functions of the DA transporter and the glutamate transporter, both extracellular enzymes, are inhibited by exposure to DA quinone [Berman et al., 1996; Berman and Hastings, 1997]. Tyrosine hydroxylase, the rate-limiting enzyme in catecholamine biosynthesis, is also modified by DA quinone, inactivating the protein and potentially altering it to a redox-cycling quinoprotein [Xu et al, 1998; Kuhn et al, 1999]. Creatine kinase, a cysteine-dependent enzyme, can also be inhibited by exposure to DA, [Maker et al., 1986; Miura et al., 1999; Jiang et al., 2002] potentially contributing to bioenergetic defects in the cell as discussed above.

DA toxicity can also contribute to impairment of mitochondrial function. Exposure of isolated rat brain mitochondria to DA quinone has been shown to alter mitochondrial respiration [Berman and Hastings, 1999]. Another factor in mitochondrial dysfunction linked to DA exposure and oxidative stress is the opening of the permeability transition pore (PTP). The PTP is a nonselective pore formed by a complex of mitochondrial proteins between the inner and outer mitochondrial membrane [reviewed in Brdiczka et al., 1998; reviewed in Vyssokikh and Brdiczka, 2003]. Opening of the PTP dramatically decreases membrane potential of the mitochondria, and if prolonged can lead to a loss of mitochondrial function, membrane integrity, and the release of apoptotic factors [reviewed in Olanow and Tatton, 1999; reviewed in Hastings and Berman, 2000]. Opening of the pore is known to be triggered by oxidative stress and exposure to DA quinones [Berman and Hastings, 1999; Dolder et al., 2001].

As previously mentioned, mitochondrial dysfunction is known to lead to ROS production. The added oxidative stress may contribute to DA oxidation, which contributes to further oxidative damage in the mitochondria and the cell. The two pathways are thus intertwined, setting up a vicious cycle of increasing oxidative damage and toxicity in the cell. Therefore, dopaminergic neurons may be especially vulnerable to propagating oxidative stress, leading to increased cellular damage, and perhaps leading to the selective dopaminergic PD pathogenesis.

Genetic Links for PD and Oxidative Stress

Though the vast majority of PD cases are considered sporadic, rare cases of PD are familial. Several genetically-linked forms of PD are known, and are described as PARK1 through PARK11. Of the genetic links that have been identified thus far, several have been associated with oxidative stress in PD pathogenesis [Zhang et al., 2000; Dawson and Dawson, 2003; Greenamyre and Hastings, 2004], as discussed below.

Alpha-Synuclein

Mutations in the *alpha-synuclein* gene, encoding a presynaptic nerve terminal-associated protein, were found to be linked to an autosomal dominant form of PD, and were the first direct genetic links to PD discovered [Polymeropoulos et al., 1997]. The alpha-synuclein mutations, although important, were soon revealed to account for only a small portion of familial PD cases [Vaughan et al., 1998; Wang et al., 1998]. Even without mutations, an overabundance of the normal form of alpha-synuclein has also been shown to cause PD, as evidenced by a genetic form of the disease in which afflicted family members carry a triplication of the *alpha-synuclein* gene [Singleton et al., 2003; Farrer et al., 2004]. The exact role of alpha-synuclein in PD is uncertain, though the pathway may involve protein oxidation and aggregation.

Evidence indicates that alpha-synuclein is prone to aggregation, which is augmented by the genetic variations associated with PD, and by oxidative and nitrative modifications [Conway et al., 1998; Giasson et al., 2000; Paxinou et al., 2001]. Alpha-synuclein aggregates form nonfibrillar protofibrils, thought to be pathogenic [Goldberg and Lansbury, 2000; Rochet et al., 2004], and fibrillar deposits, which are major components of Lewy bodies and Lewy neurites, pathogenic hallmarks of PD [Baba et al., 1998; Spillantini et al., 1998]. *In vitro* study shows that nitrative and oxidative stress leads to the formation of stable dimers and oligomers of alpha-synuclein [Souza et al., 2000]. The oxidation-induced oligomerization may be the limiting step in alpha-synuclein aggregation [Krishnan et al., 2003]. One study also demonstrated that Complex I inhibition by rotenone causes alpha-synuclein aggregation, associating mitochondrial dysfunction, oxidation, and alpha-synuclein aggregation [Sherer et al, 2002]. Alpha-synuclein aggregation and oxidative DA modification have also been linked, showing that covalent modification by DA stabilized and promoted the aggregation of alpha-synuclein protofibrils [Conway et al., 2001; Rochet et al., 2004]. The discovery of the DA—alpha-synuclein protein adduct suggests a relationship between oxidative stress, DA, and alpha-synuclein that could lead to the promotion of PD pathogenesis [Rochet et al., 2004].

Parkin and a Role in Oxidative Stress

Mutations in the *parkin* gene were first linked to a form of Parkinson's disease known as autosomal recessive juvenile parkinsonism in 1998 [Kitada et al., 1998]. Parkin, an E3 ubiquitin ligase protein, plays a role in the ubiquitin-dependent proteasome pathway, and mutations of parkin have been associated with dysfunction in the protein degradation pathway [McNaught and Olanow, 2003]. One study also showed that proteasome inhibition alone can increase oxidative stress [Lee et al., 2001]. Recent studies have linked the role of parkin to oxidative stress and mitochondrial dysfunction, as discussed below.

A proteomic study in parkin-deficient mice found changes in proteins involved with mitochondrial respiration, including Complex I, and with the cellular oxidative stress response, as compared to wildtype mice [Palacino et al., 2004]. In addition, the study found that the parkin deficient mice had reduced striatal mitochondrial respiration and decreased serum antioxidant capacity [Palacino et al., 2004]. These results suggest that mitochondrial dysfunction and oxidative stress may play a role in the parkin-linked familial form of PD. In another study, cell lines transfected with wildtype or mutated forms of parkin found that cells expressing mutant parkin exhibited increased markers of oxidative and nitrative stress, while cells expressing wildtype parkin showed lower levels of these markers [Hyun et. al, 2002]. These results are consistent with a study utilizing overexpression of wildtype parkin in the human dopaminergic neuroblastoma cell line, SH-SY5Y, which protected against DA-induced toxicity and oxidative stress [Jiang et. al, 2004]. However, expression of known mutant forms of parkin in this study did not protect against the DA-induced toxicity [Jiang et. al, 2004]. Parkin has also been shown to suppress toxicity induced by the expression of mutant alpha-synuclein and over expression of wildtype alpha-synuclein [Petrucelli et al., 2002; Oluwatosin-Chigbu et al., 2003], implicating parkin as a protective agent against protein oxidation and aggregation associated with alpha-synuclein. These studies suggest that parkin is likely involved in protection against oxidative stress and mitochondrial dysfunction, further linking the role of parkin to oxidative stress in PD.

DJ-1 and Antioxidant Properties

The gene for the ubiquitous, conserved protein DJ-1 has been linked to familial PARK7 [Bonifati et al., 2003]. A study in separate Italian and Dutch families revealed mutations resulting in loss of DJ-1 protein stability and protein production, respectively, leading to the development of PD [Bonifati et al., 2003]. While the specific function of DJ-1 remains unknown, various studies have implicated this protein in oxidative stress response and antioxidant mechanisms.

The antioxidant properties of DJ-1 may be attributed to the protein's own susceptibility to oxidation. Crystallography studies on the structure of DJ-1 revealed that the protein has one exceptionally reactive cysteine residue at Cys-106 [Wilson et al., 2003]. In cell culture studies, DJ-1 prevented H_2O_2-induced toxicity by oxidizing itself as indicated by a shift in the isoelectric point of the protein [Taira et al., 2004]. Further study revealed that the residue Cys-106 is selectively modified to a cysteinyl sulfinic acid in the oxidized form of DJ-1 [Canet-Aviles et al., 2004]. In addition, siRNA-induced downregulation of DJ-1 expression resulted in an increased susceptibility to H_2O_2 and ER stress in cells [Yokota et al., 2003; Takahashi-Niki et al., 2004]. Thus DJ-1 appears to be a strong candidate as an antioxidant protector.

PINK1, a Mitochondrial Protein

The gene for the PTEN-induced putative kinase 1 protein, *PINK1*, is the most recent genetic connection made to PD [Valente et al., 2004]. The genetic mutation was found in separate Spanish and Italian families expressing familial PARK6 associated PD. PINK1 is a nuclear encoded mitochondrial protein, and is the first genetic mutation discovered involving a mitochondrial protein associated directly and unambiguously with PD [Greenamyre and Hastings, 2004].

Currently, little is known about the mechanisms underlying mutated PINK1 induced PD, but the discovery of this genetic mutation may elicit new avenues of PD research [Greenamyre and Hastings, 2004]. Valente *et al.* (2004) hypothesized that PINK1 may be responsible for phosphorylation of mitochondrial proteins in times of cellular stress. Interestingly, the functions of several important mitochondrial proteins are modulated by phosphorylation. Complex I, whose reduced function has been related to PD, has many components that are reversibly phosphorylated [Schulenberg et al., 2003; Chen et al., 2004]. Thus, PINK1 could have a role in preventing mitochondrial dysfunction in times of cellular stress [Greenamyre and Hastings, 2004].

Although few cases of PD have been determined to be genetically associated, studies have suggested that multiple gene defects or possibly the interactions between environmental and genetic factors could account for many cases of PD, both familial and sporadic [Gasser 2001; Greenamyre et al, 2003]. Nonetheless, studies of the genetic causes of PD may lead to a better understanding of PD pathology, and result in more effective treatments for the sporadic form of the disease, as well.

Conclusion

Observations from PD patients and laboratory PD models have been directly related to oxidative stress. Some indications of increased oxidative stress appear to occur early in PD, such as GSH depletion [Jenner and Olanow, 1996], suggesting a role for oxidative stress early in the disease pathogenesis. In addition, pathogenic pathways involving mitochondrial dysfunction and oxidative protein modification have been suggested, but whether these pathways have primary or secondary roles in PD pathogenesis remains to be elucidated. Collective evidence suggests the possibility that interactions between environmental and genetic factors could account for many cases of PD [Greenamyre et al, 2003]. Nevertheless, many pathological characteristics of PD appear to be either downstream of or the direct result of oxidative stress in the cells. Thus, oxidative stress appears as a likely culprit of PD pathogenesis and therefore a promising target of PD treatment therapies [Beal, 2003].

References

Alam ZI, Daniel SE, Lees AJ, Marsden DC, Jenner P, Halliwell B. A generalised increase in protein carbonyls in the brain in Parkinson's but not incidental Lewy body disease. *J Neurochem.* 1997a;69(3):1326-9.

Alam ZI, Jenner A, Daniel SE, Lees AJ, Cairns N, Marsden CD, Jenner P, Halliwell B. Oxidative DNA damage in the parkinsonian brain: an apparent selective increase in 8-hydroxyguanine levels in substantia nigra. *J Neurochem.* 1997b;69(3):1196-203.

Ambani LM, Van Woert MH, Murphy S. Brain peroxidase and catalase in Parkinson disease. *Arch Neurol.* 1975;32:114-118.

Aoyama K, Matsubara K, Fujikawa Y, Nagahiro Y, Shimizu K, Umegae N, Hayase N, Shiono H, Kobayashi S. Nitration of manganese superoxide dismutase in cerebrospinal fluids is a marker for peroxynitrite-mediated oxidative stress in neurodegenerative diseases. *Ann Neurol.* 2000;47(4):524-7.

Baba M, Nakajo S, Tu PH, Tomita T, Nakaya K, Lee VM, Trojanowski JQ, Iwatsubo T. Aggregation of alpha-synuclein in Lewy bodies of sporadic Parkinson's disease and dementia with Lewy bodies. *Am J Pathol.* 1998;152(4):879-84.

Bajaj NP, Shaw C, Warner T, RayChaudhuri K. The genetics of Parkinson's disease and parkinsonian syndromes. *J Neurol.* 1998;245(10):625-33.

Barnham KJ, Masters CL, Bush AI. Neurodegenerative diseases and oxidative stress. *Nat Rev Drug Discov.* 2004;3(3):205-14.

Baynes JW. Role of oxidative stress in development of complications in diabetes. *Diabetes.* 1991; 40(4):405-12.

Beal MF. Mitochondria, oxidative damage, and inflammation in Parkinson's disease.*Ann N Y Acad Sci.* 2003;991:120-31.

Beal MF. Oxidatively modified proteins in aging and disease. *Free Radic Biol Med.* 2002;32(9):797-803.

Beal MF. Experimental models of Parkinson's disease. *Nat Rev Neurosci.* 2001;2(5):325-34.

Ben-Shachar D, Riederer P, Youdim MB. Iron-melanin interaction and lipid peroxidation: implications for Parkinson's disease. *J Neurochem.* 1991;57(5):1609-14.

Berman SB, Zigmond MJ, Hastings TG. Modification of dopamine transporter function: effect of reactive oxygen species and dopamine. *J Neurochem.* 1996; 67;593-600.

Berman SB, Hastings TG. Inhibition of glutamate transport in synaptosomes by dopamine oxidation and reactive oxygen species. *J Neurochem.* 1997;69:1185-1195.

Berman SB, Hastings TG. Dopamine oxidation alters mitochondrial respiration and induces permeability transition in brain mitochondria: implications for Parkinson's disease. *J Neurochem.* 1999;73:1127-1137.

Betarbet R, Sherer TB, Di Monte DA, Greenamyre JT. Mechanistic approaches to Parkinson's disease pathogenesis. *Brain Pathol.* 2002;12(4):499-510.

Betarbet R, Sherer TB, MacKenzie G, Garcia-Osuna M, Panov AV, Greenamyre JT. Chronic systemic pesticide exposure reproduces features of Parkinson's disease. *Nat Neurosci.* 2000;3(12):1301-6.

Bindoli A, Rigobello MP, Deeble DJ. Biochemical and toxicological properties of the oxidation products of catecholamines. *Free Radic Biol Med.* 1992;13(4):391-405.

Blandini F, Nappi G, Greenamyre JT. Quantitative study of mitochondrial complex I in platelets of parkinsonian patients. *Mov Disord.* 1998; 13(1):11-5.

Bonifati V, Rizzu P, van Baren MJ, Schaap O, Breedveld GJ, Krieger E, Dekker MC, Squitieri F, Ibanez P, Joosse M, van Dongen JW, Vanacore N, van Swieten JC, Brice A, Meco G, van Duijn CM, Oostra BA, Heutink P. Mutations in the DJ-1 gene associated with autosomal recessive early-onset parkinsonism. *Science.* 2003;299(5604):256-9.

Brdiczka D, Beutner G, Ruck A, Dolder M, Wallimann T. The molecular structure of mitochondrial contact sites. Their role in regulation of energy metabolism and permeability transition. *Biofactors.* 1998;8(3-4):235-42.

Browne SE, Bowling AC, MacGarvey U, Baik MJ, Berger SC, Muqit MM, Bird ED, Beal MF. Oxidative damage and metabolic dysfunction in Huntington's disease: selective vulnerability of the basal ganglia. *Ann Neurol.* 1997;41(5):646-53.

Butterfield DA, Castegna A, Lauderback CM, Drake J. Evidence that amyloid beta-peptide-induced lipid peroxidation and its sequelae in Alzheimer's disease brain contribute to neuronal death. *Neurobiol Aging.* 2002;23(5):655-64.

Butterfield DA, Drake J, Pocernich C, Castegna A. Evidence of oxidative damage in Alzheimer's disease brain: central role for amyloid beta-peptide. *Trends Mol Med.* 2001;7(12):548-54.

Butterfield DA, Kanski J. Brain protein oxidation in age-related neurodegenerative disorders that are associated with aggregated proteins. *Mech Ageing Dev.* 2001;122(9):945-62

Canet-Aviles RM, Wilson MA, Miller DW, Ahmad R, McLendon C, Bandyopadhyay S, Baptista MJ, Ringe D, Petsko GA, Cookson MR. The Parkinson's disease protein DJ-1 is neuroprotective due to cysteine-sulfinic acid-driven mitochondrial localization. *Proc Natl Acad Sci U S A.* 2004;101(24):9103-8.

Carri MT, Ferri A, Cozzolino M, Calabrese L, Rotilio G. Neurodegeneration in amyotrophic lateral sclerosis: the role of oxidative stress and altered homeostasis of metals. *Brain Res Bull.* 2003;61(4):365-74.

Castellani R, Smith MA, Richey PL, Perry G. Glycoxidation and oxidative stress in Parkinson disease and diffuse Lewy body disease. *Brain Res.* 1996;737(1-2):195-200.

Chen R, Fearnley IM, Peak-Chew SY, Walker JE. The phosphorylation of subunits of complex I from bovine heart mitochondria. *J Biol Chem.* 2004;279(25):26036-45.

Conway KA, Harper JD, Lansbury PT. Accelerated in vitro fibril formation by a mutant alpha-synuclein linked to early-onset Parkinson disease. *Nat Med.* 1998;4(11):1318-20.

Conway KA, Rochet JC, Bieganski RM, Lansbury PT Jr. Kinetic stabilization of the alpha-synuclein protofibril by a dopamine-alpha-synuclein adduct. *Science.* 2001; 294 (5545): 1346-9

Cooper JR, Bloom FE, Roth RH. *The Biochemical Basis of Neuropharmachology.* New York, NY: Oxford University Press; 1991.

Cornelio AR, Rodrigues V Jr, de Souza Wyse AT, Dutra-Filho CS, Wajner M, Wannmacher CM. Tryptophan reduces creatine kinase activity in the brain cortex of rats. *Int J Dev Neurosci.* 2004;22(2):95-101.

Dauer W, Przedborski S. Parkinson's disease: mechanisms and models. *Neuron.* 2003;39(6):889-909.

Dawson TM, Dawson VL. Molecular pathways of neurodegeneration in Parkinson's disease. *Science.* 2003;302(5646):819-22 da Silva CG, Bueno AR, Schuck PF, Leipnitz G, Ribeiro CA, Rosa RB, Dutra Filho CS, Wyse AT, Wannmacher CM, Wajner M. Inhibition of creatine kinase activity from rat cerebral cortex by D-2-hydroxyglutaric acid in vitro. *Neurochem Int.* 2004;44(1):45-52.

Degli Esposti M. Inhibitors of NADH-ubiquinone reductase: an overview. *Biochim Biophys Acta.* 1998;1364(2):222-35.

Deumens R, Blokland A, Prickaerts J. Modeling Parkinson's disease in rats: an evaluation of 6-OHDA lesions of the nigrostriatal pathway. *Exp Neurol.* 2002;175(2):303-17.

Dexter DT, Carayon A, Javoy-Agid F, Agid Y, Wells FR, Daniel SE, Lees AJ, Jenner P, Marsden CD. Alterations in the levels of iron, ferritin and other trace metals in Parkinson's disease and other neurodegenerative diseases affecting the basal ganglia. *Brain.* 1991;114 (Pt 4):1953-75.

Dexter DT, Carter CJ, Wells FR, Javoy-Agid F, Agid Y, Lees A, Jenner P, Marsden CD.
Basal lipid peroxidation in substantia nigra is increased in Parkinson's disease. *J Neurochem.* 1989a;52(2):381-9.

Dexter DT, Holley AE, Flitter WD, Slater TF, Wells FR, Daniel SE, Lees AJ, Jenner P, Marsden CD. Increased levels of lipid hydroperoxides in the parkinsonian substantia nigra: an HPLC and ESR study. *Mov Disord.* 1994;9(1):92-7.

Dexter DT, Wells FR, Lees AJ, Agid F, Agid Y, Jenner P, Marsden CD. Increased nigral iron content and alterations in other metal ions occurring in brain in Parkinson's disease. *J Neurochem.* 1989b;52(6):1830-6.

Dexter DT, Ward RJ, Wells FR, Daniel SE, Lees AJ, Peters TJ, Jenner P, Marsden CD. Alpha-tocopherol levels in brain are not altered in Parkinson's disease. *Ann Neurol.* 1992;32(4):591-3.

Dolder M, Wendt S, Wallimann T. Mitochondrial creatine kinase in contact sites: interaction with porin and adenine nucleotide translocase, role in permeability transition and sensitivity to oxidative damage. *Biol Signals Recept.* 2001;10(1-2):93-111.

Farooqui AA, Horrocks LA. Lipid peroxides in the free radical pathophysiology of brain diseases. *Cell Mol Neurobiol.* 1998;18(6):599-608.

Farrer M, Kachergus J, Forno L, Lincoln S, Wang DS, Hulihan M, Maraganore D, Gwinn-Hardy K, Wszolek Z, Dickson D, Langston JW. Comparison of kindreds with parkinsonism and alpha-synuclein genomic multiplications. Ann Neurol. 2004; 55(2):174-9

Ferrante RJ, Browne SE, Shinobu LA, Bowling AC, Baik MJ, MacGarvey U, Kowall NW, Brown RH Jr, Beal MF. Evidence of increased oxidative damage in both sporadic and familial amyotrophic lateral sclerosis. *J Neurochem.* 1997;69(5):2064-74.

Fiskum G, Starkov A, Polster BM, Chinopoulos C. Mitochondrial mechanisms of neural cell death and neuroprotective interventions in Parkinson's disease. *Ann N Y Acad Sci.* 2003;991:111-9.

Floor E, Wetzel MG. Increased protein oxidation in human substantia nigra pars compacta in comparison with basal ganglia and prefrontal cortex measured with an improved dinitrophenylhydrazine assay. *J Neurochem.* 1998;70(1):268-75.

Fornstedt B, Bergh I, Rosengren E, Carlsson A. An improved HPLC-electrochemical detection method for measuring brain levels of 5-S-cysteinyldopamine, 5-S-cysteinyl-3,4-dihydroxyphenylalanine, and 5-S-cysteinyl-3,4-dihydroxyphenylacetic acid. *J Neurochem.* 1990; 54:578-586.

Fornstedt B, Brun A, Rosengren E, Carlsson A. The apparent autoxidation rate of catechols in dopamine-rich regions of human brains increases with the degree of depigmentation of substantia nigra. *J Neural Transm Park Dis Dement Sect.* 1989;1(4):279-95.

Gabbita SP, Lovell MA, Markesbery WR. Increased nuclear DNA oxidation in the brain in Alzheimer's disease. *J Neurochem.* 1998;71(5):2034-40.

Gasser T. Genetics of Parkinson's disease. *J Neurol.* 2001;248(10):833-40.

Giasson BI, Duda JE, Murray IV, Chen Q, Souza JM, Hurtig HI, Ischiropoulos H, Trojanowski JQ, Lee VM. Oxidative damage linked to neurodegeneration by selective alpha-synuclein nitration in synucleinopathy lesions. *Science.* 2000;290(5493):985-9.

Goldberg MS, Lansbury PT Jr. Is there a cause-and-effect relationship between alpha-synuclein fibrillization and Parkinson's disease? *Nat Cell Biol.* 2000;2(7):E115-9.

Good PF, Hsu A, Werner P, Perl DP, Olanow CW. Protein nitration in Parkinson's disease. *J Neuropathol Exp Neurol.* 1998;57(4):338-42.

Graham DG. Oxidative pathways for catecholamines in the genesis of neuromelanin and cytotoxic quinones. *Mol Pharmacol.* 1978;14:633-643.

Greenamyre JT, Betarbet R, Sherer TB. The rotenone model of Parkinson's disease: genes, environment and mitochondria. *Parkinsonism Relat Disord.* 2003;9 Suppl 2:S59-64.

Greenamyre JT, Hastings TG. Biomedicine. Parkinson's--divergent causes, convergent mechanisms. *Science.* 2004 May 21;304(5674):1120-2.

Halliwell B. Reactive oxygen species and the central nervous system. *J Neurochem.* 1992;59(5):1609-23

Halliwell B. Role of free radicals in the neurodegenerative diseases: therapeutic implications for antioxidant treatment. *Drugs Aging.* 2001;18(9):685-716.

Halliwell B and Gutteridge JMC. *Free Radicals in Biology and Medicine*, Third Edition. New York, NY: Oxford University Press; 1999.

Hastings TG, Lewis DA, Zigmond MJ. Role of oxidation in the neurotoxic effects of intrastriatal dopamine injections. *Proc Natl Acad Sci U S A.* 1996;93(5):1956-61.

Hastings TG, Berman SB. Dopamine-Induced Toxicity and Quinone Modification of Protein: Implications for Parkinson's Disease. In: Creveling CR, ed. *Role of Catechol Quinone Species in Cellular Toxicity.* Johnson City, TN: FP Graham Publishing; 2000:69-89.

Hastings TG, Zigmond MJ. Identification of catechol-protein conjugates in neostriatal slices incubated with [3H]dopamine: impact of ascorbic acid and glutathione. *J Neurochem.* 1994; 63:126-1132.

Heales SJ, Bolanos JP, Stewart VC, Brookes PS, Land JM, Clark JB. Nitric oxide, mitochondria and neurological disease. *Biochim Biophys Acta.* 1999;1410(2):215-28.

Hirsch EC, Brandel JP, Galle P, Javoy-Agid F, Agid Y. Iron and aluminum increase in the substantia nigra of patients with Parkinson's disease: an X-ray microanalysis. *J Neurochem.* 1991;56(2):446-51.

Hirsch EC, Faucheux BA. Iron metabolism and Parkinson's disease. *Mov Disord.* 1998;13 Suppl 1:39-45

Hirsch E, Graybiel AM, Agid YA. Melanized dopaminergic neurons are differentially susceptible to degeneration in Parkinson's disease. *Nature.* 1988;334(6180):345-8.

Honda K, Casadesus G, Petersen RB, Perry G, Smith MA. Oxidative stress and redox-active iron in Alzheimer's disease. *Ann N Y Acad Sci.* 2004;1012:179-82.

Hoyt KR, Reynolds IJ, Hastings TG. Mechanisms of dopamine-induced cell death in cultured rat forebrain neurons: interactions with and differences from glutamate-induced cell death. *Exp Neurol.* 1997;143(2):269-81.

Hyun DH, Lee M, Hattori N, Kubo S, Mizuno Y, Halliwell B, Jenner P.Effect of wild-type or mutant Parkin on oxidative damage, nitric oxide, antioxidant defenses, and the proteasome. *J Biol Chem.* 2002;277(32):28572-7.

Janetzky B, Hauck S, Youdim MB, Riederer P, Jellinger K, Pantucek F, Zochling R, Boissl KW, Reichmann H. Unaltered aconitase activity, but decreased complex I activity in substantia nigra pars compacta of patients with Parkinson's disease. *Neurosci Lett.* 1994;169(1-2):126-8.

Jenner P. Oxidative stress in Parkinson's disease. *Ann Neurol.* 2003;53 Suppl 3:S26-36.

Jenner P, Olanow CW. Oxidative stress and the pathogenesis of Parkinson's disease. *Neurology.* 1996;47(6 Suppl 3):S161-70.

Jiang H, Ren Y, Zhao J, Feng J. Parkin protects human dopaminergic neuroblastoma cells against dopamine-induced apoptosis. *Hum Mol Genet.* 2004;13(16):1745-54

Jiang P, Van Laar VS, and Hastings TG. Modification of mitochondrial creatine kinase by dopamine quinone. *Society for Neuroscience 32nd Annual Meeting* 2002; Abstract 595.13

Jones DC, Gunasekar PG, Borowitz JL, Isom GE. Dopamine-induced apoptosis is mediated by oxidative stress and Is enhanced by cyanide in differentiated PC12 cells. *J Neurochem.* 2000;74(6):2296-304.

Kish SJ, Morito C, Hornykiewicz O. Glutathione peroxidase activity in Parkinson's disease brain. *Neurosci Lett.* 1985;58:343-346.

Kitada T, Asakawa S, Hattori N, Matsumine H, Yamamura Y, Minoshima S, Yokochi M, Mizuno Y, Shimizu N. Mutations in the parkin gene cause autosomal recessive juvenile parkinsonism. *Nature.* 1998;392(6676):605-8.

Koshimura K, Tanaka J, Murakami Y, Kato Y. Effects of dopamine and L-DOPA on survival of PC12 cells. *J Neurosci Res.* 2000;62(1):112-9.

Koufen P, Ruck A, Brdiczka D, Wendt S, Wallimann T, Stark G. Free radical-induced inactivation of creatine kinase: influence on the octameric and dimeric states of the mitochondrial enzyme (Mib-CK). *Biochem J.* 1999;344 Pt 2:413-7.

Krishnan S, Chi EY, Wood SJ, Kendrick BS, Li C, Garzon-Rodriguez W, Wypych J, Randolph TW, Narhi LO, Biere AL, Citron M, Carpenter JF. Oxidative dimer formation is the critical rate-limiting step for Parkinson's disease alpha-synuclein fibrillogenesis. *Biochemistry.* 2003;42(3):829-37.

Kuhn DM, Arthur RE Jr, Thomas DM, Elferink LA. Tyrosine hydroxylase is inactivated by catechol-quinones and converted to a redox-cycling quinoprotein: possible relevance to Parkinson's disease. *J Neurochem.* 1999;73(3):1309-17.

Kumar MJ, Nicholls DG, Andersen JK. Oxidative alpha-ketoglutarate dehydrogenase inhibition via subtle elevations in monoamine oxidase B levels results in loss of spare respiratory capacity: implications for Parkinson's disease. *J Biol Chem.* 2003;278(47):46432-9

Lang AE, Lozano AM. Parkinson's disease. First of two parts. *N Engl J Med.* 1998a;339(15):1044-53.

Lang AE, Lozano AM. Parkinson's disease. Second of two parts. *N Engl J Med.* 1998b;339(16):1130-43.

Langston JW, Ballard P, Tetrud JW, Irwin I. Chronic Parkinsonism in humans due to a product of meperidine-analog synthesis. *Science*. 1983;219(4587):979-80.

Lee MH, Hyun DH, Jenner P, Halliwell B. Effect of proteasome inhibition on cellular oxidative damage, antioxidant defences and nitric oxide production. *J Neurochem*. 2001;78(1):32-41.

Lenaz G, Bovina C, D'Aurelio M, Fato R, Formiggini G, Genova ML, Giuliano G, Merlo Pich M, Paolucci U, Parenti Castelli G, Ventura B. Role of mitochondria in oxidative stress and aging. *Ann N Y Acad Sci*. 2002;959:199-213.

Liu W, Kato M, Akhand AA, Hayakawa A, Suzuki H, Miyata T, Kurokawa K, Hotta Y, Ishikawa N, Nakashima I. 4-hydroxynonenal induces a cellular redox status-related activation of the caspase cascade for apoptotic cell death. *J Cell Sci*. 2000;113 (Pt 4):635-41.

Lotharius J, O'Malley KL. The parkinsonism-inducing drug 1-methyl-4-phenylpyridinium triggers intracellular dopamine oxidation. A novel mechanism of toxicity. *J Biol Chem*. 2000;275(49):38581-8

Maker HS, Weiss C, Brannan TS. Amine-mediated toxicity. The effects of dopamine, norepinephrine, 5-hydroxytryptamine, 6-hydroxydopamine, ascorbate, glutathione and peroxide on the in vitro activities of creatine and adenylate kinases in the brain of the rat. *Neuropharmacology*. 1986;25(1):25-32.

Maker HS, Weiss C, Silides DJ, Cohen G. Coupling of dopamine oxidation (monoamine oxidase activity) to glutathione oxidation via the generation of hydrogen peroxide in rat brain homogenates. *J Neurochem*. 1981;36(2):589-93.

Marttila RJ, Lorentz H, Rinne UK. Oxygen toxicity protecting enzymes in Parkinson's disease: increase of superoxide dismutase-like activity in the substantia nigra and basal nucleus. *J Neurol Sci*. 1988;86:321-331.

McNaught KS, Olanow CW. Proteolytic stress: a unifying concept for the etiopathogenesis of Parkinson's disease. *Ann Neurol*. 2003;53 Suppl 3:S73-84.

Miura T, Muraoka S, Fujimoto Y. Inactivation of creatine kinase induced by dopa and dopamine in the presence of ferrylmyoglobin. *Chem Biol Interact*. 1999;123(1):51-61.

Mizuno Y, Saitoh T, Sone N. Inhibition of mitochondrial NADH-ubiquinone oxidoreductase activity by 1-methyl-4-phenylpyridinium ion. *Biochem Biophys Res Commun*. 1987;143(1):294-9.

Mizuno Y, Matuda S, Yoshino H, Mori H, Hattori N, Ikebe S. An immunohistochemical study on alpha-ketoglutarate dehydrogenase complex in Parkinson's disease. *Ann Neurol*. 1994;35(2):204-10

Monks TJ, Hanzlik RP, Cohen GM, Ross D, Graham DG. Quinone chemistry and toxicity. *Toxicol Appl Pharmacol*. 1992;112(1):2-16.

Mytilineou C, Han SK, Cohen G. Toxic and protective effects of L-dopa on mesencephalic cell cultures. *J Neurochem*. 1993;61(4):1470-8.

Nicklas WJ, Vyas I, Heikkila RE. Inhibition of NADH-linked oxidation in brain mitochondria by 1-methyl-4-phenyl-pyridine, a metabolite of the neurotoxin, 1-methyl-4-phenyl-1,2,5,6-tetrahydropyridine. *Life Sci*. 1985;36(26):2503-8.

Nulton-Persson AC, Starke DW, Mieyal JJ, Szweda LI. Reversible inactivation of alpha-ketoglutarate dehydrogenase in response to alterations in the mitochondrial glutathione status. *Biochemistry*. 2003;42(14):4235-42.

Nulton-Persson AC, Szweda LI. Modulation of mitochondrial function by hydrogen peroxide. *J Biol Chem.* 2001;276(26):23357-61.

Nussbaum RL, Polymeropoulos MH. Genetics of Parkinson's disease. *Hum Mol Genet.* 1997;6(10):1687-91.

Offen D, Ziv I, Gorodin S, Barzilai A, Malik Z, Melamed E. Dopamine-induced programmed cell death in mouse thymocytes. *Biochim Biophys Acta.* 1995;1268(2):171-7.

Olanow CW, Tatton WG. Etiology and pathogenesis of Parkinson's disease. *Annual Rev Neurosci.* 1999;22:123-44.

Oluwatosin-Chigbu Y, Robbins A, Scott CW, Arriza JL, Reid JD, Zysk JR. Parkin suppresses wild-type alpha-synuclein-induced toxicity in SHSY-5Y cells. *Biochem Biophys Res Commun.* 2003;309(3):679-84

Orth M, Schapira AH. Mitochondria and degenerative disorders. *Am J Med Genet.* 2001;106(1):27-36.

Orth M, Schapira AH. Mitochondrial involvement in Parkinson's disease. *Neurochem Int.* 2002;40(6):533-41

Palacino JJ, Sagi D, Goldberg MS, Krauss S, Motz C, Wacker M, Klose J, Shen J. Mitochondrial dysfunction and oxidative damage in parkin-deficient mice. *J Biol Chem.* 2004;279(18):18614-22.

Passi S, Picardo M, Nazzaro-Porro M. Comparative cytotoxicity of phenols in vitro. *Biochem J.* 1987;245(2):537-42.

Park SW, Kim SH, Park KH, Kim SD, Kim JY, Baek SY, Chung BS, Kang CD. Preventive effect of antioxidants in MPTP-induced mouse model of Parkinson's disease. *eurosci Lett.* 2004;363(3):243-6

Paxinou E, Chen Q, Weisse M, Giasson BI, Norris EH, Rueter SM, Trojanowski JQ, Lee VM, Ischiropoulos H. Induction of alpha-synuclein aggregation by intracellular nitrative insult. *J Neurosci.* 2001;21(20):8053-61.

Pedersen WA, Fu W, Keller JN, Markesbery WR, Appel S, Smith RG, Kasarskis E, Mattson MP. Protein modification by the lipid peroxidation product 4-hydroxynonenal in the spinal cords of amyotrophic lateral sclerosis patients. *Ann Neurol.* 1998;44(5):819-24.

Perry TL, Godin DV, Hansen S. Parkinson's disease: a disorder due to nigral glutathione deficiency? *Neurosci Lett.* 1982;33(3):305-10.

Perry TL, Yong VW. Idiopathic Parkinson's disease, progressive supranuclear palsy and glutathione metabolism in the substantia nigra of patients. *Neurosci Lett.* 1986; 67(3):269-74.

Petrucelli L, O'Farrell C, Lockhart PJ, Baptista M, Kehoe K, Vink L, Choi P, Wolozin B, Farrer M, Hardy J, Cookson MR. Parkin protects against the toxicity associated with mutant alpha-synuclein: proteasome dysfunction selectively affects catecholaminergic neurons. *Neuron.* 2002;36(6):1007-19

Pitkanen S and Robinson BH. Mitochondrial complex I deficiency leads to increased production of superoxide radicals and induction of superoxide dismutase. *J Clin Invest.* 1996;98:345-351.

Polymeropoulos MH, Lavedan C, Leroy E, Ide SE, Dehejia A, Dutra A, Pike B, Root H, Rubenstein J, Boyer R, Stenroos ES, Chandrasekharappa S, Athanassiadou A, Papapetropoulos T, Johnson WG, Lazzarini AM, Duvoisin RC, Di Iorio G, Golbe LI, Nussbaum RL. Mutation in the alpha-synuclein gene identified in families with Parkinson's disease. *Science.* 1997;276(5321):2045-7.

Przedborski S, Vila M. The 1-methyl-4-phenyl-1,2,3,6-tetrahydropyridine mouse model: a tool to explore the pathogenesis of Parkinson's disease. *Ann N Y Acad Sci.* 2003;991:189-98

Quijano C, Alvarez B, Gatti RM, Augusto O, Radi R. Pathways of peroxynitrite oxidation of thiol groups. *Biochem J.* 1997;322(1):167-173.

Rabinovic AD, Hastings TG. Role of endogenous glutathione in the oxidation of dopamine. *J Neurochem.* 1998;71(5):2071-8.

Rabinovic AD, Lewis DA, Hastings TG. Role of oxidative changes in the degeneration of dopamine terminals after injection of neurotoxic levels of dopamine. *Neuroscience.* 2000;101(1):67-76.

Riederer P, Sofic E, Rausch WD, Schmidt B, Reynolds GP, Jellinger K, Youdim MB. Transition metals, ferritin, glutathione, and ascorbic acid in parkinsonian brains. *J Neurochem.* 1989;52(2):515-20.

Rochet JC, Outeiro TF, Conway KA, Ding TT, Volles MJ, Lashuel HA, Bieganski RM, Lindquist SL, Lansbury PT. Interactions among alpha-synuclein, dopamine, and biomembranes: some clues for understanding neurodegeneration in Parkinson's disease. *J Mol Neurosci.* 2004;23(1-2):23-34.

Saggu H, Cooksey J, Dexter DT, et al. A selective increase in particulate superoxide dismutase activity in parkinsonian substantia nigra. *J Neurochem.* 1989;53:692-697.

Sanchez-Ramos JR, Overvik E, Ames BN. A marker of oxyradical-mediated DNA damage (8-hydroxy-2' deoxyguanosine) is increased in nigro-striatum of Parkinson's disease brain. *Neurodegeneration.* 1994; 3:197-204.

Samii A, Nutt JG, Ransom BR. Parkinson's disease. *Lancet.* 2004;363(9423):1783-93.

Schapira AH, Cooper JM, Dexter D, Clark JB, Jenner P, Marsden CD. Mitochondrial complex I deficiency in Parkinson's disease. *J Neurochem.* 1990;54(3):823-7

Schulenberg B, Aggeler R, Beechem JM, Capaldi RA, Patton WF. Analysis of steady-state protein phosphorylation in mitochondria using a novel fluorescent phosphosensor dye. *J Biol Chem.* 2003;278(29):27251-5.

Schulz JB, Henshaw DR, Matthews RT, Beal MF. Coenzyme Q10 and nicotinamide and a free radical spin trap protect against MPTP neurotoxicity. *Exp Neurol.* 1995;132(2): 279-83.

Scotcher KP, Irwin I, DeLanney LE, Langston JW, Di Monte D. Effects of 1-methyl-4-phenyl-1,2,3,6-tetrahydropyridine and 1-methyl-4-phenylpyridinium ion on ATP levels of mouse brain synaptosomes. *J Neurochem.* 1990;54(4):1295-301.

Selley ML. (E)-4-hydroxy-2-nonenal may be involved in the pathogenesis of Parkinson's disease. *Free Radic Biol Med.* 1998; 25(2):169-74.

Sherer TB, Betarbet R, Stout AK, Lund S, Baptista M, Panov AV, Cookson MR, Greenamyre JT. An in vitro model of Parkinson's disease: linking mitochondrial impairment to altered alpha-synuclein metabolism and oxidative damage. *J Neurosci.* 2002;22(16):7006-15.

Sherer TB, Kim JH, Betarbet R, Greenamyre JT. Subcutaneous rotenone exposure causes highly selective dopaminergic degeneration and alpha-synuclein aggregation. *Exp Neurol.* 2003a;179(1):9-16.

Sherer TB, Betarbet R, Testa CM, Seo BB, Richardson JR, Kim JH, Miller GW, Yagi T, Matsuno-Yagi A, Greenamyre JT. Mechanism of toxicity in rotenone models of Parkinson's disease. *J Neurosci.* 2003b;23(34):10756-64.

Shimura-Miura H, Hattori N, Kang D, Miyako K, Nakabeppu Y, Mizuno Y. Increased 8-oxo-dGTPase in the mitochondria of substantia nigral neurons in Parkinson's disease. *Ann Neurol.* 1999;46(6):920-4.

Sheehan JP, Swerdlow RH, Parker WD, Miller SW, Davis RE, Tuttle JB. Altered calcium homeostasis in cells transformed by mitochondria from individuals with Parkinson's disease. *J Neurochem.* 1997;68(3):1221-33.

Shoffner JM, Watts RL, Juncos JL, Torroni A, Wallace DC. Mitochondrial oxidative phosphorylation defects in Parkinson's disease. *Ann Neurol.* 1991; 30(3):332-9.

Sian J, Dexter DT, Lees AJ, Daniel S, Agid Y, Javoy-Agid F, Jenner P, Marsden CD. Alterations in glutathione levels in Parkinson's disease and other neurodegenerative disorders affecting basal ganglia. *Ann Neurol.* 1994a;36(3):348-55.

Sian J, Dexter DT, Lees AJ, Daniel S, Jenner P, Marsden CD. Glutathione-related enzymes in brain in Parkinson's disease. *Ann Neurol.* 1994b;36(3):356-61.

Singleton AB, Farrer M, Johnson J, Singleton A, Hague S, Kachergus J, Hulihan M, Peuralinna T, Dutra A, Nussbaum R, Lincoln S, Crawley A, Hanson M, Maraganore D, Adler C, Cookson MR, Muenter M, Baptista M, Miller D, Blancato J, Hardy J, Gwinn-Hardy K. alpha-Synuclein locus triplication causes Parkinson's disease. *Science.* 2003;302(5646):841

Sofic E, Lange KW, Jellinger K, Riederer P. Reduced and oxidized glutathione in the substantia nigra of patients with Parkinson's disease. *Neurosci Lett.* 1992;142(2):128-30.

Sofic E, Paulus W, Jellinger K, Riederer P, Youdim MB. Selective increase of iron in substantia nigra zona compacta of parkinsonian brains. *J Neurochem.* 1991;56(3):978-82.

Souza JM, Giasson BI, Chen Q, Lee VM, Ischiropoulos H. Dityrosine cross-linking promotes formation of stable alpha -synuclein polymers. Implication of nitrative and oxidative stress in the pathogenesis of neurodegenerative synucleinopathies. *J Biol Chem.* 2000;275(24):18344-9.

Spencer JP, Jenner P, Daniel SE, Lees AJ, Marsden DC, Halliwell B. Conjugates of catecholamines with cysteine and GSH in Parkinson's disease: possible mechanisms of formation involving reactive oxygen species. *J Neurochem.* 1998;71(5):2112-22.

Spillantini MG, Crowther RA, Jakes R, Hasegawa M, Goedert M. alpha-Synuclein in filamentous inclusions of Lewy bodies from Parkinson's disease and dementia with lewy bodies. *Proc Natl Acad Sci U S A.* 1998;95(11):6469-73.

Swerdlow RH, Parks JK, Miller SW, Tuttle JB, Trimmer PA, Sheehan JP, Bennett JP Jr, Davis RE, Parker WD Jr. Origin and functional consequences of the complex I defect in Parkinson's disease. *Ann Neurol.* 1996;40(4):663-71.

Taira T, Saito Y, Niki T, Iguchi-Ariga SM, Takahashi K, Ariga H. DJ-1 has a role in antioxidative stress to prevent cell death. *EMBO Rep.* 2004;5(2):213-8.

Takahashi-Niki K, Niki T, Taira T, Iguchi-Ariga SM, Ariga H. Reduced anti-oxidative stress activities of DJ-1 mutants found in Parkinson's disease patients. *Biochem Biophys Res Commun.* 2004;320(2):389-97

Tretter L, Sipos I, Adam-Vizi V. Initiation of neuronal damage by complex I deficiency and oxidative stress in Parkinson's disease. *Neurochem Res.* 2004;29(3):569-77.

Tse DC, McCreery RL, Adams RN. Potential oxidative pathways of brain catecholamines. *J Med Chem.* 1976; 19, 37-40.

Turrens JF. Mitochondrial formation of reactive oxygen species. *J Physiol.* 2003;552 (Pt 2):335-44.

Valente EM, Abou-Sleiman PM, Caputo V, Muqit MM, Harvey K, Gispert S, Ali Z, Del Turco D, Bentivoglio AR, Healy DG, Albanese A, Nussbaum R, Gonzalez-Maldonado R, Deller T, Salvi S, Cortelli P, Gilks WP, Latchman DS, Harvey RJ, Dallapiccola B, Auburger G, Wood NW. Hereditary early-onset Parkinson's disease caused by mutations in PINK1. *Science.* 2004;304(5674):1158-60.

van Bladeren PJ. Glutathione conjugation as a bioactivation reaction. *Chem Biol Interact.* 2000; 129(1-2):61-76.

van der Walt JM, Nicodemus KK, Martin ER, Scott WK, Nance MA, Watts RL, Hubble JP, Haines JL, Koller WC, Lyons K, Pahwa R, Stern MB, Colcher A, Hiner BC, Jankovic J, Ondo WG, Allen FH Jr, Goetz CG, Small GW, Mastaglia F, Stajich JM, McLaurin AC, Middleton LT, Scott BL, Schmechel DE, Pericak-Vance MA, Vance JM. Mitochondrial polymorphisms significantly reduce the risk of Parkinson disease. *Am J Hum Genet.* 2003;72(4):804-11

Vaughan JR, Farrer MJ, Wszolek ZK, Gasser T, Durr A, Agid Y, Bonifati V, DeMichele G, Volpe G, Lincoln S, Breteler M, Meco G, Brice A, Marsden CD, Hardy J, Wood NW. Sequencing of the alpha-synuclein gene in a large series of cases of familial Parkinson's disease fails to reveal any further mutations. The European Consortium on Genetic Susceptibility in Parkinson's Disease (GSPD). *Hum Mol Genet.* 1998;7(4):751-3.

Votyakova TV, Reynolds IJ. DeltaPsi(m)-Dependent and -independent production of reactive oxygen species by rat brain mitochondria. *J Neurochem.* 2001;79:266-277.

Vyssokikh MY, Brdiczka D. The function of complexes between the outer mitochondrial membrane pore (VDAC) and the adenine nucleotide translocase in regulation of energy metabolism and apoptosis. *Acta Biochim Pol.* 2003;50(2):389-404.

Wakamatsu K, Fujikawa K, Zucca FA, Zecca L, Ito S. The structure of neuromelanin as studied by chemical degradative methods. *J Neurochem.* 2003;86(4):1015-23.

Wang WW, Khajavi M, Patel BJ, Beach J, Jankovic J, Ashizawa T. The G209A mutation in the alpha-synuclein gene is not detected in familial cases of Parkinson disease in non-Greek and/or Italian populations. *Arch Neurol.* 1998;55(12):1521-3.

Wilson MA, Collins JL, Hod Y, Ringe D, Petsko GA. The 1.1-A resolution crystal structure of DJ-1, the protein mutated in autosomal recessive early onset Parkinson's disease. *Proc Natl Acad Sci U S A.* 2003;100(16):9256-61.

Xu Y, Stokes AH, Roskoski R Jr, Vrana KE. Dopamine, in the presence of tyrosinase, covalently modifies and inactivates tyrosine hydroxylase. *J Neurosci Res.* 1998;54(5):691-7.

Yokota T, Sugawara K, Ito K, Takahashi R, Ariga H, Mizusawa H.Down regulation of DJ-1 enhances cell death by oxidative stress, ER stress, and proteasome inhibition. *Biochem Biophys Res Commun.* 2003;312(4):1342-8.

Yoritaka A, Hattori N, Uchida K, Tanaka M, Stadtman ER, Mizuno Y. imunohistochemical detection of 4-hydroxynonenal protein adducts in Parkinson disease. *Proc Natl Acad Sci U S A.* 1996;93(7):2696-701.

Yoshida E, Mokuno K, Aoki S-I, et al. Cerebrospinal fluid levels of superoxide dismutases in neurological diseases detected by sensitive enzyme immunoassays. *J Neurol Sci.* 1994;124:25-31.

Youdim MB, Ben Shachar D, Riederer P. Is Parkinson's disease a progressive siderosis of substantia nigra resulting in iron and melanin induced neurodegeneration? Acta Neurol Scand Suppl 1989; 126: 47-54.

Zecca L, Zucca FA, Wilms H, Sulzer D. Neuromelanin of the substantia nigra: a neuronal black hole with protective and toxic characteristics. *Trends Neurosci.* 2003;26(11): 578-80.

Zhang Y, Dawson VL, Dawson TM. Oxidative stress and genetics in the pathogenesis of Parkinson's disease. *Neurobiol Dis.* 2000; 7(4):240-50.

Zhang J, Perry G, Smith MA, Robertson D, Olson SJ, Graham DG, Montine TJ. Parkinson's disease is associated with oxidative damage to cytoplasmic DNA and RNA in substantia nigra neurons. *Am J Pathol.* 1999;154(5):1423-9.

In: Fatty Acids and Oxidative Stress...
Editors: R. Reddy and J.K. Yao, pp. 77-97
ISBN: 978-1-60021-677-0

Chapter VI

Oxidative Stress and Altered Sphingolipid Metabolism in Alzheimer's Disease

Ward A. Pedersen and Mark P. Mattson

Introduction

It would be impossible to find someone that could seriously argue against the involvement of oxidative stress in Alzheimer's disease (AD). Oxidative stress is now as much a part of the core of AD pathogenesis as insulin resistance is to type 2 diabetes or human immunodeficiency virus is to AIDS. With more than one thousand publications on the topic, an extensive list of markers of oxidative damage and lipid peroxidation in AD tissues have been analyzed. These include increased levels of the lipid peroxidation products acrolein (Calingasan et al., 1999; Lovell et al., 2001) and 4-hydroxy-2-nonenal (HNE; Lovell et al., 1997; Montine et al., 1997; Sayre et al., 1997; Markesbery and Lovell, 1998), protein nitrotyrosine and dityrosine (Hensley et al., 1998), carbonyls and advanced glycation endproducts (Picklo et al., 2002), hydroperoxyacids generated from 12/15-lipoxygenase (Pratico et al., 2004), isoprostanes and neuroprostanes (Pratico et al., 1998; Reich et al., 2001; Tuppo et al., 2001; Yao et al., 2003), and 8-hydroxyguanosine (Nunomura et al., 1999; Abe et al., 2002). Increased oxidative modification of lipids, proteins, and DNA/RNA have been well documented in vulnerable regions of AD brain, well beyond that which occurs in normal aging. There is also evidence from animal models of AD (Pratico et al., 2001) and from studies of human AD tissue (Nunomura et al., 2001) that oxidative stress and lipid peroxidation occur as early events in the course of AD. Thus, oxidative stress does not appear to be merely a consequence of disease processes, but rather may have a causative role in AD pathogenesis.

Deposition of the amyloid β-peptide (Aβ) is a neuropathological hallmark of AD, and a wealth of evidence exists supporting a central role for Aβ in oxidative stress in AD brain. It was shown more than 10 years ago that Aβ and its fragments are capable of producing reactive peptide free radicals and reactive oxygen species (ROS), and free radical-induced oxidative damage to lipids, proteins, and DNA/RNA occur extensively in regions of Aβ accumulation (Butterfield et al., 2001). Prominent substrates of Aβ -induced oxidation are lipoproteins, such as apolipoprotein E (ApoE)-containing lipoproteins (Bassett and Montine, 2003). Indeed, ApoE has been shown to protect against oxidative stress and lipid peroxidation in an isoform-specific manner (Miyata and Smith, 1996; Pedersen et al., 2000). ApoE functions as the major cholesterol carrier protein in the brain, and the role of cholesterol in Aβ production and clearance has generated considerable interest in the AD field (Puglielli et al., 2003b). It has been shown that there is a correlation between Aβ production and the levels of cholesterol esters, generated by activation of acyl-coenzyme A:cholesterol acyltransferase (ACAT) (Wolozin et al., 2004). This enzyme may be indirectly regulated by ceramide, which is generated by *de novo* sphingolipid biosynthesis and by the turnover of complex sphingolipids. Thus, altered sphingolipid metabolism can have effects on cholesterol metabolism and the generation of Aβ, which, in turn, promotes oxidative stress. Sphingolipid metabolism is disrupted by increased oxidative stress, resulting in a vicious cycle that culminates in profound impairments in neuronal function and ultimately neuronal degeneration. Here, we review the evidence for a relationship between oxidative stress and altered sphingolipid metabolism in the pathogenesis of AD. Potential new targets for AD therapy have emerged from recent studies in this field.

Sphingolipid Metabolism and Alterations in AD

Our understanding of the roles of sphingolipids have expanded greatly beyond that of simply maintaining cell structure, to include the regulation of cell proliferation, differentiation, and survival (Cutler and Mattson, 2001). The functions of complex sphingolipids, i.e., sphingomyelins, cerebrosides, and gangliosides, have been studied extensively for many years in both neuronal and non-neuronal cells. However, more recent attention has been given to simple sphingolipids, such as ceramide and sphingosine-1-phosphate (S1P). The abundance of sphingolipids in cells of neuronal origin underscores the important roles they play in the nervous system, with ceramide, for example, established as an important second messenger in signaling pathways that control axonal outgrowth and neuronal apoptosis during development (Buccoliero and Futerman, 2003).

Ceramides can be produced *de novo* or by sphingomyelin hydrolysis through the activities of sphingomyelinases (SMases; Fig. 1; Cutler and Mattson, 2001; Menaldino et al., 2003). In *de novo* synthesis, serine palmitoyltransferase (SPT) catalyses the condensation of palmitoyl-CoA and serine to form dehydrosphingosine, which is then converted into sphingosine; fatty acids are then attached to sphingosine via N-acylation to form ceramide, from which the terminal hydroxyl group is substituted to form sphingomylein, cerebroside, and gangliosides. The extensive diversity of sphingolipids arises from the various long-chain sphingoid bases, with sphingosines, sphinganines, and 4-hydroxysphinganines being the most common in mammalian cells, from differences in the amide-linked fatty acids of the ceramide moiety, with typical chain lengths ranging from 16-24 carbon atoms, and from numerous polar head groups (Mendalino et al., 2003).

All organisms that synthesize sphingolipids require SPT, but its activity is also critical in maintaining sphingolipid levels such that an accumulation of metabolites, particularly ceramide, do not reach toxic concentrations. The critical role for SPT in cellular sphingolipid homeostasis is demonstrated by the fact that its activity is regulated by transcriptional and post-transcriptional mechanisms in response to a variety of stimuli (Perry, 2002; Hanada, 2003). In primary cultures of cerebellar neurons, for example, the addition of sphingoid bases causes a down-regulation of SPT activity in the absence of effects on certain other enzymes of lipid metabolism (Mandon et al., 1991). Stimuli that up-regulate SPT activity include ultraviolet light B, endotoxins, interleukins, tumor necrosis factor-α (TNFα), the DNA-damaging agent etoposide, fatty acids, and deficiency of ApoE. Thus, while SPT activity can be down-regulated as an important homeostatic control mechanism, many cellular responses to injury involve increased SPT activity and the consequent effects on sphingolipid biosynthesis.

The activation of SMases, namely the lysosomal acidic and neutral SMases, also appear to play important roles in regulating sphingolipid levels in response to cellular injury. In non-neuronal cells, both acidic and neutral SMases are activated by a variety of extracellular stimuli, including cytokines, growth factors, differentiating agents/hormones, and stress inducers (Levade and Jaffrézou, 1999). While less is know about SMase activation in neuronal cells, there is evidence that increased ceramide production mediates nerve growth factor (NGF)-induced death of cultured hippocampal neurons through activation of neutral SMase (Brann et al., 2002) and excitotoxic (glutamate)-induced death of cultured cortical neurons through activation of acidic SMase (Yu et al., 2001). While the initial view of the subcellular localization of SMases considered acidic SMase as being confined to lysosomes and related acidic compartments, recent findings suggest that acidic SMase translocates to the plasma membrane in response to stimuli such as Fas ligand and oxidative stress (Cremesti et al., 2003). It has been proposed that the translocation of acidic SMase to the plasma membrane involves fusion of acidic SMase-rich vesicles with the membrane (Gulbins and Kolesnick, 2003). Studies of non-neuronal cells from patients with acidic SMase deficiency and from acidic SMase-deficient mice have shown that this enzyme is essential for ceramide production and apoptosis induced by radiation (Santana et al., 1996; Garcia-Barros et al., 2003). Our own studies have shown that acidic SMase is required for the generation of ceramide in neurons in response to ischemic and excitotoxic insults (Yu et al., 2001). The latter study further showed that ceramide generated by acidic SMase activation is required for the production of pro-inflammatory cytokines and neuronal damage induced by a stroke. It therefore appears that acidic SMase plays a major role in ceramide production in response to the activation of death receptors and oxidative/metabolic stress. The role of neutral SMase in these settings remains to be determined.

The extensive amount of data regarding the involvement of ceramides in the regulation of cell death processes and the considerable evidence that apoptosis occurs in AD, Parkinson's disease (PD), and amyotrophic lateral sclersosis (ALS), has led to the proposal that altered sphingolipid metabolism is a common pathophysiological event in neurodegenerative disorders (Ariga et al., 1998). While this proposal has validity, it has not yet been established to what extent altered sphingolipid metabolism contributes to the pathogenesis of neurodegenerative disorders versus to what extent this results from the neurodegenerative

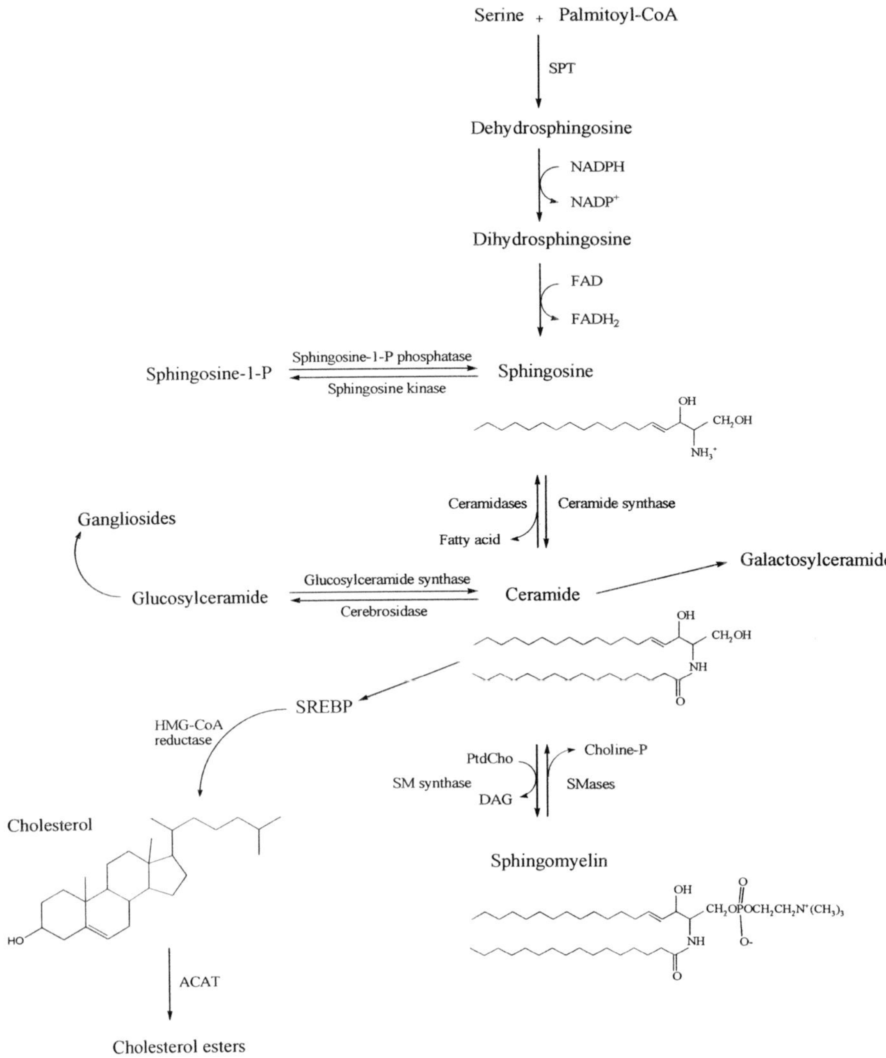

Figure 1. Sphingolipid biosynthetic and metabolic pathways. The main pathways leading to ceramide accumulation involve activation of SPT with sphingosine as an intermediate (*de novo* synthesis) or SM hydrolysis following activation of SMases. Ceramide causes maturation of SREBP, thereby affecting the expression of genes in the cholesterol biosynthetic pathway. The complexity of sphingolipid pathways affords the opportunity to manipulate ceramide levels in a number of ways. Not shown is the conversion of ceramide to ceramide-1-P catalyzed by ceramide kinase, the conversion of sphingosine-1-P to ethanolamine-P and fatty aldehyde catalyzed by lyase, or the reverse reactions catalyzed by phosphatases (see Hannun and Luberto, 2000; Mendalino et al., 2003; and references therein). Abbreviations: DAG, diacylglycerol; PtdCho, phosphatidylcholine; choline-P, phosphocholine; SREBP, sterol regulatory element-binding protein; other abbreviations as defined in the text.

process. However, that alterations in sphingolipid metabolism have a causal role in diseases characterized by neuronal degeneration is evidenced by pathogenetic studies of hereditary sensory neuropathy type 1 (HSN1). This disorder results from mutations in the gene encoding the long chain base subunit-1 of SPT, which is associated with increased *de novo* synthesis of the glucosylated derivative of ceramide, the cerebroside glucosyl-ceramide (Dawkins et al.,

2001). Interestingly, chemical inhibition of glucocerebrosidase in cultured hippocampal neurons, resulting in accumulation of glycosylceramide, caused a greater rise in intracellular calcium concentration and sensitivity to cell death in response to glutamate (Korkotian et al., 1999). These results suggest a mechanistic link between a genetic mutation that disrupts sphingolipid metabolism and the degeneration of dorsal root ganglia and motor neurons characteristic of HSN1.

Recent findings suggest that altered sphingolipid metabolism has a prominent role in the pathogenesis of other neurodegenerative disorders, including ALS (Cutler et al., 2002), AD (Han et al., 2002; Cutler et al., 2004), and dementia caused by HIV (Haughey et al., 2004). Moreover, evidence inherent to these studies suggests that altered sphingolipid metabolism may actually contribute to disease processes. In our studies, we used electrospray ionization tandem mass spectrometry to measure the levels of SMs and ceramides of various chain lengths and double bond contents in post-mortem brain samples from AD patients and neurologically normal control subjects (Cutler et al., 2004). We found that there was a marked increase in the levels of ceramide C24:0 and galactosylceramide C24:0, accompanied by reduced levels of SM C24:0, in samples of the middle frontal gyrus from AD brain relative to control brain. Samples from the cerebellum showed no differences in the levels of these lipids between AD and control cases, suggesting that certain aspects of sphingolipid metabolism may be altered selectively in vulnerable regions of AD brain. One explanation for the reduced levels of SM is that phosphatidylcholine, the phospholipid used in the synthesis of SM, is also reduced in AD frontal cortex (Nitsch et al., 1992). Alternatively, or additionally, SM hydrolysis results from abnormal activation of SMases. Activation of SM hydrolysis by neutral SMase and increased ceramide levels cause maturation of the sterol regulatory element-binding protein-1 (Lawler et al., 1998), which translocates to the nucleus and activates genes involved in the cholesterol biosynthetic pathway, such as HMG-CoA reductase (Fig. 1; Brown and Goldstein, 1999). Indeed, we observed increased levels of free cholesterol in the middle frontal gyrus of AD brain concomitant with elevated ceramide and reduced SM levels (Cutler et al., 2004). The extent of increased membrane-associated ceramides and free cholesterol levels was correlated with the degree of dementia severity of the patients, inviting the possibility that altered sphingolipid metabolism plays a role in mechanisms underlying cognitive impairment in AD.

Altered sphingolipid metabolism occurs in white matter as well as in gray matter of AD brain. White matter axons serve important roles in cognitive functions (Filley, 2001), and white matter degeneration may contribute significantly to the pathogenesis of AD. In one study, the concentration of gangliosides was found to be markedly reduced in gray matter from frontal cortex, temporal cortex, caudate nucleus and hippocampus, and in white matter from frontal cortex of AD brain relative to controls (Svennerholm and Gottfries, 1994). In addition, the results of this study showed a 38% reduction in sulfatide levels in frontal cortex white matter of AD brain relative to control brain. The sulfatides are a class of sulfated galactocerebrosides that are synthesized almost exclusively by oligodendrocytes in the CNS. More recently, sulfatide depletions of up to 93% in gray matter and up to 58% in white matter were reported, which did not appear to be due to a deficiency in biosynthesis as indicated by no differences in galactocerebroside sulfotransferase activities (Han et al., 2002). As sulfatide deficiencies were found in samples from AD patients with very mild dementia, i.e., clinical dementia ratings of 0.5 (Han et al., 2002; 2003b), the results suggest an involvement of this sphingolipid abnormality in early cognitive processes in AD. Increased ceramide levels relative to controls, specifically the molecular species C18:0 and C24:1, were also observed in white matter samples from AD brain relative to control brain (Han et al., 2002). Ceramides can result from the degradation of sulfatides, and the increased content of ceramides was correlated with the observed sulfatide deficiency in white matter. Notably, the increased ceramide content was found in cortical regions as well as in the cerebellum, with the magnitude of the increase being similar for both regions, indicating that certain sphingolipid abnormalities are widespread in AD brain. Overall, white matter alterations in sphingolipid metabolism may have a prominent role in the pathogenesis of AD, possibly contributing to axonal demyelination and dysfunction.

Oxidative Stress in the Alterations of Sphingolipid Metabolism in AD

Despite the wealth of evidence for oxidative damage in AD brain, the precise mechanisms by which increased levels of oxidative stress exert pathophysiological effects in AD remain rather obscure. However, the evidence now exists that altered sphingolipid metabolism may be a key component of the mechanisms underlying oxidative stress-induced pathology in AD and certain other neurodegenerative disorders. In contrast to HSN1, where mutations in a gene encoding an enzyme directly involved in sphingolipid biosynthetic pathways occur, altered sphingolipid metabolism in AD and certain other neurodegenerative disorders may occur in response to cellular oxidative stress. Perhaps the best evidence for a cause-and-effect relationship between oxidative stress and altered sphingolipid metabolism is that obtained from studies of transgenic ALS mice. Mutations in the gene encoding the antioxidant enzyme Cu/Zn-superoxide dismutase (SOD1) account for about 1-2% of the total cases of ALS, and, as in humans, transgenic mice overexpressing mutant human SOD1 develop progressive muscle weakness and ultimately paralysis due to the degeneration of spinal cord motor neurons (Cleveland and Rothstein, 2001). Mutations in SOD1 alter the chemistry of the enzyme resulting in the use of aberrant substrates, such as hydrogen peroxide, thus increasing hydroxyl radical formation, and peroxynitrite, which leads to protein nitration. In mice overexpressing mutant human SOD1, we observed an increase in membrane lipid peroxidation concomitant with higher levels of ceramides in lumbar spinal cord samples (Cutler et al., 2002). Importantly, these increases could be detected prior to the onset of symptoms, suggesting not only that oxidative stress and membrane lipid peroxidation result in ceramide accumulation, but that oxidative stress-induced alterations in sphingolipid metabolism may trigger events that ultimately precipitate the expression of full-blown ALS.

There is currently no direct evidence that increased levels of oxidative stress, due to Aβ burden, alter sphingolipid metabolism in AD brain. Nonetheless, the results of several studies have demonstrated a relationship between oxidative stress and increased ceramide production in non-neuronal cells (Andrieu-Abadie et al., 2001). For example, in cultures of microcapillary glomerular endothelial cells, exposure to superoxide-generating agents caused a marked increase in the levels of ceramides, effects that were prevented by pretreatment with the antioxidants glutathione and *N*-acetylcysteine (Huwiler et al., 2001). An important implication of this study is that oxidative stress acts upstream of ceramide production. Similar mechanisms are now being revealed in neuronal cells. In primary cultures of hippocampal neurons from rat, treatment with Aβ 1-42 increased the levels of ceramides C18:0 and C24:0, and free cholesterol (Cutler et al., 2004). In conjunction, treatment with Aβ peptides caused membrane lipid peroxidation, as indicated by the levels of free HNE and the adducts lysine-HNE and histidine-HNE. All of these effects of Aβ were found to be attenuated by a pretreatment with α -tocopherol (vitamin E). It was also recently shown that Aβ(25-35)-induced ceramide production in C6 glioma cells is attenuated by vitamin E (Ayasolla et al., 2004). While the latter studies were carried out in cell culture, it was recently reported that, in rats, a single intracerebral injection of Aβ (25-35) peptide caused increased SMase activity and ceramide production selectively in the hippocampus (Alessenko et al., 2004). These effects were preceded by increased levels of lipid peroxidation products (conjugated dienes and ketodienes).

There are several potential ways by which oxidative stress-induced ceramide accumulation can occur. These include increased synthesis by activation of SPT, hydrolysis of sphingomyelins by SMases, or defective transport to the Golgi apparatus where ceramide is used as a substrate in the synthesis of complex sphingolipids. In regards to the latter, oxidative stress induced in non-neuronal cells by a photosensitizing agent caused an increase in *de novo* ceramide accumulation in the absence of effects on SPT activity or sphingomyelin degradation (Dolgachev et al., 2004). Moreover, ATP-dependent transport of ceramide from the endoplasmic reticulum (ER) to the Golgi is inhibited following hypoxic injury in oligodendrocytes (Kendler and Dawson, 1992). Whether such effects occur in AD brain are unknown at present. However, SPT and SMases are activated in response to various insults that involve oxidative stress as a component of their cytotoxic effects (Levade and Jaffrézou, 1999; Andrieu-Abadie and Levade, 2002; Perry, 2002; Hanada, 2003). Superoxide-generating agents increase not only ceramide levels in glomerular endothelial cells, but also the activities of neutral and acidic SMases, without effects on neutral or acidic ceramidase activities (Huwiler et al., 2001). In human breast cancer cells, overexpression of glutathione peroxidase-1 prevented ROS generation, activation of neutral SMase, SM hydrolysis and ceramide production in response to a chemotherapeutic agent (Gouaze et al., 2001). We found that Aβ 1-42 treatment of primary hippocampal neurons decreased SM C24:0 levels (Cutler et al., 2004), suggesting SM hydrolysis. Yet, we did not measure SPT or SMase activities upon Aβ treatment, and we cannot rule out a contribution from impaired synthesis of complex sphingolipids due to defective ceramide transport.

It is not yet known which aspects of sphingolipid metabolism are affected by Aβ peptides to result in ceramide accumulation. At present, we cannot rule out any of the above-mentioned mechanisms, and all may contribute to ceramide accumulation in neuronal cells (Fig. 1). It is even less clear how Aβ affects the activities of enzymes involved in sphingolipid metabolism. One possibility is that SMases, in particular, are sensitive to the redox state of the cell. For instance, there is evidence that a magnesium-dependent neutral SMase purified from rat brain is coupled to redox state (Sawai and Hannun, 1999; Martin et al., 2003) and can be inhibited by reduced glutathione but not oxidized glutathione (Liu and Hannun, 1997; Huwiler et al., 2001). In human neuroblastoma cells, Aβ peptides, in the presence of advanced glycation endproducts, increase the ratio of oxidized to reduced glutathione subsequent to superoxide and hydrogen peroxide production (Deuther-Conrad et al., 2001). The generation of ROS follows Aβ reduction of redox-active Cu^{2+} and Fe^{3+} (Huang et al., 2004), which in turn promotes aggregation of Aβ (Mattson, 2004). Indeed, markedly increased levels of Cu, Fe and Zn occur in the neuropil of AD brain relative to controls, and are concentrated in plaques (Lovell et al., 1998). In AD brain, therefore, ROS formed as a result of the interaction between certain biometals and Aβ, may change the cellular redox state and decrease the levels of reduced glutathione, thereby activating SMases and promoting ceramide accumulation (Fig. 2). There is currently no direct evidence that Aβ can affect SPT activity. Yet, in primary cultures of hippocampal neurons, we found that the SPT inhibitor ISP-1 attenuated Aβ -induced increases in ceramide and free cholesterol levels, similar to what was observed with α-tocopherol (Cutler et al., 2004). Thus, Aβ -induced oxidative stress may alter the activities of a number of enzymes involved in sphingolipid metabolism and the regulation of ceramide levels.

A potentially important consideration in the relationship between Aβ -induced oxidative stress and altered sphingolipid metabolism in AD is that of the role of ApoE. Of the three alleles of *APOE*, encoding the isoforms E2, E3 and E4, inheritance of the □4 allele increases the risk of AD. ApoE is up-regulated in response to neuronal injury, with differential effects of the three isoforms on repair mechanisms involving neuronal sprouting (Teter, 2004). Moreover, there are differential effects of ApoE isoforms on metal-induced aggregation of Aβ and on the ability of cells to cope with Aβ -induced oxidative stress and membrane lipid peroxidation. In particular, ApoE4 shows the greatest ability to promote Aβ aggregation in the presence of zinc or copper (Moir et al., 1999), and has the least ability to prevent Aβ -induced oxidative damage to mouse synaptosomes (Lauderback et al., 2002) or to prevent Aβ /HNE-induced cell death of cultured neurons (Pedersen et al., 2000). Notably, as shown in mice, ApoE may modulate brain sulfatide content (Han et al., 2003a). Particularly noteworthy is that mice expressing a transgene encoding human ApoE4 have lower brain sulfatide levels than mice expressing a transgene for human ApoE3, with the levels in both groups of mice being lower than in their wild-type counterparts. Thus, ApoE genotype may influence sphingolipid metabolism in AD, possibly an indirect effect related to the differential abilities of the isoforms to regulate cellular oxidative stress levels and the consequent effects on the activities of enzymes involved in sphingolipid pathways.

Potential Roles of Ceramide in the Pathogenesis of AD

Any investigation of the involvement of altered sphingolipid metabolism in the pathogenesis of AD will inevitably focus on the role of ceramide, as it has been the most studied sphingolipid mediator of cell death and survival. Ceramide is now recognized as a key regulator of neuronal growth, differentiation, and cell death by apoptosis during development (Futerman et al., 1998; Mitoma et al., 1998; Herget et al., 2000; Bieberich et al., 2001; Buccoliero and Futerman, 2003). The balance between survival and death is governed by the levels of ceramide reached within a cell, with high concentrations leading to apoptosis. The classical *in vitro* paradigm of neuronal apoptosis during development is that of NGF withdrawal from cultured sympathetic neurons, and exogenously added ceramide can prevent apoptosis in this model (Ito and Horigome, 1995; Nair et al., 2000). However, NGF causes a switch from axonal outgrowth to apoptosis in developing hippocampal neurons in culture, an effect mediated by increased expression of the p75 NGF receptor ($p75^{NGFR}$) and increased ceramide levels arising from SM hydrolysis (Buccoliero and Futerman, 2003). It has been proposed that many aspects of the neurodegenerative process in the adult brain recapitulate regulated cell death during development, and the possibility of ceramide-induced apoptosis in AD brain is consistent with this proposal.

Ceramide has now been shown to induce apoptosis in various neuronal cell types, including hippocampal neurons (Brann et al., 2002), cortical neurons (Yu et al., 2001; Toman et al., 2002), cerebellar granule cells (Taniwaki et al., 1999; Toman et al., 2002), mesenchephalic neurons (Brugg et al., 1996), sensory neurons from dorsal root ganglia (Ping and Barrett, 1998), and spinal cord motor neurons (Irie and Hirabayashi, 1998). The ability of ceramide to induce apoptosis also extends to glial cells. For instance, ceramide generated from neutral SMase mediates oliogdendrocyte cell death upon exposure to Aβ, which could be prevented by glutathione (Lee et al., 2004). Interestingly, oligodendrocytes appear to be more sensitive than either neurons or astrocytes to ceramide-induced apoptosis (Casaccia-Bonnefil et al., 1996), but the reasons for these differential sensitivities are unclear. Ceramide may also be involved in AD pathology arising from defects in non-neuronal cells. In rat brain, ceramides potently induce alterations such as constriction of cortical venules, increased venular wall permeability, postcapillary venule rupture, and often microhemorrhaging (Altura et al., 2002). While speculative at present, it is possible that ceramide plays a role in cerebral amyloid angiopathy of AD, which occurs in about 80% of cases (Greenberg, 1998).

Glial cells may have a critical role in the alterations of sphingolipid metabolism in neurons in AD brain. It is well established that Aβ-activated microglia release cytokines, such as TNF□, responsible for the inflammatory processes that occur in AD brain. TNF□ can activate both SPT and neutral SMase (Andrieu-Abadie and Levade, 2002; Hanada, 2003), and it may potentiate Aβ neurotoxicity mediated by the death domain of $p75^{NGFR}$ (Perini et al., 2002). Moreover, it was recently shown that TNF receptor type I signaling is required for Aβ-induced apoptosis in cultured neurons (Li et al., 2004). These results suggest that, in AD brain, TNFα released from activated microglia acts synergistically with Aβ to increase ceramide production and induce apoptosis in neurons (Fig. 2). On the other hand, findings from several studies suggest that activation of TNF□ receptors in neurons can protect against death induced by Aβ and oxidative stress (Mattson and Camandola, 2001). In addition, ceramide may mediate the cell survival-promoting actions of some neurotrophic factors (DeFreitas et al., 2001). It therefore appears that the roles of ceramide in neuronal survival and neuronal degeneration are complex and dependent upon the cell type(s) involved and the particular physiological or pathological condition.

Much insight has been gained into the mechanisms by which ceramide induces neuronal apoptosis, and these mechanisms overlap with those shown to mediate the neurotoxic effects of Aβ. For instance, it has been shown in a number of studies that the c-Jun N-terminal kinase (JNK), the p38 mitogen-activated protein kinase (MAPK), p53, caspases, and NF-kB are activated by Aβ (Kuner et al., 1998; Okazawa and Estus, 2002; Wei et al., 2002; Tamagno et al., 2003) and ceramide (Mattson et al., 1997; Buccoliero and Futerman, 2003; Ruvolo, 2003; Willaime-Morawek et al., 2003). Moreover, ceramide induces apoptosis only in neural progenitor cells that also express prostate apoptosis response-4 (Bieberich et al., 2003), a potential mediator of Aβ-induced cell death (Guo et al., 1998). It is now apparent that, in neurons, ceramide evokes stress activated protein kinase-dependent pathways and "classical" apoptotic events involving Bcl-2 family members and mitochondrial alterations. For example, ceramide increases p53 expression in cortical neurons, likely as a result of JNK activation, which, in turn, activates the pro-apoptotic Bcl-2 family member Bax. Ceramide also causes dephosphorylation and inactivation of Akt (Stoica et al., 2003), which normally phosphorylates and inactivates Bad, thus initiating mitochondrial depolarization, formation of the apoptosome (cytochrome C/Apaf-1/procaspase-9), and activation of caspase-3. Caspase-8 also appears to be involved, which cleaves Bid, further promoting mitochondrial alterations (Darios et al., 2003). Consistent with these pathways, ceramide-induced apoptosis can be blocked by over-expression of Bcl-2 (Srinivasan et al., 1996). Interestingly, impaired neuronal endosomal/lysosomal functions, which occur early in AD, have been shown to cause SM hydrolysis and ceramide accumulation within endosomal/lysosomal vesicles, associated with activation of procaspases-8 and -9 and caspase-3 (Ditaranto-Desimone et al., 2003).

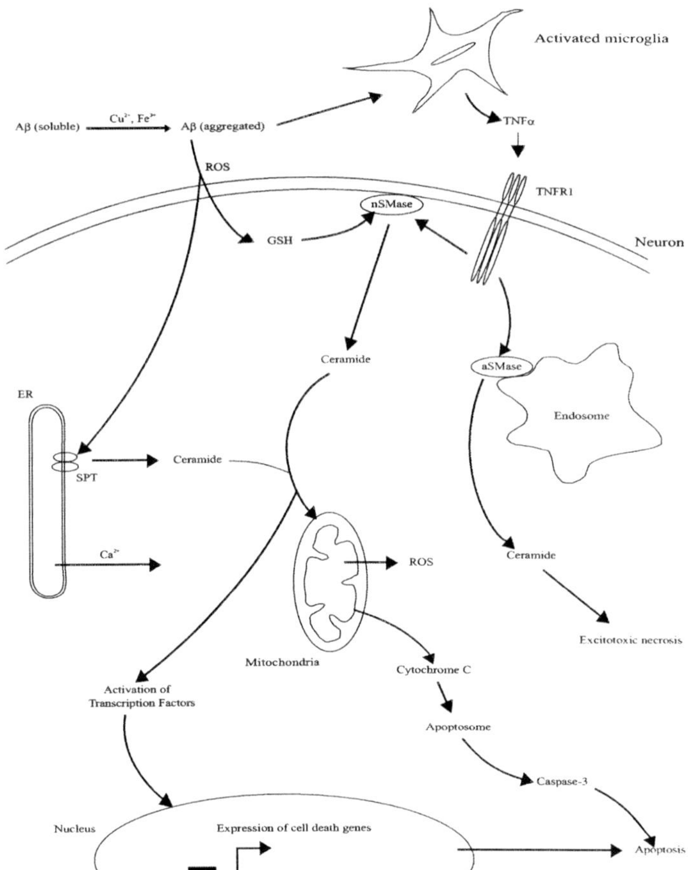

Figure 2. Potential mechanisms relating aggregation of Aβ to ceramide accumulation in neurodegeneration in AD brain. Aggregated Aβ may increase ceramide production by three distinct pathways: (1) activation of microglia, which release TNFα, resulting in TNF receptor type I (TNFR1)-mediated activation of neutral SMase (nSMase) and acidic SMase (aSMase); (2) ROS-mediated reduction in reduced glutathione (GSH) resulting in activation of nSMase; and (3) ROS-mediated activation of SPT in the ER. Ceramides generated in the aSMase pathway appear to promote excitotoxic necrosis, whereas ceramides generated in the nSMase and SPT pathways induce apoptosis. Ceramide also causes the release of calcium from ER-sensitive stores, increasing both cytosolic and mitochondrial calcium concentrations, thereby enhancing oxidative stress through ROS generated by the mitochondria.

Ceramide has been shown to affect the activities of a number of kinases, including JNK, Akt, PKC, ceramide-activated protein kinase, which activates MAPK, and the Ca^{2+}/calmodulin-regulated death-associated protein kinase (Buccoliero and Futerman, 2003; Ruvolo, 2003). Additionally, ceramide activates serine-threonine protein phosphatases, such as PPA1, which acts on cell cycle regulatory proteins, and PPA2, which inhibits the cell survival-promoting effects of Akt, PKCα, and Bcl-2 (Hannun and Luberto, 2000; Ruvolo, 2003). The cumulative effect is activation of transcription factors that increase the expression of cell death genes, at the expense of expression of genes that affect cell survival. Ceramide is also capable of potentiating cell death pathways. While ceramide levels are increased by cellular oxidative stress, ceramide can further promote oxidative stress by increasing mitochondrial ROS generation through complex I of the electron transport chain (France-Lanord et al., 1997) and by up-regulating expression of inducible nitric oxide synthase (Ayasolla et al., 2004; Won et al., 2004), leading to formation of nitric oxide which reacts with superoxide to produce peroxynitrite. Moreover, expression of the neuron-specific damage-induced neuronal endopeptidase in COS cells prevents ceramide-induced apoptosis, possibly through proteolytic cleavage and activation of antioxidant enzymes such as Cu/Zn-SOD (SOD1), Mn-SOD, and glutathione peroxidase (Kiryu-Seo et al., 2000). Ceramide causes the release of calcium from ER-sensitive stores, resulting in increased cytosolic and mitochondrial calcium levels (Darios et al., 2003) and activation of calpains (Moore et al., 2002). Disruption in calcium homeostasis further promotes ROS production, which impair ion-motive ATPases, voltage-dependent calcium channels, and glucose and glutamate membrane transport proteins (Mattson, 2002). Excess levels of glutamate can result in excitoxicity, and ceramides, generated from acidic SMase, may sensitize neurons to excitotoxic necrosis (Yu et al., 2001; 2002). Overall, one can envision a scenario wherein Aβ-induced oxidative stress causes accumulation of ceramides, which further promote ROS generation and calcium dyshomeostasis, resulting in a vicious cycle that disrupts ion homeostasis and neuronal function, and culminating in neuronal apoptosis (initiated through neutral SMase activation) and excitotoxic necrosis (initiated through acidic SMase activation) (Fig. 2).

Many of the adverse effects of ceramides in AD brain may be due to mechanisms other than direct activation of cell death pathways, many of which may be potentiated by increased levels of cholesterol. Such effects involve changes in the composition of lipid rafts, plasma membrane microdomains enriched in cholesterol and sphingolipids and sites where signaling molecules are concentrated. Thus, the finding of higher levels of ceramides and free cholesterol in membranes prepared from AD brain (Cutler et al., 2004) has implications for perturbed signaling mechanisms. For example, proper expression and functionality of the nicotinic acetylcholine receptor, critical for memory function and levels of which are reduced in the hippocampus of AD brain (Maelicke et al., 2000), appear to be very sensitive to alterations in the sphingolipid compositions of membranes (Roccamo et al., 1999). It has also now been shown that there is an interaction between Aβ and SM, between Aβ and galactosylceramide, and between Aβ and gangliosides (Ariga et al., 2001; Kakio et al., 2002; Mahfoud et al., 2002), and increased concentrations of sphingolipids in lipid rafts may have an impact on Aβ conformation and its resultant toxicity. Furthermore, increased ceramide and cholesterol levels may affect the production of Aβ. Ceramide generated from neutral SMase appears to stabilize the β-site amyloid precursor protein (APP)-cleaving enzyme, BACE1

(Puglielli et al., 2003a), required for amyloidogenic processing of APP. As ACAT is known to modulate Aβ production by exerting a tight control over the equilibrium between free cholesterol and cholesterol esters (Puglielli et al., 2003b), increased cholesterol levels may affect this balance and the levels of Aβ. Again, this increased production of Aβ would serve to perpetuate a vicious cycle of oxidative stress and altered sphingolipid metabolism, until the cell finally succumbs to death.

Future Directions: Inhibition of Ceramide and Cholesterol Production in the Treatment of AD

It is becoming clear that oxidative stress and altered sphingolipid metabolism can be integrated into a mechanistic framework that not only forms the basis for important future studies, but also which provides new approaches for therapeutic interventions in AD. We reviewed the evidence that altered sphingolipid metabolism, and accumulation of ceramides, in particular, mediates oxidative stress-induced neuronal apoptosis. Because increased oxidative stress is now recognized as invariant occurrence in AD brain, these novel insights have important implications for AD pathology and for AD therapy, where any strategy aimed at reducing cellular oxidative stress levels may inherently target altered sphingolipid metabolism and ceramide accumulation.

While ceramides can induce apoptosis in virtually every neuronal cell type, it may be the case that local increases in ceramide levels, perhaps due to Aβ accumulation, partly account for the selective vulnerability of certain types of neurons in AD brain. Altered sphingolipid metabolism may occur simultaneously in neurons and glial cells, with TNF□ being a key mediator of these effects. However, an important issue that remains to be determined is what are the relative contributions of altered sphingolipid metabolism in neurons versus glial cells to the pathogenesis of AD. In addition to gray matter, the white matter of AD brain is also burdened with increased levels of Aβ (Roher et al., 2002). Given that Aβ induces apoptosis in cultured oligodendrocytes through neutral SMase-ceramide pathways, altered sphingolipid metabolism may play a role in white matter as well as gray matter degeneration in AD brain. It should be pointed out that Aβ treatment of cultured oligodendrocytes did not result in a sulfatide deficiency as was found in AD brain (Cheng et al., 2003), suggesting that not all sphingolipid abnormalities in AD brain are directly related to effects of Aβ.

In terms of therapeutic approaches, a critical aspect of the findings of altered sphingolipid metabolism in AD is that this may be an early feature of the disease and may therefore have a causal role in its pathogenesis. This offers attractive, and potentially very effective, targets for delaying the progression of AD. Several targets are available given the complex nature of sphingolipid pathways. However, for obvious reasons, the focus will inevitably be on ceramides.

Compounds that disrupt the balance between ceramide and S1P, the so-called "sphingolipid rheostat", have been considered for some time now in the treatment of cancer (Kester and Kolesnick, 2003). The rationale for the use of such compounds in cancer treatment is to increase ceramide levels and induce apoptosis, but clearly the opposite effect would be needed for AD therapy. A reduction in endogenous ceramide levels could potentially be accomplished by conversion to other sphingolipid metabolites through the activation of ceramide kinase, SM synthase, or ceramidases, or by preventing sphingolipid metabolism through inhibition of SMases, cerebrosidase, or ceramide synthase (Fig.1). Activation of ceramide kinase or inhibition of ceramide synthase would increase sphingosine production, which *in vitro* has been shown to promote neuronal cell survival (Ping and Barrett, 1998). However, sphingosine would be converted to S1P, and while S1P can exert neuroprotective and proliferative effects, chronic exposure induces neuronal apoptosis (Moore et al., 1999). This likely explains the cytotoxicity caused by fumonisins, naturally occurring inhibitors of ceramide synthase, in certain cell types (Mendalino et al., 2003), and is the basis for the use of the sphingosine kinase inhibitor, safingol, in cancer treatment (Kester and Kolesnick, 2003). The lessons learned are that caution must be exercised when manipulating any one component of sphingolipid biosynthetic and metabolic pathways. For AD therapy, the most useful approach may lie in the dual inhibition of SPT, to prevent *de novo* synthesis of ceramides, and SMase, to prevent hydrolysis of SM, as both effects appear to occur in response to Aβ-induced oxidative stress.

There are currently no clinical trials with AD patients involving the use of drugs that target sphingolipid pathways. However, clinical trials have begun to determine if cholesterol-lowering drugs, such as the statins which inhibit HMG-CoA reductase, will be useful in the treatment of AD. This therapeutic strategy is related to altered sphingolipid metabolism in that ceramide may indirectly alter the expression of genes encoding HMG-CoA reductase and other enzymes of cholesterol biosynthesis. The benefits of lowering neuronal cholesterol levels were demonstrated by the finding that simvastatin reduces Aβ 1-40 and Aβ 1-42 levels in neuronal cells expressing mutant APP and in the CSF and brains of guinea pigs (Fassbender et al., 2001). As discussed in the previous section, a possible mechanism is that the cholesterol content of lipid rafts may affect enzymes involved in amyloidogenic processing of APP. Related to sphingolipid metabolism is the finding that simvastatin reduces cytokine expression, including TNF□, in the brains of rats subjected to hypoxic-ischemic injury (Balduini et al., 2003). This indicates an additional benefit of using statins in the treatment of AD.

The brain synthesizes most, if not all, of the cholesterol it requires, and steady-state levels are achieved by the conversion of cholesterol to, and efflux of, 24*S*-hydroxycholesterol. Increased plasma levels of 24*S*-hydroxycholesterol are an indication of neuronal injury, and noteworthy is the report that plasma 24*S*-hydroxycholesterol levels are reduced by statin treatment in AD patients (Vega et al., 2003). Is there also a benefit to reducing plasma cholesterol levels? The answer to this question appears to be yes, because diet-induced hypercholesterolemia increases brain lipid peroxidation in animals, perhaps due to increased passage of iron into the brain (Ong and Halliwell, 2004). This, in turn, may promote alterations in sphingolipid metabolism in the brain. In summary, the benefits of using cholesterol-reducing drugs in AD seem promising. These beneficial effects may be enhanced through the use of drugs that target sphingolipid metabolism, and this will undoubtedly be an exciting area of future AD research.

References

Abe T, Tohgi H, Isobe C, Murata T, Sato C. Remarkable increase in the concentration of 8-hydroxyguanosine in cerebrospinal fluid from patients with Alzheimer's disease. *J Neurosci Res.* 2002;70:447-450.

Alessenko AV, Bugrova AE, Dudnik LB. Connection of lipid peroxide oxidation with the sphingomyelin pathway in the development of Alzheimer's disease. *Biochem Soc Trans.* 2004;32:144-146.

Altura BM, Gebrewold A, Zheng T, Altura BT. Sphingomyelinase and ceramide analogs induce vasoconstriction and leukocyte-endothelial interactions in cerebral venules in the intact rat brain: insight into mechanisms and possible relation to brain injury and stroke. *Brain Res Bull.* 2002;58:271-278.

Andrieu-Abadie N, Gouaze V, Salvayre R, Levade T. Ceramide in apoptosis signaling: relationship with oxidative stress. *Free Radic Biol Med.* 2001;31:717-728.

Andrieu-Abadie N, Levade T. Sphingomyelin hydrolysis during apoptosis. *Biochim Biophys Acta.* 2002;1585:126-134.

Ariga T, Jarvis WD, Yu RK. Role of sphingolipid-mediated cell death in neurodegenerative diseases. *J Lipid Res.* 1998;39:1-16.

Ariga T, Kobayashi K, Hasegawa A, Kiso M, Ishida H, Miyatake T. Characterization of high-affinity binding between gangliosides and amyloid β-protein. *Arch Biochem Biophys.* 2001;388:225-230.

Ayasolla K, Khan M, Singh AK, Singh I. Inflammatory mediator and β-amyloid (25-35)-induced ceramide generation and iNOS expression are inhibited by vitamin E. *Free Radic Biol Med.* 2004;37:325-338.

Balduini W, Mazzoni E, Carloni S, De Simoni MG, Perego C, Sironi L, Cimino M. Prophylactic but not delayed administration of simvastatin protects against long-lasting cognitive and morphological consequences of neonatal hypoxic-ischemic brain injury, reduces interleukin-1β and tumor necrosis factor-α mRNA induction, and does not affect endothelial nitric oxide synthase expression. *Stroke.* 2003;34:2007-2012.

Bassett CN, Montine TJ. Lipoproteins and lipid peroxidation in Alzheimer's disease. *J Nutr Health Aging.* 2003;7:24-29.

Bieberich E, MacKinnon S, Silva J, Noggle S, Condie BG. Regulation of cell death in mitotic neural progenitor cells by asymmetric distribution of prostate apoptosis response 4 (PAR-4) and simultaneous elevation of endogenous ceramide. *J Cell Biol.* 2003;162:469-479.

Bieberich E, MacKinnon S, Silva J, Yu RK. Regulation of apoptosis during neuronal differentiation by ceramide and b-series complex gangliosides. *J Biol Chem.* 2001;276:44396-44404.

Brann AB, Tcherpakov M, Williams IM, Futerman AH, Fainzilber M. Nerve growth factor-induced p75-mediated death of cultured hippocampal neurons is age-dependent and transduced through ceramide generated by neutral sphingomyelinase. *J Biol Chem.* 2002;277:9812-9818.

Brown MS, Goldstein JL. A proteolytic pathway that controls the cholesterol content of membranes, cells, and blood. *Proc Natl Acad Sci USA.* 1999;96:11041-11048.

Brugg B, Michel PP, Agid Y, Ruberg M. Ceramide induces apoptosis in cultured mesencephalic neurons. *J Neurochem.* 1996;66:733-739.

Buccoliero R, Futerman AH. The roles of ceramide and complex sphingolipids in neuronal cell function. *Pharmacol Res.* 2003;47:409-419.

Butterfield DA, Drake J, Pocernich C, Castegna A. Evidence of oxidative damage in Alzheimer's disease brain: central role for amyloid □-peptide. *Trends Mol Med.* 2001;7:548-554.

Calingasan NY, Uchida K, Gibson GE. Protein-bound acrolein: a novel marker of oxidative stress in Alzheimer's disease. *J Neurochem.* 1999;72:751-756.

Casaccia-Bonnefil P, Aibel L, Chao MV. Central glial and neuronal populations display differential sensitivity to ceramide-dependent cell death. *J Neurosci Res.* 1996;43: 382-389.

Cheng H, Xu J, McKeel DW Jr, Han X. Specificity and potential mechanism of sulfatide deficiency in Alzheimer's disease: an electrospray ionization mass spectrometric study. *Cell Mol Biol.* 2003;49:809-818.

Cleveland DW, Rothstein JD. From Charcot to Lou Gehrig: deciphering selective motor neuron death in ALS. *Nat Rev Neurosci.* 2001;2:806-819.

Cutler RG, Mattson MP. Sphingomyelin and ceramide as regulators of development and lifespan. *Mech Ageing Dev.* 2001;122:895-908.

Cutler RG, Kelly J, Storie K, Pedersen WA, Tammara A, Hatanpaa K, Troncoso JC, Mattson MP. Involvement of oxidative stress-induced abnormalities in ceramide and cholesterol metabolism in brain aging and Alzheimer's disease. *Proc Natl Acad Sci USA.* 2004;101: 2070-2075.

Cutler RG, Pedersen WA, Camandola S, Rothstein JD, Mattson MP. Evidence that accumulation of ceramides and cholesterol esters mediates oxidative stress-induced death of motor neurons in amyotrophic lateral sclerosis. *Ann Neurol.* 2002;52:448-457.

Darios F, Lambeng N, Troadec JD, Michel PP, Ruberg M. Ceramide increases mitochondrial free calcium levels via caspase 8 and Bid: role in initiation of cell death. *J Neurochem.* 2003;84:643-654.

Dawkins JL, Hulme DJ, Brahmbhatt SB, Auer-Grumbach M, Nicholson GA. Mutations in *SPTLC1*, encoding serine palmitoyltransferase, long chain base subunit-1, cause hereditary sensory neuropathy type I. *Nat Genet.* 2001;27:309-312.

DeFreitas MF, McQuillen PS, Shatz CJ. A novel p75NTR signaling pathway promotes survival, not death, of immunopurified neocortical subplate neurons. *J Neurosci.* 2001;21:5121-5129.

Deuther-Conrad W, Loske C, Schinzel R, Dringen R, Riederer P, Munch G. Advanced glycation endproducts change glutathione redox status in SH-SY5Y human neuroblastoma cells by a hydrogen peroxide dependent mechanism. *Neurosci Lett.* 2001;312:29-32.

Ditaranto-Desimone K, Saito M, Tekirian TL, Saito M, Berg M, Dubowchik G, Soreghan B, Thomas S, Marks N, Yang AJ. Neuronal endosomal/lysosomal membrane destabilization activates caspases and induces abnormal accumulation of the lipid secondary messenger ceramide. *Brain Res Bull.* 2003;59:523-531.

Dolgachev V, Farooqui MS, Kulaeva OI, Tainsky MA, Nagy B, Hanada K, Separovic D. De novo ceramide accumulation due to inhibition of its conversion to complex sphingolipids in apoptotic photosensitized cells. *J Biol Chem.* 2004;279:23238-23249.

Fanzo JC, Lynch MP, Phee H, Hyer M, Cremesti A, Grassme H, Norris JS, Coggeshall KM, Rueda BR, Pernis AB, Kolesnick R, Gulbins E. CD95 rapidly clusters in cells of diverse origins. *Cancer Biol Ther.* 2003;2:392-395.

Fassbender K, Simons M, Bergmann C, Stroick M, Lütjohann D, Keller P, Runz H, Kühl S, Bertsch T, von Bergmann K, Hennerici M, Beyreuther K, Hartmann T. Simvastatin strongly reduces levels of Alzheimer's disease β-amyloid peptides Aβ42 and Aβ40 in vitro and in vivo. *Proc Natl Acad Sci USA.* 2001;98:5856-5861.

Filley CM. *The Behavioral Neurology of White Matter.* New York: Oxford; 2001.

France-Lanord V, Brugg B, Michel PP, Agid Y, Ruberg M. Mitochondrial free radical signal in ceramide-dependent apoptosis: a putative mechanism for neuronal death in Parkinson's disease. *J Neurochem.* 1997;69:1612-1621.

Futerman AH, Boldin S, Brann AB, Schwarz A, Zisling R. Regulatory roles for sphingolipids in the growth of polarized neurons. *Ann NY Acad Sci.* 1998;845:176-187.

Garcia-Barros M, Paris F, Cordon-Cardo C, Lyden D, Rafti S, Haimovitz-Friedman A, Fuks Z, Kolesnick R. Tumor response to radiotherapy regulated by endothelial cell apoptosis. *Science* 2003;300:1155-1159.

Gouaze V, Mirault ME, Carpentier S, Salvayre R, Levade T, Andrieu-Abadie N. Glutathione peroxidase-1 overexpression prevents ceramide production and partially inhibits apoptosis in doxorubicin-treated human breast carcinoma cells. *Mol Pharmacol.* 2001;60:488-496.

Greenberg, SM. Cerebral amyloid angiopathy: prospects for clinical diagnosis and treatment. *Neurology* 1998;51:690-694.

Gulbins E, Kolesnick R. Raft ceramide in molecular medicine. *Oncogene* 2003;22: 7070-7077.

Guo Q, Fu W, Xie J, Luo H, Sells SF, Geddes JW, Bondada V, Rangnekar VM, Mattson MP. Par-4 is a mediator of neuronal degeneration associated with the pathogenesis of Alzheimer disease. *Nat Med.* 1998;4:957-962.

Han X, Cheng H, Fryer JD, Fagan AM, Holtzman DM. Novel role for apolipoprotein E in the central nervous system: modulation of sulfatide content. *J Biol Chem.* 2003a;278: 8043-8051.

Han X, Fagan AM, Cheng H, Morris JC, Xiong C, Holtzman DM. Cerebrospinal fluid sulfatide is decreased in subjects with incipient dementia. *Ann Neurol.* 2003b;54: 115-119.

Han X, Holtzman DM, McKeel DW Jr, Kelley J, Morris JC. Substantial sulfatide deficiency and ceramide elevation in very early Alzheimer's disease: potential role in disease pathogenesis. *J Neurochem.* 2002;82:809-818.

Hanada K. Serine palmitoyltransferase, a key enzyme of sphingolipid metabolism. *Biochim Biophys Acta.* 2003;1632:16-30.

Hannun YA, Luberto C. Ceramide in the eukaryotic stress response. *Trends Cell Biol.* 2000;10:73-80.

Haughey NJ, Cutler RG, Tamara A, McArthur JC, Vargas DL, Pardo CA, Turchan J, Nath A, Mattson MP. Perturbation of sphingolipid metabolism and ceramide production in HIV-dementia. *Ann Neurol.* 2004;55:257-267.

Hensley K, Maidt ML, Yu Z, Sang H, Markesbery WR, Floyd RA. Eletrochemical analysis of protein nitrotyrosine and dityrosine in the Alzheimer brain indicates region-specific accumulation. *J Neurosci.* 1998;18:8126-8132.

Herget T, Esdar C, Oehrlein SA, Heinrich M, Schütze S, Maelicke A, van Echten-Deckert G. Production of ceramides causes apoptosis during early neural differentiation in vitro. *J Biol Chem.* 2000;275:30344-30354.

Huang X, Moir RD, Tanzi RE, Bush AI, Rogers JT. Redox-active metals, oxidative stress, and Alzheimer's disease pathology. *Ann NY Acad Sci.* 2004;1012:153-163.

Huwiler A, Böddinghaus B, Pautz A, Dorsch S, Briner VA, Brade V, Pfeilschifter. Superoxide potently induces ceramide formation in glomerular endothelial cells. *Biochem Biophys Res Commun.* 2001;284:404-410.

Irie F, Hirabayashi Y. Application of exogenous ceramide to cultured rat spinal motoneurons promotes survival or death by regulation of apoptosis depending on its concentration.*J Neurosci Res.* 1998;54:475-485.

Ito A, Horigome K. Ceramide prevents neuronal programmed cell death induced by nerve growth factor deprivation. *J Neurochem.* 1995;65:463-466.

Kakio A, Nishimoto S, Yanagisawa K, Kozutsumi Y, Matsuzaki K. Interactions of amyloid β–protein with various gangliosides in raft-like membranes: importance of GM1 ganglioside-bound form as an endogenous seed for Alzheimer amyloid. *Biochemistry.* 2002;41:7385-7390.

Kendler A, Dawson G. Hypoxic injury to oligodendrocytes: reversible inhibiton of ATP-dependent transport of ceramide from the endoplasmic reticulum to the Golgi. *J Neurosci Res.* 1992;31:205-211.

Kester M, Kolesnick R. Sphingolipids as therapeutics. *Pharmacol Res.* 2003;47:365-371.

Kiryu-Seo S, Sasaki M, Yokohama H, Nakagomi S, Hirayama T, Aoki S, Wada K, Kiyama H. Damage-induced neuronal endopeptidase (DINE) is a unique metallopeptidase expressed in response to neuronal damage and activates superoxide scavengers. *Proc Natl Acad Sci USA.* 2000;97:4345-4350.

Korkotian E, Schwarz A, Pelled D, Schwarzmann G, Segal M, Futerman AH. Elevation of intracellular glucosylceramide levels results in an increase in endoplasmic reticulum density and in functional calcium stores in cultured neurons. *J Biol Chem.* 1999;274:21673-21678.

Kuner P, Schubenel R, Hertel C. β-amyloid binds to p75NTR and activates NFκB in human neuroblastoma cells. *J Neurosci Res.* 1998;54:798-804.

Lauderback CM, Kanski J, Hackett JM, Maeda, Kindy MS, Butterfield DA. Apolipoprotein E modulates Alzheimer's Aβ(1-42)-induced oxidative damage to synaptosomes in an allele-specific manner. *Brain Res.* 2002;924:90-97.

Lawler JF Jr, Yin M, Diehl AM, Roberts E, Chatterjee S. Tumor necrosis factor-α stimulates the maturation of sterol regulatory binding element binding protein-1 in human hepatocytes through the action of neutral sphingomyelinase. *J Biol Chem.* 1998;273:5053-5059.

Lee JT, Xu J, Lee JM, Ku G, Han X, Yang DI, Chen S, Hsu CY. Amyloid-β peptide induces oligodendrocyte death by activating the neutral sphingomyelinase-ceramide pathway. *J Cell Biol.* 2004;164:123-131.

Levade T, Jaffrézou J-P. Signalling sphingomyelinases: which, where, how and why? *Biochim Biophys Acta* 1999;1438:1-17.

Li R, Yang L, Lindholm K, Konishi Y, Yue X, Hampel H, Zhang D, Shen Y. Tumor necrosis factor death receptor signaling cascade is required for amyloid-β protein-induced neuron death. *J Neurosci.* 2004;24:1760-1771.

Liu B, Hannun YA. Inhibition of the neutral magnesium-dependent sphingomyelinase by glutathione. *J Biol Chem.* 1997;272:16281-16287.

Lovell MA, Ehmann WD, Mattson MP, Markesbery WR. Elevated 4-hydroxynonenal in ventricular fluid in Alzheimer's disease. *Neurobiol Aging.* 1997;18:457-461.

Lovell MA, Xie C, Markesbery WR. Acrolein is increased in Alzheimer's disease brain and is toxic to primary hippocampal cultures. *Neurobiol Aging.* 2001;22:187-194.

Lovell MA, Robertson JD, Teesdale WJ, Campbell JL, Markesbery WR. Copper, iron and zinc in Alzheimer's disease senile plaques. *J Neurol Sci.* 1998;158:47-52.

Maelicke A, Schrattenholz A, Samochocki M, Radina M, Albuquerque EX. Allosterically potentiating ligands of nicotinic receptors as a treatment strategy for Alzheimer's disease. *Behav Brain Res.* 2000;113:199-206.

Mahfoud R, Garmy N, Maresca M, Yahi N, Puigserver A, Fantini J. Identification of a common sphingolipid-binding domain in Alzheimer, prion, and HIV-1 proteins. *J Biol Chem.* 2002;277:11292-11296.

Mandon EC, van Echten G, Birk R, Schmidt RR, Sandoff K. Sphingolipid biosynthesis in cultured neurons. Down-regulation of serine palmitoyltransferase by sphingoid bases. *Eur J Biochem.* 1991;198:667-674.

Markesbery WR, Lovell MA. Four-hydroxynonenal, a product of lipid peroxidation, is increased in the brain in Alzheimer's disease. *Neurobiol Aging.* 1998;19:33-36.

Martin SF, Gomez-Diaz C, Bello RI, Navas P, Villalba JM. Inhibition of neutral Mg^{2+}-dependent sphingomyelinase by ubiquinol-mediated plasma membrane electron transport. Protoplasma 2003;221:109-116.

Mattson MP. Contributions of mitochondrial alterations, resulting from bad genes and a hostile environment, to the pathogenesis of Alzheimer's disease. *Int Rev Neurobiol.* 2002;53:387-409.

Mattson MP. Metal-catalyzed disruption of membrane protein and lipid signaling in the pathogenesis of neurodegenerative disorders. *Ann NY Acad Sci.* 2004;1012:37-50.

Mattson MP, Camandola S. NF-□B in neuronal plasticity and neurodegenerative disorders. *J Clin Invest.* 2001;107:247-254.

Mattson MP, Goodman Y, Luo H, Fu W, Furukawa K. Activation of NF-κB protects hippocampal neurons against oxidative stress-induced apoptosis: evidence for induction of manganese superoxide dismutase and suppression of peroxynitrite production and protein tyrosine nitration. *J Neurosci Res.* 1997;49:681-697.

Menaldino DS, Bushnev A, Sun A, Liotta DC, Symolon H, Desai K, Dillehay DL, Peng Q, Wang E, Allegood J, Trotman-Pruett S, Sullards MC, Merrill AH Jr. Sphingoid bases and de novo ceramide synthesis: enzymes involved, pharmacology and mechanisms of action. *Pharmacol Res.* 2003;47:373-381.

Mitoma J, Ito M, Furuya S, Hirabayashi Y. Bipotential roles of ceramide in the growth of hippocampal neurons: promotion of cell survival and dendritic outgrowth in dose- and developmental stage-dependent manners. *J Neurosci Res.* 1998;51:712-722.

Miyata M, Smith JD. Apolipoprotein E allele-specific antioxidant activity and effects on cytotoxicity by oxidative insults and □-amyloid peptides. *Nat Genet.* 1996;15:55-61.

Moir RD, Atwood CS, Romano DM, Laurans MH, Huang X, Bush AI, Smith JD, Tanzi RE. Differential effects of apolipoprotein E isoforms on metal-induced aggregation of Aβ using physiological concentrations. *Biochemistry.* 1999;38:4595-4603.

Montine KS, Olson SJ, Amarnath V, Whetsell WO Jr, Graham DG, Montine TJ. Immunohistochemical detection of 4-hydroxy-2-nonenal adducts in Alzheimer's disease is associated with inheritance of APOE4. *Am J Pathol.* 1997;150:437-443.

Moore AN, Kampfl AW, Zhao X, Hayes RL, Dash PK. Sphingosine-1-phosphate induces apoptosis of cultured hippocampal neurons that requires protein phosphatases and activator protein-1 complexes. *Neuroscience.* 1999;94:405-415.

Moore JD, Rothwell NJ, Gibson RM. Involvement of caspases and calpains in cerebrocortical neuronal cell death is stimulus-dependent. *Br J Pharmacol.* 2002;135:1069-1077.

Nair P, Tammariello SP, Estus S. Ceramide selectively inhibits apoptosis-associated events in NGF-deprived sympathetic neurons. *Cell Death Differ.* 2000;7:207-214.

Nitsch RM, Blusztajn JK, Pittas AG, Slack BE, Growdon JH, Wurtman RJ. Evidence for a membrane defect in Alzheimer disease brain. *Proc Natl Acad Sci USA.* 1992;89:1671-1675.

Nunomura, A, Perry G, Aliev G, Hirai K, Takeda A, Balraj EK, Jones PK, Ghanbari H, Wataya T, Shimohama S, Chiba S, Atwood CS, Petersen RB, Smith MA. Oxidative damage is the earliest event in Alzheimer disease. *J Neuropathol Exp Neurol.* 2001;60:759-767.

Nunomura A, Perry G, Pappolla MA, Wade R, Hirai K, Chiba S, Smith MA. RNA oxidation is a prominent feature of vulnerable neurons in Alzheimer's disease. *J Neurosci.* 1999;19:1959-1964.

Okazawa H, Estus S. The JNK/c-Jun cascade and Alzheimer's disease. *Am J Alzheimers Dis Other Demen.* 2002;17:79-88.

Ong WY, Halliwell B. Iron, atherosclerosis, and neurodegeneration: a key role for cholesterol in promoting iron-dependent oxidative damage? *Ann NY Acad Sci.* 2004;1012:51-64.

Pedersen WA, Chan SL, Mattson MP. A mechanism for the neuroprotective effect of apolipoprotein E: isoform-specific modification by the lipid peroxidation product 4-hydroxynonenal. *J Neurochem.* 2000;74:1426-1433.

Perini G, Della-Bianca V, Politi V, Della Valle G, Dal-Pra I, Rossi F, Armato U. Role of p75 neurotrophin receptor in the neurotoxicity by □-amyloid peptides and synergistic effect of inflammatory cytokines. *J Exp Med.* 2002;195:907-918.

Perry DK. Serine palmitoyltransferase: role in apoptotic de novo ceramide synthesis and other stress responses. *Biochim Biophys Acta.* 2002;1585:146-152.

Picklo MJ, Montine TJ, Amarnath V, Neely MD. Carbonyl toxicology and Alzheimer's disease. *Toxicol Appl Pharmacol.* 2002;184:187-197.

Ping SE, Barrett GL. Ceramide can induce cell death in sensory neurons, whereas ceramide analogues and sphingosine promote survival. *J Neurosci Res.* 1998;54:206-213.

Pratico D, Lee VM, Trojanowski JQ, Rokach J, Fitzgerald GA. Increased F_2-isoprostanes in Alzheimer's disease: evidence for enhanced lipid peroxidation in vivo. *FASEB J.* 1998; 12:1777-1783.

Pratico D, Uryu K, Leight S, Trojanowski JQ, Lee VM. Increased lipid peroxidation precedes plaque formation in an animal model of Alzheimer amyloidosis. *J Neurosci.* 2001;21:4183-4187.

Pratico D, Zhukareva V, Yao Y, Uryu K, Funk CD, Lawson JA, Trojanowski JQ, Lee VM. 12/15-lipoxygenase is increased in Alzheimer's disease: possible involvement in brain oxidative stress. *Am J Pathol.* 2004;164:1655-1622.

Puglielli L, Ellis BC, Saunders AJ, Kovacs DM. Ceramide stabilizes β–site amyloid precursor protein-cleaving enzyme 1 and promotes amyloid β-peptide biogenesis. *J Biol Chem.* 2003a;278:19777-19783.

Puglielli L, Tanzi RE, Kovacs DM. Alzheimer's disease: the cholesterol connection. *Nat Neurosci.* 2003b;6:345-351.

Reich EE, Markesbery WR, Roberts LJ 2nd, Swift LL, Morrow JD, Montine TJ. Brain regional quantification of F-ring and D-/E-ring isoprostanes and neuroprostanes in Alzheimer's disease. *Am J Pathol.* 2001;158:293-297.

Roccamo AM, Pediconi MF, Aztiria E, Zanello L, Wolstenholme A, Barrantes FJ. Cells defective in sphingolipids biosynthesis express low amounts of muscle micotinic acetylcholine receptor. *Eur J Neurosci.* 1999;11:1615-1623.

Roher AE, Weiss N, Kokjohn TA, Kuo YM, Kalback W, Anthony J, Watson D, Luehrs DC, Sue L, Walker D, Emmerling M, Goux W, Beach T. Increased Aβ peptides and reduced cholesterol and myelin proteins characterize white matter degeneration in Alzheimer's disease. *Biochemistry.* 2002;41:11080-11090.

Ruvolo PP. Intracellular signal transduction pathways activated by ceramide and its metabolites. *Pharmacol Res.* 2003;47:383-392.

Santana P, Pena LA, Haimovitz-Friedman A, Martin S, Green D, McLoughlin M, Cordon-Cardo C, Schuchman EH, Fuks Z, Kolesnick R. Acid sphingomyelinase-deficient human lymphoblasts and mice are defective in radiation-induced apoptosis. *Cell* 1996;86: 189-199.

Sawai H, Hannun YA. Ceramide and sphingomyelinases in the regulation of stress responses. *Chem Phys Lipids.* 1999;102:141-147.

Sayre LM, Zelasko DA, Harris PL, Perry G, Salomon RG, Smith MA. 4-Hydroxynonenal-derived advanced lipid peroxidation end products are increased in Alzheimer's disease. *J Neurochem.* 1997;68:2092-2097.

Srinivasan A, Foster LM, Testa MP, Ord T, Keane RW, Bredesen DE, Kayalar C. Bcl-2 expression in neural cells blocks activation of ICE/CED-3 family proteases during apoptosis. *J Neurosci.* 1996;16:5654-5660.

Stoica BA, Movsesyan VA, Lea PM 4th, Faden AI. Ceramide-induced neuronal apoptosis is associated with dephosphorylation of Akt, BAD, FKHR, GSK-3β, and induction of the mitochondrial-dependent intrinsic caspase pathway. *Mol Cell Neurosci.* 2003;22:365-382.

Svennerholm L, Gottfries CG. Membrane lipids, selectively diminished in Alzheimer brains, suggest synapse loss as a primary event in early-onset form (type I) and demyelination in late-onset form (type II). *J Neurochem.* 1994;62:1039-1047.

Tamagno E, Parola M, Guglielmotto M, Santoro G, Bardini P, Marra L, Tabaton M, Danni O. Multiple signaling events in amyloid β-induced, oxidative stress-dependent neuronal apoptosis. *Free Radic Biol Med.* 2003;35:45-58.

Taniwaki T, Yamada T, Asahara H, Ohyagi Y, Kira J. Ceramide induces apoptosis to immature cerebellar granule cells in culture. *Neurochem Res.* 1999;24:685-690.

Teter B. ApoE-dependent plasticity in Alzheimer's disease. *J Mol Neurosci.* 2004;23: 167-180.

Toman RE, Movsesyan V, Murthy SK, Milstien S, Spiegel S, Faden AI. Ceramide-induced cell death in primary neuronal cultures: upregulation of ceramide levels during neuronal apoptosis. *J Neurosci Res.* 2002;68:323-330.

Tuppo EE, Forman LJ, Spur BW, Chan-Ting RE, Chopra A, Cavalieri TA. Sign of lipid peroxidation as measured in the urine of patients with probable Alzheimer's disease. *Brain Res Bull.* 2001;54:565-568.

Vega GL, Weiner MF, Lipton AM, von Bergmann K, Lütjohann D, Moore C, Svetlik D. Reduction in levels of 24S-hydroxycholesterol by statin treatment in patients with Alzheimer disease. *Arch Neurol.* 2003;60:510-515.

Wei W, Norton DD, Wang X, Kusiak JW. Aβ 17-42 in Alzheimer's disease activates JNK and caspase-8 leading to neuronal apoptosis. *Brain.* 2002;125:2036-2043.

Willaime-Morawek S, Brami-Cherrier K, Mariani J, Caboche J, Brugg B. C-Jun N-terminal kinases/c-Jun and p38 pathways cooperate in ceramide-induced neuronal apoptosis. *Neuroscience.* 2003;119:387-397.

Wolozin B, Brown J 3rd, Theisler C, Silberman S. The cellular biochemistry of cholesterol and statins: insights into the pathophysiology and therapy of Alzheimer's disease. *CNS Drug Rev.* 2004;10:127-146.

Won JS, Im YB, Khan M, Singh AK, Singh I. The role of neutral sphingomyelinase produced ceramide in lipopolysaccharide-mediated expression of inducible nitric oxide synthase. *J Neurochem.* 2004;88:583-593.

Yao Y, Zhukareva V, Sung S, Clark CM, Rokach J, Lee VM, Trojanowski JQ, Pratico D. Enhanced brain levels of 8,12-*iso*-iPF$_{2}$□-VI differentiate AD from frontotemporal dementia. *Neurology* 2003;61:436-437.

Yu Z, Cheng G, Wen X, Wu GD, Lee WT, Pleasure D. Tumor necrosis factor α increases neuronal vulnerability to excitotoxic necrosis by inducing expression of the AMPA-glutamate receptor subunit GluR1 via an acid sphingomyelinase and NF–κB-dependent mechanism. *Neurobiol Dis.* 2002;11:199-213.

Yu ZF, Nikolova-Karakashian M, Zhou D, Cheng G, Schuchman EH, Mattson MP. Pivotal role for acidic sphingomyelinase in cerebral ischemia-induced ceramide and cytokine production, and neuronal apoptosis. *J Mol Neurosci.* 2001;15:85-97.

SECTION II: FATTY ACID DEFICITS

In: Fatty Acids and Oxidative Stress...
Editors: R. Reddy and J.K. Yao, pp. 101-114
ISBN: 978-1-60021-677-0

Chapter VII

Membrane Fatty Acid Deficit in Schizophrenia and Mood Disorders

Malcolm Peet

Introduction

In recent years there has been increasing interest in the possible role of lipids in the aetiology of a variety of mental disorders including schizophrenia and depression. This has derived from increasing awareness of the importance of lipid-related messengers in neurotransmission (Bazan, 200) together with direct and indirect evidence of abnormal fatty acid and phospholipid biochemistry in patients with these mental disorders. This chapter will focus on studies of cell membrane polyunsaturated fatty acid (PUFA) levels in schizophrenia and depression.

Polyunsaturated Fatty Acids: do Peripheral Tissue Levels Reflect Those in the Brain?

Because of the obvious difficulty in obtaining brain tissue from people with mental illness, most direct studies of polyunsaturated fatty acid levels have been carried out on peripheral tissues. The tissue most commonly investigated because of the ease of acquisition is red blood cells (RBC). It may be questioned whether RBC levels are a valid reflection of brain levels of PUFA. Several studies have addressed this issue. There is evidence that in developing rodents and piglets, RBC levels of PUFA reflect those in neuronal membranes, particularly after omega-3 PUFA supplementation (Carlson et al, 1986; Lapillone et al, 2002; Suarez et al, 1996; Ward et al, 1998). More pertinent are the few studies which have been done in humans and other primates. In baboon neonates, there is evidence that RBC levels of docosahexaenoic acid (DHA) correlate significantly with levels in brain, but that arachidonic acid (AA) levels show no such correlation (Sarkadi-Nagy et al, 2003). In human infants, it has been shown that the percentage of DHA in RBC and cortex is greater in breast fed than in formula fed infants, and that the percentage of RBC DHA correlates with that in the brain

(Boehm et al, 1996; Farquharson et al, 1992; Makrides et al, 1994). These studies have focused on the developing brain because of the known importance of PUFA to early brain development, which has particular significance when considering the composition of formula feeds for human infants which in the past have lacked adequate amounts of DHA and AA (Crawford et al, 1998). However, such studies are not necessarily relevant to adults. In an important study, the fatty acid composition in RBC and cerebral cortex was measured in a group of human subjects and between 2 and 88 years (Carver et al, 2001). It was found that levels of 20:3 n-6 (dihomogamma-linolenic acid), AA, 22:4 n-6 (adrenic acid) and DHA in RBC were significant predictors of levels of those fatty acids in the cerebral cortex, but only in subjects over the age of 18. No correlation between RBC and cortex levels of any fatty acids were found in subjects aged 2 to 18 years.

In summary, there is evidence that RBC levels of PUFA correlate with levels in the cerebral cortex in the developing brain of neonates and in adults over the age of 18, but no such correlations have been demonstrated in children and adolescents.

Even where no correlation exists between peripheral and central measures of PUFA, peripheral PUFA levels can still be relevant to brain function. For example, eicosapentaenoic acid (EPA) is present in RBC membranes but is found in brain at negligible levels. However, EPA can still affect brain function through its effects on inflammatory processes, gene expression and enzyme activity (Kavanaugh et al, 2004; Kearns and Haag, 2002; Kitajka et al, 2002; Martin et al, 2002; Song et al, 2004; Yamagata et al, 2003)

Potential Confounding Factors

Findings of abnormal cell membrane levels of PUFA have led some investigators, particularly in studies of schizophrenic patients, to conclude that there is an abnormality of PUFA metabolism in these patients. However, exogenous factors which can affect PUFA levels must be taken into account. These include diet, smoking, alcohol consumption, medication, and storage artefact.

Diet

Both animal and human studies have shown that RBC fatty acid composition is an indicator of medium term dietary intake of PUFA (Katan et al, 1997). This is illustrated by a recent study (Peet et al, 2004) which included schizophrenic patients and control subjects from India and from Malaysia. Levels of n-3 PUFA were four times higher in the Malaysian subjects (both patients and controls) than in the Indian subjects because large quantities of fish are eaten in Malaysia whereas the Indian subjects were vegetarian. Furthermore, RBC membrane levels of n-3 PUFA in the ethnic Indian subjects who lived in Malaysia, reflected those of the Malaysian population rather than those of the Indian population. This shows that the variability is not related to genetically determined differences in metabolism. Because of the marked effects of diet, it is important to control the diet in studies of RBC membrane PUFA levels.

It is well established that schizophrenic patients eat an unhealthy diet which is high in sugar and saturated fat and lacking in fresh fruit and vegetables (Brown et al, 1999; Krajcovicova-Kulackova et al, 2004; McCreadie et al, 1998; Ryan et al, 2003; Stokes, 2004). As a result, schizophrenic patients consume significantly less of the antioxidant vitamins C and E (McCreadie et al, 1998) and greater amounts of sugar (Stokes, 2004). Both of these nutrients can have effects on fatty acids. Vitamin E protects against lipid peroxidation (Krajcovicova-Kulackova et al, 2004). Vitamin E treatment is associated with increased RBC membrane levels of AA and DHA acid in patients with hepatitis C virus (Ota et al, 2004), and increased brain levels of AA and DHA in diabetic rats (Celik et al, 2002). Some schizophrenic patients consume extremely large amounts of sucrose (Stokes, 2004) and it has been shown in rats that prolonged sucrose loading leads to markedly reduced levels of AA and DHA acid in the plasma lipids of rats (Fukuchi et al, 2004). Thus, aspects of diet other than fatty acid intake need to be controlled for.

Depressive patients characteristically show loss of appetite and prefer to eat snack food rather than balanced meals (Edwards and Peet, 1999). In depressive patients simultaneous assessment of diet and RBC PUFA levels have shown that there is indeed a strong correlation between dietary intake of n-3 PUFA and levels in RBC membranes. Furthermore, dietary intake correlates with severity of depressive symptoms (Edwards et al, 1998). This finding is consistent with epidemiological evidence that international variations in the prevalence of depression correlate strongly with consumption of fish in the national diet (Hibbeln, 1998; Peet, 2004).

Smoking

Smoking causes increased free radical activity (Morrow et al, 1995) and has the potential to reduce membrane fatty acid levels, particularly DHA (Hibbeln et al, 2003; Ling et al, 1994). The majority of patients who suffer from schizophrenia and depression are smokers, at a rate much higher than the general population (McCreadie, 2003). The damaging effects of smoking will be further exacerbated by a diet low in the antioxidant vitamins C and E. It is known that schizophrenic patients are under increased oxidative stress. This is shown by increased plasma levels of thiobarbituric acid reactive substances (TBARS) (Arvindakshan et al, 2003) and increased levels of alkanes in expired air (Phillips et al, 1993). Several studies have shown that the decreased levels of RBC levels of PUFAs in schizophrenic patients correlate with increased levels of TBARS (Arvindakshan et al, 2003; Peet et al, 1995). Hibbeln et al (2003) found that schizophrenic patients who smoke had significantly lower RBC membrane levels of DHA. There was also evidence of an interaction between smoking and diet: smokers consumed less linolenic acid and EPA than non-smokers

Medication

Antipsychotic drugs have complex effects on phospholipid metabolism. Thus, it is reported that phenothiazines increase fatty acid turnover in fibroblast phospholipids without altering membrane composition (Maziere et al, 1988). These drugs are also potent inhibitors of peroxisomal beta-oxidation (Vamecq, 1987; Van den Branden and Roels, 1985). Fischer et al (1992) reported a dose related reduction of AA and DHA in platelets from patients taking phenothiazines and thioxanthines, with profound depletions in the high dose patients which are similar to the low levels previously reported in medicated schizophrenic patients. In contrast, it appears that treatment with the atypical antipsychotic drugs including clozapine, may be associated with increased RBC membrane levels of AA and DHA (Arvindakshan et al, 2003; Horrobin et al, Evans et al, 2003).

Storage Artefact

As already noted, there are several studies showing that schizophrenic patients are under increased oxidative stress. This may be related to abnormal expression of the genes related to oxidative stress (Prabhakaran et al, 2004), compounded by lifestyle factors such as smoking. RBC are particularly prone to peroxidation damage which can lead to reduced levels of AA and DHA (Clemens and Waller, 1987). This is important because oxidative damage can occur during sample processing and storage. Because samples from schizophrenic patients, particularly those early in their illness, are liable to be collected over a long period of time, whereas controls are collected towards the end of the study when they can be matched with the schizophrenic patients, there is the potential for samples from patients to be stored much longer than those from control subjects. Also, the inherent difficulties of obtaining samples in psychiatric clinics and institutions can lead to delays before they are processed and stored. Storage at –20°C is not sufficient to inhibit lipid peroxidation, whereas storage at –80°C will normally preserve polyunsaturated fatty acids from oxidation for several years (Hodson et al, 2002; Zeleniuch-Jacquotte et al, 2000). Furthermore, it has been shown that samples from schizophrenic patients stored under the same conditions at –20°C as those from healthy control subjects, will show peroxidation damage more rapidly, presumably because of the increased oxidative stress which existed in the samples even before storage (Fox et al, 2003).

Membrane Fatty Acid Levels in Schizophrenic Patients

The fatty acids most commonly found to be abnormal in RBC membranes of schizophrenic patients are 18:2n-6 (LA), 20:4n-6 (AA) and 20:6 n-3 (DHA).

Studies which have investigated these fatty acids in medicated schizophrenic patients are shown in table 1. It is striking that abnormalities were much more commonly seen in early studies than in more recent studies. This can be partly explained by methodological problems in some of the early studies. For example, samples were stored at minus 20° in the study of Glen et al (1994) which would lead to depletion of AA and DHA in the samples from schizophrenic patients, because of increased oxidative stress. However, these results might also reflect changes in the type of antipsychotic medication which was used. In the earlier studies, patients were treated with older antipsychotics including phenothiazines which have been associated with depleted membrane levels of AA and DHA (Fischer et al, 1992). In the more recent studies, patients were mainly treated with newer atypical antipsychotic drugs which are associated with higher membrane levels of PUFA (Arvindakshan et al, 2003; Horrobin et al, 1997; Evans et al, 2003). Therefore only studies of unmedicated patients can provide a pointer to abnormalities which may be associated with schizophrenia itself.

Studies which included unmedicated patients are shown in table 2. In total six patient samples have been investigated. Of these, three showed reduced LA levels and three showed normal levels; three showed reduced AA levels and three showed normal levels; and three showed reduced, one normal and two elevated levels of DHA, relative to control subjects.

Table 1. RBC Membrane PUFA Levels in Medicated Schizophrenic Patients

	18:2 n-6 (LA)	20:4 n-6 (AA)	22:6 n-3 (DHA)
Obi & Nwanze (1977) [50]	-	-	-
Vaddadi et al (1986) [51]	↓	↓	↑
Yao et al (1994) [52]	↓	↓	↓
Glen et al (1994) [53]	↓	↓	↓
Peet et al (1995) [38]	↓	↓	↓
Vaddadi et al (1996) [54]	↓	-	↓
Doris et al (1998)	-	-	↓
Assies et al (2001)	-	-	-
Khan et al (2002)	↓	↓	↓
Arvindakshan et al (2003)	-	-	-
Ranjekar et al (2003)	-	-	↓
Evans et al (2003)	-	-	-

↓ decreased; ↑ increased; - no difference

AA, Arachidonic acid; DHA, Docosahexaenoic acid; LA, Linoleic acid

Table 2. RBC Membrane PUFA Levels in Medication-Free Schizophrenic Patients

	18:2 n-6 (LA)	20:4 n-6 (AA)	22:6 n-3 (DHA)
Yao et al (1994)	↓	↓	-
Khan et al (2002)	↓	↓	↓
Arvindakshan et al (2003)	-	↓	↓
Evans et al (2003)	-	-	↓
Peet et al (2004) India	↓	-	↑
Peet et al (2004) Malaysia	-	-	↑

↓ decreased; ↑ increased; - no difference

AA, Arachidonic acid; DHA, Docosahexaenoic acid; LA, Linoleic acid

In the study of Peet et al (2004), it was striking that RBC levels of n-6 PUFA, including AA, were similar in the Indian and Malaysian populations, whereas the amount of n-3 PUFA in RBC from Malaysian patients was four times higher than in the Indian patients. This reflected dietary practices, because fish is a staple in Malaysia whereas the Indian subjects were vegetarian. As a result, the n-6: n-3 ratio was around two to one in the Malaysian population but around ten to one in the Indian population. It is apparent that studies of RBC membrane PUFA in unmedicated schizophrenic patients have not produced consistent results. International variations relating to differences in national diet are much greater than differences between patients and control subjects.

Two studies have investigated PUFA levels in cerebral cortex. Such studies are particularly prone to the confounding factors discussed earlier, especially because of delays in obtaining and storing tissue and the effects of medication. Horrobin et al (1991) found a significant reduction in LA but an increase of DHA in phosphatidylethanolamine from prefrontal cortex, whereas Yao et al (2000) found reduced levels of LA and AA but not DHA, in the caudate. One study investigated cultured fibroblasts and found a significant reduction of DHA but no change in AA (Mahadik et al, 1996).

Conclusions about possible abnormalities of the metabolic pathways of PUFA cannot be drawn from studies of steady state levels. There are many studies that point to an abnormality of phospholipids in patients who suffer from schizophrenia, but there is no simple association between this and the reported changes in membrane PUFA levels. Some attempts have been made to correlate membrane PUFA levels with other indices of disturbed phospholipid and fatty acid disposition. In one study, RBC membrane levels of AA correlated with brain levels of phosphomonoesters (PME) measured by (31)P Magnetic Resonance Spectroscopy (Yao et al, 2002). PME is one of the building blocks of phospholipid. Reduced PME has previously been shown to occur in schizophrenic patients (Pettegrew et al, 1991). Evidence has recently been reviewed that the reduced skin flush response to niacin which has been shown in schizophrenic patients may be related to reduced levels of PUFA and particularly AA which is required for the flush response to occur (Messamore, 2003). Such findings are so far tentative and not well replicated.

One plausible mechanism which might cause of low RBC membrane PUFA levels in schizophrenia is increased oxidative stress leading to increased breakdown of PUFA. Increased levels of fatty acid oxidative breakdown products have been shown in blood (TBARS) (Arvindakshan et al, 2003) and in expired air (alkanes) (Phillips et al, 1993). Several investigators have shown that decreased RBC membrane levels of PUFA correlate significantly with higher plasma levels of TBARS (Peet et al, 1995; Sarkadi-Nagy et al, 2003). Fatty acids in blood samples from schizophrenic patients oxidise more rapidly during storage than those from healthy control subjects, again indicating pre-existing oxidative stress (Fox et al, 2003). A recent study of gene and protein expression combined with measurement of intermediate metabolites in the brain of schizophrenic patients, showed evidence that the effects of genes related to mitochondrial function and oxidative stress differentiated around 90% of schizophrenic patients from controls (Prabhakaran et al, 2004). It was proposed that oxidative stress is linked to the core pathology of schizophrenia. Any pre-disposition to oxidative stress would be worsened by the dietary lack of antioxidant vitamins which has been demonstrated in several studies of schizophrenic patients ((McCreadie et al, 1998), as well as by smoking. Smoking ten or more cigarettes a day in adolescence more than doubles the risk of subsequently developing schizophrenia (Weiser et al, 2004), and smoking has been proposed as an independent risk factor for schizophrenia (Kelly et al, 1999). The effects of oxidative stress on membrane fatty acid levels would be ameliorated by a diet containing adequate levels of antioxidants and PUFA, and these nutrients would be expected to have an ameliorating effect on schizophrenia.

There is one report that schizotypal personality traits correlate positively with RBC membrane levels of several PUFA including AA, EPA and DHA (Richardson et al, 2003). Positive schizotypal traits in the personality reflect a high risk of developing a schizophrenic illness. The authors proposed that the relatively high levels of these fatty acids might be protective against the development of a full schizophrenic illness, in people who are otherwise predisposed to develop such an illness. It is worthy of note that chronic schizophrenic patients in the large psychiatric hospital in Malaysia who were investigated by Peet et al (2004), were not routinely treated with antipsychotic medications and yet did not display florid symptomatology and could be managed in an open hospital environment. This may have been possible because of the high n-3 PUFA content of their diet. Within groups of schizophrenic patients, two separate studies have shown that higher consumption of PUFA (Stokes, 2004), or specifically n-3 PUFA (Mellor et al, 1996), in the normal daily diet is associated with less severe schizophrenic symptoms. International variations in the two-year outcome of schizophrenia have been shown to correlate with the ratio of saturated fat to PUFA in the national diet, such that a lower saturated to unsaturated fatty acid ratio is associated with a better outcome (Christensen and Christensen, 1998). Some, but not all, studies have found significant associations between higher levels of some PUFA in RBC membranes and less severe schizophrenic symptoms (Arvindakshan et al, 2003; Evans et al, 2003; Mellor et al, 1996; Peet et al, 2004). Furthermore, intervention studies with n-3 PUFA have shown some promising benefits on the symptomatology of schizophrenia (Peet, 2003). All of these findings support the possibility that higher levels of PUFA in the diet and RBC membranes have some protective or ameliorating effect in schizophrenia.

Membrane Fatty Acid Levels in Mood Disorder

Studies of membrane fatty acid levels in patients with mood disorder are shown in table 3. Two early studies (Ellis and Sanders, 1977; Fehily et al, 1981), both conducted in the same laboratory, reported elevated levels of n-3 PUFA in plasma and RBC membranes from depressive patients. However, one of these studies (Ellis and Sanders, 1977) did not report specific data for the RBC membrane levels, simply stating that changes were similar to those in plasma. Both studies showed methodological deficiencies in patient identification and analytical techniques, and also both lacked information about sample storage, although it is unlikely that samples were stored at the required – 80° C. Therefore, these studies can be discounted. More recent studies have consistently shown reduced levels of n-3 PUFA in depressive patients. The two recent studies shown in table 3, again from the same laboratory, both found reduction in RBC membrane levels of n-3 PUFA (Edwards et al, 1998; Peet et al, 1998). This is consistent with other studies showing reduced n-3 levels in plasma (Maes et al, 1999) and in adipose tissue (Mamalakis et al, 2002) from depressive patients, and by a reported correlation between AA to EPA ratio in RBC membranes and severity of depression (Adams et al, 1996).

Table 3. RBC Membrane PUFA Levels in Patients with Major Depression or Mood Disorders

Major Depression					
	LA	**AA**	**ALA**	**EPA**	**DHA**
Ellis & Sanders (1977)	?	-	?	?	?
Fehily et al (1981)	↓	-	?	-	↑
Peet et al (1998)	↓	-	?	-	↓
Edwards et al (1998)	-	-	-	↓	↓
Bipolar disorder					
Ranjekar et al (2003)	-	-	↓	↓	-
Chiu et al (2003)	-	↓	?	-	↓

↓ decreased; ↑ increased; - no difference

AA, Arachidonic acid; ALA, Alpha-linolenic acid; DHA, Docosahexaenoic acid; EPA, Eicosapentaenoic acid; LA, Linoleic acid

The study of Edwards et al (1998) is particularly important because, although numbers included were small, all subjects were medication free and were carefully matched for potential confounding variables including age, gender, social class, body mass index, life events, smoking habits, alcohol intake and the number of children. In addition a full dietary assessment was carried out for both patients and control subjects using seven day weighed food intake. It was found that the depressed subjects ate food which was easier to obtain and prepare, including more snack food and cold or uncooked foods. It was of interest that these dietary patterns, albeit obtained retrospectively, appear to have antedated the onset of depressive symptoms. Despite these differences, it was found that total energy intake, as well as total dietary intake of fat, carbohydrate and protein was similar between the depressive patients and the control subjects (Edwards et al, 1998). Depressive patients were found to have significantly lower levels of ALA EPA and DHA in their RBC membranes. Furthermore, n-3 PUFA levels in RBC membranes were predicted by dietary intake of these fatty acids and both RBC and dietary levels of n-3 PUFA correlated strongly with severity of depressive symptoms. This study has therefore established that depressive patients have low levels of n-3 PUFA in their RBC membranes and that this is related to n-3 PUFA consumption in the diet independently of any other potential confounding factors. This is consistent with epidemiological data showing that national rates of major depression correlate strongly with levels of fish intake in the national diet ((Hibbeln, 1998).

Table 3 shows two studies of RBC membrane PUFA in bipolar disorder, both of which have shown reduced levels of omega-3 PUFA (Chiu et al, 2003; Ranjekar et al, 2003). This is consistent with epidemiological data showing a strong correlation between national sea food consumption and international variation in rates of bipolar disorder (Noaghiul and Hibbeln, 2003). One of these studies, which included bipolar patients recovering from mania (Chiu et al, 2003), reported reduced RBC membrane levels of AA, whereas the other study in which the bipolar patients were mostly asymptomatic, showed no change in AA levels (Ranjekar et al, 2003). One study has reported increased omega-6 PUFA in plasma phospholipids of first degree relatives of bipolar patients (Sobczaks et al, 2004). However, this does not imply a genetic abnormality because dietary habits are well known to have a familial influence (Feunekes et al, 1998).

Summary and Conclusions

Most studies of patients with either schizophrenia or major depression have shown reduced RBC membrane levels of PUFA, particularly LA, AA and DHA. However, underlying factors appear to be different between the two conditions.

In schizophrenia, the biggest abnormalities were shown in early studies, many of which were confounded by the effects of medication and other methodological problems. Studies in unmedicated patients have given less consistent results. Some studies have shown reduced levels of LA, AA and DHA in schizophrenic patients, but other studies show no abnormality, or even an increase of DHA levels. There is good evidence that schizophrenic patients are under increased oxidative stress and several investigators have found that low levels of PUFA in RBC membranes correlate significantly with plasma levels of TBARS, which suggest increased oxidative breakdown of PUFA in schizophrenic patients. Other lines of evidence suggest that increased dietary intake of PUFA and consequent increases in cell membrane levels of these fatty acids, may be protective against the development of schizophrenia or against a poor long-term outcome or may alleviate symptoms in people with established schizophrenia. The relationship between RBC membrane PUFA levels and other abnormalities of phospholipid and fatty acid disposition in schizophrenia remains to be elucidated.

In depression, the situation appears to be much more straightforward. The available evidence suggests that low levels of n-3 fatty acids in the diet relate to low levels of these PUFA in RBC membranes, and that this in turn relates to the development and severity of depressive symptoms. No metabolic abnormality is required to explain these associations. Evidence supporting a beneficial effect of n-3 PUFA supplementation on depressive symptoms is particularly strong (Peet, 2003), and this supports the proposition of a direct relationship between n-3 PUFA and depression.

References

Adams PB, Lawson S, Sanigorski A, Sinclair A. Arachidonic acid to eicosapentaenoic acid ration in blood correlates positively with clinical symptoms of depression. *Lipids* 1996; 31 Suppl: S157-S161.

Arvindakshan M, Sitasawaq S, Debsikdar V, Ghate M, Evans D, Horrobin DF, Bennett C, Ranjekar PK, Mahadik SP. Essential polyunsaturated fatty acid and lipid peroxide levels in never medicated and medicated schizophrenic patients. *Biol Psychiatry* 2003; 53: 56-64.

Assies J, Lieverse R, Vreken P, Wanders R J A, Dingemans P M J A, Linszen D H. Significantly reduced docosahexaenoic acid concentrations in erythrocyte membranes from schizophrenic patients compared with a carefully matched control group. *Biol Psychiatry*. 2001; 49: 510-522.

Bazan NG. Synaptic lipid signalling: significance of polyunsaturated fatty acids and platelet – activating factor. *J Lipid Res*. 2003; 44: 2221-33.

Boehm G, Borte M, Bohles HJ, Muller H, Kohn G, Moro G. Docosahexaenoic and arachidonic acid content of serum and red blood cell membrane phospholipids of pre-term infants fed breast milk, standard formula or formula supplemented with n-3 and n-6 long-chain polyunsaturated fatty acids. *Eur J Pediatr*. 1996; 155: 410-416.

Brown S, Birtwistle LR, Thompson C. The unhealthy lifestyle of people with schizophrenia. *Psychol Med.* 1999; 29: 697-701.

Carlson SE, Carver JD, House SG. High fat diets varying in ratios of polyunsaturated to saturated fatty acid and linolenic to linolenic acid. A comparison of rat neural and red cell membrane phospholipids. *J Nutr*. 1986; 116: 718-725.

Carver JD, Benford VJ, Han B, Cantor AB. The relationship between age and the fatty acid composition of cerebral cortex and erythrocytes in human subjects. *Brain Res Bull*. 2001; 56: 79-85.

Celik S, Baydas G, Yilmaz O. Influence of vitamin E on the levels of fatty acids and MDA in some tissues of diabetic rats. *Cell Biochem Funct*. 2002; 20: 67-71.

Chiu C-C, Huang S-Y, Su K-P, Lu m-L, Huang M-C, Chan C-C, Shen WW. Polyunsaturated fatty acid deficit inpatients with bipolar mania. *Eur Neuropsychopharmacology* 2003; 13: 99-103.

Christensen O, Christensen E. Fat consumption and schizophrenia. *Acta Psychiatr Scand.* 1998; 78: 587-591.

Clemens MR and Waller HD. Lipid peroxidation in erythrocytes. *Chem Phys Lipids.* 1987; 45: 251-268.

Crawford MA, Costeloek, Ghebremeskal K, Phylactos A. The inadequacy of the essential fatty acid content of present preterm feeds. *Eur J Pediatr.* 1998; 157 Suppl 1: S23-7.

Doris A B, Wahle K, MacDonald A, Morris S, Coffey I, Muir W, Blackwood D. Red cell membrane fatty acids, cytosolic phospholipase A2 and schizophrenia. *Schizophr Res.* 1998; 31: 185-196.

Edwards R, Peet M, Shay J, Horrobin D. Omega-3 polyunsaturated fatty acid levels in the diet and red blood cell membranes of depressed patients. *J Affect Disord.* 1998; 48: 149-155.

Edwards Rh W, Peet M. Essential fatty acid intake in relation to depression. In: Peet M, Glen I, Horrobin DF, eds. *Phospholipid Spectrum Disorder in Psychiatry.* Carnforth, Lancashire: Marius Press; 1999: 211-221.

Ellis FR, Sanders TAB. Long chain polyunsaturated fatty acids in endogenous depression. *J Neurol Neurosurg Psychiatry* 1977; 40: 168-169.

Evans DR, Parikh VV, Khan MM, Coussons C, Buckley PF, Mahadik SP. Red blood cell membrane essential fatty acid metabolism in early psychotic patients following antipsychotic drug treatment. *Postaglandins Leukotr Essent Fatty Acids.* 2003; 69: 393-399.

Farquharson J, Cockburn F, Patrick WA, Jamieson EC, Logan RQ. Infant cerebral cortex phospholipid fatty acid composition and diet. *Lancet* 1992. 340 810-3.

Fehily AMA, Bowey AOM, Ellis FR, Meade BW, Dickerson JWT. Plasma and erythrocyte membrane long chain polyunsaturated fatty acids in endogenous depression. *Neurochem International* 1981; 3: 37-42.

Feunekes GL, de Graaf C, Meyboom S, Van Staveren WA. Food choice and fat intake of adolescents and adults: associations of intakes within social networks. *Prev Med* 1998; 27: 645-656.

Fischer S, Kissling W, Kuss HJ. Schizophrenic patients treated with high-dose phenothiazines or thioxanthine become deficient in polyunsaturated fatty acids in their thrombocytes. *Biochem Pharmacol.* 1992; 44: 317-323.

Fox H, Ross B, Tocher D, Horrobin D, Glen I, St Clair D. Degradation of specific polyunsaturated fatty acids in red blood cells stored at –20oc proceeds faster in patients with schizophrenia when compared with healthy controls. *Prostaglandins Leukotr Essent Fatty Acids.* 2003, 69; 291-297.

Fukuchi S, Hamaguchi, K, Seike M, Himeno K, Salata T, Yoshimatsu H. Role of fatty acid composition in the development of metabolic disorders in sucrose-induced obese rats. *Exp Biol Med* 2004; 229: 486-493.

Glen A I M, Glen E M T, Horrobin D F, Vaddadi K S, Spellman M, Morse-Fisher N, Ellis K, Skinner FS. A red blood cell membrane abnormality in a subgroup of schizophrenic patients: evidence for two diseases. *Schizophr Res.* 1994; 12: 53-61.

Hibbeln J. Fish consumption and major depression. *Lancet* 1998; 351: 1213.

Hibbeln JR, Makino KK, Martin CE, Dickerson F, Boronow J, Fenton WS. Smoking, gender and dietary influences on erythrocyte essential fatty acid composition among patients with schizophrenia or schizoaffective disorder. *Biol Psychiatry* 2003; 53: 431-441.

Hodson L, Skeaff CM, Wallace AJ, Arribas FL. Stability of plasma and erythrocyte fatty acid composition during cold storage. *Clin Chim Acta.* 2002; 321: 63-67.

Horrobin DF, Glen AIM, Cantrill RC. Clozapine elevation of membrane unsaturated fatty acids as a new mechanism of action. *Schizophr Res.* 1997; 24: 214.

Horrobin DF, Manku MS, Hillman H, Glen IAM. Fatty acid levels in the brains of schizophrenics and controls. *Biol Psychiatry* 1991; 30: 795-805.

Katan MB, Deslypere JP, van Birgelen AP, Penders M, Zegwaard M. Kinetics of the incorporation of dietary fatty acids into serum cholesteryl esters, erythrocyte membranes, and adipose tissues: an 18-month controlled study. *J Lipid Res.* 1997; 38: 2012-2022.

Kavanaugh T, Lonergan PE, Lynch MA. Eicosapentaenoic acid and gamma-linolenic acid increase hippocampal concentrations of IL-4 and IL-10 and abrogate lipopolysacchoride-induced inhibition of long-term potential. *Postaglandins Leukotr Essent Fatty Acids.* 2004; 70: 391-397.

Kearns SD, Haag M. The effect of omega-3 fatty acids on Ca-ATPase in rat cerebral cortex. *Prostaglandins Leukot Essent Fatty Acids.* 2002; 67: 303-308.

Kelly C, McCreadie RG. Smoking habits current symptoms, and pre-morbid characteristics of schizophrenic patients in Nithsdale, Scotland. *Am J Psychiatry 1999*; 156: 1751-1757.

Khan MM, Evans DR, Gunn V, Scheffer R E, Parikh VV, Mahadik SP. Reduced erythrocyte membrane essential fatty acids and increased lipid peroxides in schizophrenia at the never-medicated first episode of psychosis and after years of treatment with antipsychotics. *Schizophr. Res.* 2003; 58: 1-10.

Kitajka K, Puskas LG, Zvara A, Hackler L Jr, Barcelo-Coblijn G, Yeo YK, Farkas T. The role of n-3 polyunsaturated fatty acids in brain: modulators of rat brain gene expression by dietary n-3 fatty acids. *Proc Natl Acad Sci USA.* 2002; 99: 2619-2624.

Krajcovicova – Kudlackova M, Paukova V, Bacekova M, Dusinska M. Lipid peroxidation in relation to vitamin C and vitamin E levels. *Cent Eur J Public Health* 2004; 12: 46-48.

Lapillone A, De Mar JC, Nanegari V, Heird WC. The fatty acid profile of buccal cheek cell phospholipids is a non-invasive marker of long-chain polyunsaturated fatty acid status in piglets. *J Nutr.* 2002, 132: 2319-23.

Ling GC, Horrobin DF, Fowkes FGR, Smith FB, Lowe GDO, Donnan PT, Ellis K. Plasma essential fatty acids, cigarette smoking and dietary antioxidants in peripheral arterial disease. *Arteroscler Thromb.* 1994; 14: 471-478.

Maes M, Christophe A, Delanghe J, Altamura C, Neels H, Meltzer HY. Lowered omega-3 polyunsaturated fatty acids in serum phospholipids and cholesteryl esters of depressed patients. *Psychiatry Res.* 1999; 85: 275-291

Mahadik SP, Mukhenjee S, Horrobin DF, Jenkins K, Correnti EE, Scheffer RE. Plasma membrane phospholipid fatty acid composition of cultured skin fibroblasts from schizophrenic patients: comparison with bipolar patients and normal subjects. *Psychiatry Res.* 1996; 63: 133-142.

Makrides M, Neumann MA, Byard RW, Simmer K, Gibson RA. Fatty acid composition of brain, retina and erythrocytes in breast and formula-fed infants. *Am J Clin Nutr.* 1994, 60: 189-194.

Mamalakis G, Kiriakaksis M, Tsibinos G, Kafotos A. Depression and adipose essential polyunsaturated fatty acids. *Prostaglandins Leukotr Essent Fatty Acids* 2002; 67: 311-318.

Martin DS, Lonergan PE, Boland B, Fogarty MP, Brady M, Horrobin DF, Campbell VA, Lynch MA. Apoptotic changes in the aged brain are triggered by interleukin-1beta-induced activation of p38 and reversed by treatment with eicosapentaenoic acid. *J Biol Chem*. 2002; 277: 34239-46.

Mazière C, Mazière J C, Mora L, Auclair M, Polonovski J. Trifluperazine increases fatty acid turnover in phospholipids in cultured human fibroblasts. *Lipids* 1988; 23: 419-423.

McCreadie R, MacDonald E, Blacklock C, Tilak-Singh D, Wiles D, Halliday J, Patenson J. Dietary intake of schizophrenic patients in Nithsdale, Scotland: a case-control study. *BMJ*. 1998; 317: 784-785.

McCreadie RG. Diet, smoking and cardiovascular risk in people with schizophrenia: a descriptive study. *Br J Psychiatry* 2003; 183: 534-539.

Mellor JE, Laugharne JDE, Peet M. Omega-3 fatty acid supplementation in schizophrenic patients. *Human Psychopharmarol*. 1996; 11: 39-46.

Messamore E. Relationship between the niacin skin flush response and essential fatty acids in schizophrenia. *Prostaglandins Leukotr Essent Fatty Acids* 2003; 69: 413-419.

Morrow JD, Frei B, Longmire AW, Gaziano JM, Lynch SM, Shyr Y, Strauss WE, Oates JA, Roberts L J. Increase in circulating products of lipid peroxidation (F2-Isoprostanes) in smokers. *N Engl J Med*. 1995; 332: 1198-1203

Noaghiul S, Hibbeln JR. Cross-national comparisons of seafood consumption and rates of bipolar disorders. *Am J Psychiatry* 2003; 160: 2222-2227.

Obi F O, Nwanze EAC. Fatty acid profiles in mental disease. Part 1. Linoleate variations in schizophrenia. *J Neurol Sci*. 1979; 43: 447-454.

Ota Y, Sasagawa T, Suzuki K, Tomioka K, Nagai A, Niiyama G, Kawanaka M, Yamada G, Okita M. Vitamin E supplementation increases polyunsaturated fatty acids of RBC membrane in HCV-infected patients. *Nutrition* 2004; 20: 358-363.

Peet M, Laugharne J, Rangarajan N, Horrobin D, Reynolds G. Depleted red cell membrane essential fatty acids in drug-treated schizophrenic patients. *J Psychiatr Res*. 1995; 29: 227-232.

Peet M, Murphy B, Shay J, Horrobin D. Depletion of omega-3 fatty acid levels in red blood cell membranes of depressive patients. *Biol Psychiatry* 1998; 43: 315-319.

Peet M. Eicosapentaenoic acid in the treatment of schizophrenia and depression: rationale and preliminary double-blind trial results. *Prostaglandins Leukotr Essent Fatty Acids*. 2003; 477-485.

Peet M. International variations in the outcome of schizophrenia and the prevalence of depression in relation to national dietary practices: an ecological analysis. *Br J Psychiatry* 2004; 184: 404-408.

Peet M. Shah S, Selvam K, Ramchand CN. Polyunsaturated fatty acid levels in red cell membranes of unmedicated schizophrenic patients. *World J Biol Psychiatry*. 2004; 5: 92-99.

Pettegrew JW, Keshavan MS, Panchalingham K, Strychor S, Kaplan DB, Tetta MG, Allen M. Alterations in brain high-energy phosphate and membrane phospholipid metabolism in first-episode, drug-naive schizophrenics. A pilot study of the dorsal prefrontal cortex by in vivo phosphous 31 nuclear magnetic resonance spectroscopy. *Arch Gen Psychiatry* 1991; 48: 563-568.

Phillips M, Sabas M, Greenberg J. Increased pentane and carbon disulfide in the breath of patients with schizophrenia. *J Clin Pathol*. 1993; 46: 861-864.

Prabhakaran S, Swatton JE, Ryan MM, Huffaker SJ, Huang JT, Groffin JL, Wayland M, Freemen T, Dudbridge F, Lilley KS, Karp NA, Hesten S, Tkachev D, Mimmock ML, Yolken RH, Webster MJ, Torrey EF, Bahn S. Mitochondrial dysfunction in schizophrenia: evidence for compromised brain metabolism and oxidative stress. *Mol Psychiatry*. 2004; 9: 684-697.

Ranjekar PK, Hinge A, Hegde V, Ghate M, Kale A, Sitasawad S, Wagh UV, Dedsikdar VB, Mahadik SP. Decreased antioxidant enzymes and membrane essential polyunsaturated fatty acids in schizophrenic and bipolar mood disorder patients. *Psychiatry Res*. 2003; 121: 109-122.

Richardson AJ, Cyhlarova E, Ross MA. Omega-3 and omega-6 fatty acid concentrations in red blood cell membranes relate to schizotypal traits in healthy adults. *Prostaglandins Leukotr Essent Fatty Acids* 2003; 69: 461-466.

Ryan MCM, Collins P, Thakore JH. Impaired fasting glucose tolerance in first episode, drug-naive patients with schizophrenia. *Am J Psychiatry*. 2003; 160: 284-289.

Sarkadi-Nagy E, Wijendran V, Diau GY, Chao AC, Hsieh AT, Turpeinen A, Nathanielsz PW, Brenna JT. The influence of prematurity and long chain polyunsaturated supplementation in 4-week adjusted age baboon neonate brain and related tissues. *Pediatr Res*. 2003; 54: 244-52.

Sobczaks, Honig A, Christophe A, Maes M, Helsdingen RW, De Vriese. SA, Riedel WJ. Lower high-density lipoprotein cholesterol and increased omega-6 polyunsaturated fatty acids in first-degree relatives of bipolar patients. *Psychol Med* 2004; 34: 103-112.

Song C, Leonard BE, Horrobin DF. Dietary ethyl-eicosapentaenoic acid but not soybean oil reverses central interleukin-1-induced changes in behaviour, corticosterone and immune response in rats. *Stress*. 2004; 7: 43-54.

Stokes C. Dietary intake of sugar and polyunsaturated fatty acids predicts the severity of symptoms of schizophrenia. *Nutr Neurosci*, 2004;7: 247-9.

Suarez A, del Carmen Ramirez M, Faus MJ, Gil A. Dietary long-chain fatty acids influence tissue fatty acid composition in rats at weaning. *J Nutr*. 1996; 126: 887-97.

Vaddadi K S, Gilleard C J, Mindham RH, Butler R. A Controlled trial of prostaglandin E1 precursor in chronic neuroleptic resistant schizophrenia. *Psychopharmacology (Berl)*. 1986; 88: 362-367.

Vaddadi KS, Gilleard CJ, Soosai E, Polonowita AK, Gibson RA, Burrows GD. Schizophrenia, tardive dyskinesia and essential fatty acids. *Schizophrenia Res*. 1996; 20: 287-294.

Vamecq J. Chlorpromazine and carnitine – dependency of rat liver peroxisomal beta-oxidation of long-chain fatty acids. *Biochem J*. 1987; 241: 783-791.

Van den Branden C, Roels F. Thioridazine a selective inhibitor of peroxisomal beta-oxidation in vito. *FEBS*. 1985; 187: 331-333.

Ward GR, Huang YS, Bobik E, Xing HC, Mutsaers L, Auestad N, Montalto M, Wainwright P. Long-chain polyunsaturated fatty acid levels in formulae influence deposition of docosahexaenoic acid and arachidonic acid in brain and red blood cells of artificially reared neonatal rats. *J Nutr*. 1998; 128: 2473-87.

Weiser M, Reichenberg A, Grotto I, Yasvitzky R, Rabinowitz J, Lubin G, Nahon D, Knobler HY, Davidson M. Higher rates of cigarette smoking in male adolescents before the onset of schizophrenia: a historical-prospective cohort study. *Am J Psychiatry* 2004; 161: 1219-1223.

Yamagata K, Tagami M, Takenaea F, Yamori Y, Navia Y, Itoh S. Polyunsaturated fatty acids induce tight junctions to form in brain capillary endothelial cells. *Neuroscience*. 2003; 116: 649-656.

Yao J K, van Kammen D P, Welker JA. Red blood cell membrane dynamics in schizophrenia II. Fatty acid composition. *Schizophr Res*. 1994; 13: 217-226.

Yao JK, Leonard S, Reddy RD. Membrane phospholipid abnormalities in post-mortem brains from schizophrenic patients. *Schizophr Res*. 2000; 42: 7-17.

Yao JK, Stanley JA, Reddy R, Keshavan MS, Pettegrew JW. Correlations between peripheral polyunsaturated fatty acid content and in vivo membrane phospholipid metabolites. *Biol Psychiatry* 2002; 52: 823-830.

Zeleniuch-Jacquotte A, Chajes V, Van Kappel AL, Riboli E, Toniolo P. Reliability of fatty acid composition in human serum phospholipids. *Eur J Clin Nutr*. 2000; 54: 367-372.

In: Fatty Acids and Oxidative Stress...
Editors: R. Reddy and J.K. Yao, pp. 115-128
ISBN: 978-1-60021-677-0

Chapter VIII

Essential Fatty Acids and Neurodevelopmental Disorders

Antonio Y. Hardan and Jeff A. Stanley

Introduction

There is a growing interest in the investigation of essential fatty acids (EFA) metabolism in neurodevelopmental disorders (NDD) due to the key role that these compounds play in cell physiology and during periods of development (Neuringer et al., 1988; Neuringer et al., 1994; Yoshida et al., 1998). EFA are polyunsaturated fatty acids (PUFA), which cannot be synthesize de novo in the body and therefore, must rely on dietary consumption and its ability to metabolize within normal levels in the brain. EFA can influence processes of brain development including neuronal migration, axonal and dendritic growth, remodeling and pruning of synaptic connections (Kaufmann et al. 1996; Smalheiser et al. 1996; Hughes et al. 1996; Crawford, 1992; Horrobin, 1998). The importance of certain highly PUFA for brain development and function has been documented with evidence of cognitive, learning and visual deficits accompanying alterations of PUFA (Neuringer et al., 1994; Yoshida et al., 1998). Therefore, it is not surprising that neuropsychiatric symptoms can be observed secondary to small changes in key PUFA such as AA and DHA at critical periods of brain development leading to a broad range of membrane dysfunctions involving receptor bindings, neurotransmission, signal transduction, and prostaglandin synthesis (Horrobin, 1998). A cascade of events can potentially occur when abnormalities of the antioxidant system and EFA are suspected which may underlie many biological, physiological and clinical manifestation of childhood developmental disorders such as autistic disorder (AutD), attention deficit hyperactivity disorder (ADHD), dyslexia and dyspraxia.

The potential involvement of EFA deficits in the pathophysiology of NDD is multifaceted. The importance of an adequate dietary supply of highly PUFA in early development has been the subject of considerable research (Dobbing, 1997; Williatts and Forsyth, 2000). Careful studies of the effects of highly PUFA supplementation on the cognitive developments of infants have been conducted emphasizing the critical role that these compounds play in early development with evidence of favorable influence of long-chain PUFA on measures of visual attention and problem solving (Williatts and Forsyth, 2000). Additionally, theoretical models have been proposed on the ways in which EFA alterations could account for the abnormalities in NDD such as the visual magnocellular deficits observed in dyslexia (Taylor and Richardson, 2000). Dietary deficits may not be the sole mechanism by which EFA can be incriminated, phospholipase A_2 (PL A_2) and other phospholipid enzymes abnormalities can precipitate the onset of these disorders early in childhood. These enzymes are good candidates for contributing not only to the genetic predisposition to NDD but to their onset because they are strongly expressed during brain growth and development (Bennet and Horrobin, 2000). However, gene-environment interactions should be kept in mind since the metabolism of cell membranes is dependent on the dietary intake of EFA (Taylor and Richardson, 2000).

While interest in the role of EFA in the pathophysiology of several psychiatric disorders such schizophrenia is increasing, the role of these compounds in early onset neuropsychiatric disorders including AutD and ADHD has been largely neglected. This is particularly unfortunate since the early onset and the course of NDD is consistent with theoretical model of EFA deficits altering the physical environment of cell membranes and/or the functioning of neurotransmitters and receptors and consequently the course of brain development. In fact, EFA deficits are more likely to manifest themselves early in life than in late adolescence or adulthood. Additionally, the absence of gross anatomic and metabolic alterations in AutD, ADHD and other similar disorders and the presence of diffuse brain anomalies support the potential involvement of altered processes associated with membrane phospholipids in their pathophysiology. Finally, emerging evidence though limited on the benefits of dietary supplementation of EFA in some disorders is encouraging and should stimulate more interest in the investigation of its effect on the metabolites of membrane phospholipids in NDD (e.g., as measured with *in vivo* ^{31}P spectroscopy).

EFA in Neurodevelopmental Disorders

Over the last decade, a number of investigations have assessed peripheral lipid content of red blood cells (RBC) and platelets in several NDD and have reported disturbances in the content of EFA and antioxidant defense system in AutD, ADHD and dyslexia (Table 1). Most of these disorders have been associated with EFA deficits ranging from genetic abnormalities in the enzymes involved in phospholipid metabolism to symptoms reportedly to have improved following dietary supplementation with long chain fatty acid (Ward, 2000). These observations have led some experts to suggest that these conditions may fall within a phospholipid spectrum of disorders in light of the overlap between them with the high degree of comorbidity and their aggregation within families (Richardson and Ross, 2000). While much more research is needed, the available evidence is encouraging and will be discussed below.

Table 1. Evidence of Phospholipid Metabolism Abnormalities in Specific Neurodevelopmental Disorders

Disorder	Evidence
Autistic Disorder	+
Attention Deficit Hyperactivity Disorder	+
Apraxia	+
Asperger's Disorder	+
Childhood Disintegrative disorder	0
Down's Syndrome	+
Dyslexia or Reading Disorder	+
Expressive language disorder	0
Fetal Alcohol Syndrome	+
Fragile X	0
Long-chain 3-hydroxyacyl-CoA dehydrogenase deficiency	++
Motor tic disorders	0
Phytanic acid storage disease	++
Rett's disorder	+
Stuttering	0

0 = No studies; - = negative findings; + = some evidence; ++ = Strong evidence

Autism

Autistic disorder is a pervasive developmental disorder characterized by impairment in communication skills, reciprocal social interaction and imaginative activity. The etiology of autistic disorder remains unknown. Although the disorder was first considered to be psychosocial or psychodynamic in origin, much evidence has accumulated to support a biological substrate. Several etiological factors have been implicated in the pathogenesis of autism including genetics, biochemical, immunological, and environmental. Phospholipid metabolism and the antioxidant system which play a key role in altering the course of normal brain development have recently been incriminated in the pathophysiology of pervasive developmental disorders including AutD. This effort is supported by evidence of abnormalities of these systems in several developmental disorders such as Down Syndrome (Brugge et al., 1999).

EFA in Autism

Growing interest has been observed in fatty aacid metabolism in autism as indicated by an anecdotal surge of EFA supplementation in children with pervasive developmental disorders. While strong and convincing scientific evidence is still lacking, some reports exist in the literature pointing to phospholipid metabolism abnormalities in AutD. Vancassel et al. (2001) reported the presence of abnormalities in the PUFA levels in the plasma of autistic children, compared to a mentally retarded group. The study population consisted of 15 children with autism (4 girls, 11 boys) mean age 8 years 4 months, and 18 mentally retarded children (5 girls, 13 boys) mean age 8 years 8 months. All children were in excellent physical health, and none had a history of endocrine or systemic disease. Levels of AA, DHA and total (n-3) PUFA levels were reduced by approximately 20% in autistic children. A tendency of reduced EPA levels was also observed. In another study, Bell et al. (2000) investigated RBC EFA compositions in a patient with autistic spectrum disorder (ASD) and reported that the levels of PUFA, particularly DHA and AA, were reduced in comparison to control subjects. They suggested that the existence of PUFA instability in this patient with ASD and suggested that other NDD may bes caused by increased phospholipase activity, perhaps in conjunction with auto-oxidative stress.

Antioxidant System in Autism

The direct evidence of abnormal antioxidant system and phospholipid metabolism in AutD is limited. Several investigations have examined the antioxidants system and reported conflicting results but all pointed to the existence of increased oxidative stress leading to free radical accumulation and potentially contributing to the pathosphyiology of AutD (Golse et al., 1978; Yorbik et al., 2002; Sogut et al., 2003). In 1978, Golse et al., (1978) reported on abnormalities in the platelets' activities of SOD I, and GSH-Px in children with infantile autism. Recently, Yorbik et al., (2002) investigated the plasma levels of GSH-Px, and SOD, and erythrocyte levels of GSH-Px in 45 autistic children (39 boys and 6 girls) and 41 healthy matched controls (35 boys and 6 girls). The age range of autistic group was 4 to 12 years, and the normal control group was between 4 and 12 years. Activity levels of erythrocyte SOD, erythrocyte and plasma GSH-Px in the autistic children were significantly lower than in the control group. In contrast, Sogut et al., (2003) compared SOD, GSH-Px and thiobarbituric acid-reactive substances (TBARS) levels in plasma as well as nitric oxide (NO) levels in red blood cells (RBC) in 27 patients with autism and 30 age- and sex- matched controls. In the autistic group, increased RBC NO levels and plasma GSP-Px activity were observed when compared with controls with no differences in plasma TBARS, and SOD activity between the two groups (Sogut et al., 2003). More recently, Zoroglu et al., (2003) examined RBC levels of TBARS SOD, CAT, adenosine deaminase (ADA), and xanthine oxidase (XO) activities in 27 patients with autism and 26 age- and sex-matched controls. The activity level of TBARS, XO, and SOD was increased and CAT was decreased in the patient group while no difference in ADA levels was found when comparing the two groups. Differences in results between studies may be related to methodological limitations such as the method of diagnosis since neither the autism diagnostic interview or the autism diagnostic observation schedule have been used. Nevertheless, altered enzymatic oxidation have been reported by all these studies which warrant further research in more rigorous and well-designed investigations.

While some literature is available on the enzymatic oxidation, there is limited data about non-enzymatic molecules in autism. Dolske et al. (1993) reported a 30-week double-blind, placebo controlled trial exploring the effectiveness of ascorbic acid as a supplemental treatment for autistic children in a residential facility. Ascorbic acid is a biological antioxidant that acts as a chain-breaking scavenger for peroxyradicals and acts synergistically with alpha-tocopherol (Yao et al., 2001). Ascorbic acid was significantly more beneficial than placebo in improving autistic symptoms as measured by the Ritvo-Freeman Scale. In another study of non-enzymatic antioxidants molecules, uric acid was measured in urine and serum in 22 autistic patients and 22 matched controls. No difference in the blood levels of uric acid was observed between the two groups (Launay et al., 1988). In contrast, several other studies have documented increased urinary excretion of uric acid in at least a subgroup of children with autism (Rosenberger-Debiesse J et al., 1986; Coleman M, et al., 1976). Studies undertaken to determine the percentage of the autistic population which overproduces uric acid arrived at a figure of 11-28% as compared to 0-3% of non-autistic children (Page and Coleman, 2000).

Pathophysiology of Autism

Neuropathological studies have provided evidence in autism of structural abnormalities of developmental origin in the cerebral cortex, portions of the limbic system, cerebellum and inferior olivary nuclei (Kemper and Bauman, 1998). These findings suggest abnormalities in neuronal migration, alignment and the elaboration, pruning, and selective elimination of neuronal processes in cerebral cortex, the limbic system and the cerebellum (Kemper and Bauman, 1998; Minshew et al., 1996; Bailey et al 1998). Since the underlying pathophysiology of these abnormalities is unknown, it is possible that dysfunction of the antioxidant defense system and EFA deficits could contribute to these changes in brain development. The decrease in Purkinje's cell number observed in autopsy studies could be related to the accumulation of free radicals reported to be cell toxic and possibly leading to neuronal cell death (Mahadik and Mukherjee, 1996). Furthermore, low SOD can contribute to superoxide-mediated DNA damage and has been suggested to affect neuronal growth and migration (Mahadik and Mukherjee, 1996), and therefore could explain the abnormal brain development and neuronal migration observed in autism.

Antioxidant defense system has been incriminated in several features associated with autism such EEG abnormalities and seizures, abnormalities of pain thresholds, mental retardation, polydipsia and attentional deficits. Electroencephalogram (EEG) abnormalities and epileptic seizures are often observed in autistic children (Gillberg, 1991). Accumulation of free radical leading to abnormalities of neuronal cell membrane has been suggested to increase susceptibility for seizure activity (Vreugdenhil et al., 1996). AA is important in pain physiology and its depletion could diminish pain responses (Horrobin 1996), which is one of the characteristics of autistic children. While about 70% of autistic disorder has mental retardation, the EFA-rich phospholipids may be related to intelligence (Horrobin et al., 1998). Polydipsia, which is considered as one of symptoms of EFA deficiency (Burr and Burr, 1930), has also been described in autism (Terai et al., 1999). Whereas attention problems are frequently observed in autistic children (Noterdaeme et al., 2001), a reduction in AA and DHA levels has been observed in the serum of children with ADHD (Burgess et al., 2000).

Finally, the metabolism of neuronal membrane fatty acids appears to be closely related to neurotrophic factors. Neurotrophic growth facto (NGF) was found to induce ethanolamine glycerophospholipid synthesis which is suppressed by AA but not DHA (Ikemeto et al., 1997). Recently, Li and Wurtman (1999) reported that NGF did increase phospholipase A2 activity, and increased (by 50-330%) the incorporation of monosaturated (oleic acid) and polyunsaturated (linoleic acid or AA) fatty acids into diacylglycerol and by 15-70% into phospatidylcholine. Furthermore, in some abnormal neuronal migration disorders such as the cerebro-hepato-renal (Zellenger) syndrome, the incorporation of abnormal fatty acids into neuronal membrane leads to unresponsiveness to neurotrophic factors necessary for normal function and survival or to increased permeability of calcium channels and cell deaths (Powers, 2001). These observations support the critical role that these different systems including neurotrophic factors play in brain growth and in developmental disorders such as autism. This particularly relevant in light of the recent report of elevated neurotrophin 4 (NT4), vasoactive intestinal peptide (VIP), brain derived neurotrophic factors (BDNF), and calcitonin-related gene peptide in children with autism when controlled with healthy controls (Nelson et al., 2000).

Summary

While significant progress has been made to elucidate the etiology of autism, the cause of this severe developmental disorder remains unknown. Chemical, anatomic, genetic and, neuroimaging studies have provided more insight into the pathophysiology of AutD but no comprehensive study to date has examined the antioxidant system, neuronal membrane phospholipid metabolism, and neurotrophic factors as possible contributors to the emergence and the development of this disorder. Therefore, it is essential to develop investigations that will attempt to replicate previous findings and to expand our understanding by further examining membrane lipid metabolism and antioxidant defense system in individuals with autism and wide range of intellectual abilities and age range. An increase in our understanding of the these systems will be instrumental in the identification of effective and safe treatment but the development of preventive strategies such as EFA and antioxidant supplementation to pregnant women, and children at risk, as well as toddlers with a recent diagnosis of autism.

Attention Deficit Hyperactivity Disorder

The neurodevelopmental disorder, attention-deficit/hyperactive disorder (ADHD), is one of the most prevalent childhood behavioral disorders, affecting approximately 3% to 9% of the population (Swanson et al. 1995; Greenhill 1998; Todd 2000). ADHD is first diagnosed in children and is characterized by the inability to focus attention, control impulses and refrain from extraneous motor activity. The majority of ADHD children treated with psychostimulants (primarily methylphenidate and dextroamphetamine) show improvements in ADHD symptoms; however, approximately 25% of ADHD children do not respond or partially respond to psychostimulant treatment (Simeon & Wiggins 1993; Swanson et al. 1993; Findling & Dogin 1998). Neuroimaging findings have implicated the basal ganglia and prefrontal regions, which are regions associated with the neural networks of attention; however the molecular/biochemical underpinning(s) causing ADHD has yet to be fully understood. The pathophysiology of ADHD is complex and different etiological factors have been incriminated and recent research has focused on several hypotheses including the existence of phospholipid metabolism disturbances.

Fatty Acid Abnormalities in ADHD

Based on genetic, molecular and neurochemical studies and the clinical improvement with psychostimulants suggest ADHD is a "receptor disorder" involving monoamines (e.g., dopamine and norepinephrine); however, there is increasing evidence implicating the metabolism of EFAs in ADHD. As first hypothesized by Colquhoun (1981), a lack of EFAs may play a role in the cause of ADHD.

Abnormalities of fatty acid metabolism have been reported in ADHD. Decreased EFA content (specifically EPA, DHA, AA and DGLA) in plasma and RBC have been reported in ADHD children (Mitchell et al. 1987; Stevens et al. 1995; Chen et al. 2004) and ADHD adults (Young et al. 2004). Furthermore, an association between greater AA and DHA plasma deficits and increased frequency of EFA deficiency symptoms (polydipsia, polyuria, dry skin and dry hair) in ADHD children, and between total plasma n-6 and n-3 fatty acid deficits and severity of behavioral scores also have been reported (Burgess et al., 2000; Richardson & Puri, 2000). Other less direct evidence of a EFAs deficiency in ADHD include: 1) a greater incident of ADHD in males than females (Gaub & Carlson 1997) and males have a greater requirement for EFAs (Pudelkewicz et al. 1968; Huang & Horrobin 1987; Marra & de Alaniz 1989; Huang et al. 1990); 2) ADHD children tend to have difficulties sleeping (Kaplan et al. 1987) and EFAs are involved in the regulation of sleep (Fagioli et al. 1989; Yehuda et al. 1998); and 3) increased symptoms of eczema, asthma and other allergies in ADHD children (Hartsough & Lambert 1985; Szatmari et al. 1989) and eczema is associated with a deficit of EFAs (Horrobin 2000).

Neuroimaging Investigations in ADHD

One approach to examine the role of phospholipid metabolism in the pathophysiology of ADHD is the application of novel neuroimaging methodologies such ^{31}P MRS which provides information on the synthesis and degradation of membrane phospholipids (MPL) by quantifying the phosphomonoester [PME; precursors of membrane phospholipids] and phosphodiester [PDE; breakdown products of membrane phospholipids] resonances. The role of PMEs reflecting membrane synthesis has been demonstrated in an entorhinal cortical lesioned rat model of neuronal degeneration/regeneration where higher PME levels are observed at the time and site of neuritic sprouting (Geddes et al. 1997). Preliminary results using a multi-voxel, *in vivo* ^{31}P spectroscopy approach showed lower PML precursor levels in the prefrontal (more pronounced on the right side) and in the basal ganglia (with a greater effect on the left side) of ADHD children compared to age-matched control children (Stanley et al. 2004). The implicated regions, prefrontal and basal ganglia, are consistent with MRI structural studies in ADHD (Zametkin & Liotta 1998; Eliez & Reiss 2000) and are associated with the neural networks of attention (Posner & Petersen 1990; Mirsky et al. 1999). When compared to metabolite levels of normal developing brains, the results suggest an alteration in the pruning of overproduced cells in the prefrontal, and an underdevelopment of neuronal processes and synapses in the basal ganglia (Stanley et al. 2004). Consequently, this provides additional support implicating phospholipid abnormalities in the pathophysiology of this disorder.

Treatment Studies in ADHD

Medication treatment, which primarily involves psychostimulants [e.g., methylphenidate (Ritalin®) and dextroamphetamine (Dexedrine® or Adderall®)](Barkley, 1998), is effective in improving symptoms and hence, academic performance in the majority of children with ADHD (Simeon & Wiggins 1993; Swanson *et al.* 1993; Findling & Dogin 1998). The downside of psychostimulants includes adverse side effects (e.g., headaches, anorexia, abdominal pain and insomia) (Scarnati 1986), increased hyperactivity as the drug wears off (Ahmann et al. 1993), potential cardiotoxicity, and potential addictive abuse and unclear of long term effects since psychostimulants have similar behavior as cocaine (Scarnati 1986; Findling & Dogin 1998). Consequently, in recent years much effort has focused on alternatives for medical management of children with ADHD (Arnold 2001; Chan 2002) such as EFA supplementation studies. In a double-blind, placebo-controlled study, no behavioral/symptoms improvements were observed after 2 (Hirayama et al. 2004) and 4 months (Voigt et al. 2001) of DHA supplementation in ADHD children; however, the total plasma n-3 fatty acid levels were significantly increased while plasma AA levels were significantly reduced (Voigt et al. 2001). In contrast, significant improvements in ADHD-related symptoms were reported in children with specific learning difficulties after 3 months of supplementation with a mixture of n-6 and n-3 HUFAs; the study did not include peripheral measurements (Richardson & Puri 2002). The lack of a positive outcome in the former two studies may be attributed to the choice of which or what combination of EFAs to use for supplementation. Recently, Horrobin and colleagues (2002) have shown clinical improvements with EPA supplementation at doses that coincided with increased RBC AA levels, a finding that was not observed in the Voigt et al. study (2001).

Summary

The examination of fatty acid metabolism in ADHD provides an opportunity to understand the neurobiologic abnormalities observed in this disorder. Taking in consideration the heterogeneity of ADHD it is possible that the metabolism of phospholopid may be present in some but not all clinical or biological subgroups. Hence, the children and adolescents with ADHD who are not responsive to available and traditional treatment such as psychostimulants and atomoxetine may benefit from alternative strategies including EFA supplementation. However, despite the attractiveness of such strategies, final conclusions should not be drawn before additional and more rigorous research has been conducted and converging evidence has emerged.

Dyslexia

Dyslexia or reading disorder is defined as reading achievement that is below the expected level for a child's age, education, and intelligence (APA, 1994). Additionally impairment interferes with academic success or the daily activities that involve reading (APA, 1994). It is characterized by an impaired ability to recognize word, slow inaccurate reading and poor comprehension (APA, 1994). No specific etiology has been identified, but in light of the many associated developmental disorders such as language difficulties, ADHD and learning disorders, dyslexia probably has multifactorial causes. Fatty acid abnormalities have been described in this disorder with evidence indicating improvement of some dyslexic adults treated with omega-3 fatty acids (Stordy, 1995). More direct evidence come from an *in vivo* cerebral ^{31}P MRS study which revealed elevated PME peak area in 12 dyslexic when compared to 10 non-dyslexic adults. These findings reflect a reduced incorporation of phospholipids into cell membranes and suggest that membrane phospholipid is abnormal in dyslexia (Richardson et al., 1997).

Conclusion

Despite the complexity of NDD, the investigation of fatty acids in these disorders is quite promising with the potential of identifying the contribution of phospholipid metabolism abnormalities to their pathophysiology. While some evidence is emerging, more research is needed in the field of brain development to increase our understanding of the role of EFA during this critical period. This step will be instrumental in yielding better insight into any observed abnormalities in children and adolescents with NDD and will help in elucidating the neurobiologic underpinnings of these disorders. Well-designed studies, using rigorous methodologies such as gold standard diagnostic and assessment tools, are desperately needed to examine further the role of phospholipids in childhood onset neuropsychiatric disorders. The application of newer imaging techniques such as *in vivo* cerebral ^{31}P MRS will be key to illuminate the link between peripheral findings and brain pathology and will help in providing important clues to neuronal membrane functioning. Future neurobiologic investigations should also be complemented with well-controlled EFA supplementation studies that will offer final answers and more importantly make available alternative strategies to clinicians to treat NDD.

References

Ahmann PA, Waltonen SJ, Olson KA, Theye FW, Van Erem AJ and LaPlant RJ. Placebo-controlled evaluation of Ritalin side effects. *Pediatrics*, 91:1101-6, 1993.

American Psychiatric Association. *Diagnostic and Statistical Manual of Mental Disorders*, 4th edition (DSM-IV). Washington, DC: American Psychiatric Association; 1994.

Arnold LE. Alternative treatments for adults with attention-deficit hyperactivity disorder (ADHD). *Ann. N. Y. Acad. Sci.*, 931:310-41, 2001.

Barkley RA. Attention-deficit hyperactivity disorder : a handbook for diagnosis and treatment. New York: Guilford Press, 1998.

Bennett CN, and Horrobin DF. Gene targets related to phospholipid and fatty acid metabolism in schizophrenia and other psychiatric disorders: an update. *Prostaglandins Leukot Essent Fatty Acids.*, 63 : 47–59, 2000.

Bailey A, Luthert P, Dean A, Harding B, Janota I, Montgomery M, Rutter M, Lantos P. A clinicopathological study of autism. *Brain*, 121:889-905,1998.

Bell JG, Sargent JR, Tocher DR, Dick JR. Red blood cell fatty acid compositions in a patient with autistic spectrum disorder: a characteristic abnormality in neurodevelopmental disorders? *Prostaglandins Leukot Essent Fatty Acids*, 63:21-25, 2000.

Bourre JM, Francois M, Youyou A, Dumont O, Piciotti M, Pascal G, Durand G. The effects of dietary alpha-linolenic acid on the composition of nerve membranes, enzymatic activity, amplitude of electrophysiological parameters, resistance to poisons and performance of learning tasks in rats. *J Nutr*, 1989;119:1880-1892.

Brugge K, Nichols S, Saitoh T, Trauner D. Correlations of glutathione peroxidase activity with memory impairment in adults with Down syndrome. *Biol Psychiatry*, 46:1682-1689, 1999.

Burgess JR, Stevens L, Zhang W, Peck L. Long-chain polyunsaturated fatty acids in children with attention-deficit hyperactivity disorder. *Am J Clin Nutr*, 71:327S-30S, 2000.

Burr GO, Burr M. On the nature and role of the fatty acids in essential in nutrition. *J Biol Chem*, 86: 587-621, 1930.

Chan PH, Fishman RA, Longar S, Chen S, Yu A. Cellular and molecular effects of polyunsaturated fatty acids in brain ischemia and injury. *Prog Brain Res*, 63:227-235, 1985.

Chan PH, Fishman RA. Transient formation of superoxide radicals in polyunsaturated fatty acid-induced brain swelling. *J Neurochem*, 35:1004-1007, 1980.

Chan E. The role of complementary and alternative medicine in attention-deficit hyperactivity disorder. *J. Dev. Behav. Pediatr*, 23:S37-45, 2002.

Chen J, Hsu S, Hsu C, Hwang L and Yang S. Dietary patterns and blood fatty acid composition in children with attention-deficit hyperactivity disorder in Taiwan. *J Nutr Biochem*, 15:467-72, 2004.

Coleman M, Landgrebe MA, Landgrebe AR. Purine autism: Hyperuricosuria in autistic children: does this identify a subgroup of autism? In: Coleman M, ed. *The Autistic Syndromes*. Elsevier, New York; 183-214, 1976.

Colquhoun I, Bunday S., A lack of essential fatty acids as a possible cause of hyperactivity in children. *Med Hypotheses*, 7:673–679, 1987.

Crawford MA. Essential fatty acids and neurodevelopmental disorder. In: Bazán NG, Murphy MG, and Toffano G, eds. Neurobiology of essential fatty acids. New York: Plenum Press, pp. 307-14, 1992.

Dobbing J. *Developing brain and behaviour: the role of lipids in infant formula*. San Diego: Academic Press;1997.

Dolske MC, Spollen J, McKay S, Lancashire E, Tolbert L. A preliminary trial of ascorbic acid as supplemental therapy for autism. *Prog Neuropsychopharmacol Biol Psychiatry*, 17:765-774, 1993.

Eliez S and Reiss AL. MRI neuroimaging of childhood psychiatric disorders: A selective review. *Journal of Child Psychology & Psychiatry & Allied Disciplines*, 41:US: Cambridge Univ Press, 2000.

Ernstein L. Lipid peroxidation in biological membranes mechanisms and implications. In: Yagi K, ed. *Active oxygens, lipid peroxides, and antioxidants*. Tokyo: CRC Press; 11-38; 1993.

Fagioli I, Baroncini P, Ricour C and Salzarulo P. Decrease of slow-wave sleep in children with prolonged absence of essential lipids intake. *Sleep*, 12:495-9, 1989.

Findling RL and Dogin JW. Psychopharmacology of ADHD: children and adolescents. *J. Clin. Psychiatry*, 59:42-9, 1998.

Gaub M and Carlson CL. Behavioral characteristics of DSM-IV ADHD subtypes in a school-based population. *J. Abnorm. Child Psychol*, 25:103-11, 1997.

Geddes JW, Pandey GN, Keller JN and Pettegrew JW. Elevated phosphocholine and phosphatidyl choline following rat entorhinal cortex lesions. *Neurobiol. Aging*, 18:305-8, 1997.

Gillberg C. The treatment of epilepsy in autism. *J Autism Dev Disord*, 21:61-77, 1991.

Golse B, Debray-Ritzen P, Durosay P, Puget K, Michelson AM. Alterations in two enzymes: superoxide dismutase and glutathion peroxidase in developmental infantile psychosis (infantile autism). *Rev Neurol*, 134:699-705, 1978.

Halliwell B, Gutteridge JMC. Oxygen radicals and the central nervous sytem. *Treds Neurosci*, 8: 22-66, 1984.

Hartsough CS and Lambert NM. Medical factors in hyperactive and normal children: prenatal, developmental, and health history findings. *Am. J. Orthopsychiatry*, 55:190-201, 1985.

Hirayama S, Hamazaki T and Terasawa K. Effect of docosahexaenoic acid-containing food administration on symptoms of attention-deficit/hyperactivity disorder - a placebo-controlled double-blind study. *Eur. J. Clin. Nutr*, 58:467-73, 2004.

Horrobin DF. The membrane phospholipid hypothesis as a biochemical basis for the neurodevelopmental concept of schizophrenia. *Schizophr Res,* 30:193-208, 1998.

Huang YS and Horrobin DF. Sex differences in n-3 and n-6 fatty acid metabolism in EFA-depleted rats. *Proc. Soc. Exp. Biol. Med.*, 185:291-6, 1987.

Horrobin DF.Schizophrenia as a membrane lipid disorder which is expressed throughout the body. *Prostaglandins Leukot Essent Fatty Acids*, 55:3-7, 1996.

Horrobin DF. Essential fatty acid metabolism and its modification in atopic eczema. *Am. J. Clin. Nutr*, 71:367S-72S, 2000.

Horrobin DF, Jenkins K, Bennett CN and Christie WW. Eicosapentaenoic acid and arachidonic acid: collaboration and not antagonism is the key to biological understanding. *Prostaglandins Leukot. Essent. Fatty Acids*, 66:83-90, 2002.

Huang YS, Horrobin DF, Watanabe Y, Bartlett ME and Simmons VA. Effects of dietary linoleic acid on growth and liver phospholipid fatty acid composition in intact and gonadectomized rats. *Biochem Arch,* 6:47-54, 1990.

Hughes DA, Southon S and Pinder AC. (n-3) Polyunsaturated fatty acids modulate the expression of functionally associated molecules on human monocytes in vitro. *J. Nutr,* 126:603-10, 1996.

Ikemoto A. Kobayashi T. Watanabe S. Okuyama H. Membrane fatty acid modifications of PC12 cells by arachidonate or docosahexaenoate affect neurite outgrowth but not norepinephrine release. *Neurochemical Research*, 22:671-678, 1997.

Kalyanaraman B. Free radicals from catecholamine hormones, neuromelanins, and neurotoxins. In: Miquel J, Quintanilha AT, Weber H, eds. *Handbook of free 107 radicals and antioxidants in biomedicine.* Boca Raton: CRP Press; 147-59, 1989.

Kaplan BJ, McNicol J, Conte RA and Moghadam HK. Sleep disturbance in preschool-aged hyperactive and nonhyperactive children. *Pediatrics*, 80:839-44, 1987.

Kaufmann WE, Worley PF, Pegg J, Bremer M and Isakson P. COX-2, a synaptically induced enzyme, is expressed by excitatory neurons at postsynaptic sites in rat cerebral cortex. *Proc Natl Acad Sci U S A,* 93:2317-21, 1996.

Kemper TL, Bauman M. Neuropathology of infantile autism. *J Neuropathol Exp Neurol*, 57:645-52, 1998.

Launay JM, Ferrari P, Haimart M, Bursztejn C, Tabuteau F, Braconnier A, Pasques-Bondoux D, Luong C, Dreux C. Serotonin metabolism and other biochemical parameters in infantile autism. A controlled study of 22 autistic children. *Neuropsychobiology*, 20:1-11, 1988.

Li J. Wurtman RJ. Mechanisms whereby nerve growth factor increases diacylglycerol levels in differentiating PC12 cells. *Brain Research*, 818:252-259, 1999.

Mahadik SP, Mukherjee S. Free radical pathology and antioxidant defense in schizophrenia: a review. *Schizophr Res*, 19:1-17, 1996.

Marra CA and de Alaniz MJ. Influence of testosterone administration on the biosynthesis of unsaturated fatty acids in male and female rats. *Lipids*, 24:1014-9, 1989.

Mirsky AF, Pascualvaca DM, Dunçan CC and French LM. A model of attention and its relation to ADHD. *Mental Retardation & Developmental Disabilities Research Reviews,* 5:169-76, 1999.

Minshew NJ, Goldstein G, Dombrowski SM, Panchalingam K, Pettegrew JW. A preliminary 31P MRS study of autism: evidence for undersynthesis and increased degradation of brain membranes. *Biol Psychiatry*, 33:762-73, 1993.

Minshew NJ. (1996). Brain mechanisms in autism: Functional and structural abnormalities. *J Autism Dev Disord*, 26:205-209, 1996.

Mitchell EA, Aman MG, Turbott SH and Manku M. Clinical characteristics and serum essential fatty acid levels in hyperactive children. *Clin Pediatr (Phila)*, 26:406-11, 1987.

Murphy MG, Byczko Z. Effects of membrane polyunsaturated fatty acids on adenosine receptor function in intact N1E-115 neuroblastoma cells. *Biochem Cell Biol*, 68:392-395, 1990.

Nelson KB, Grether JK, Croen LA, Dambrosia JM, Dickens BF, Jelliffe LL, Hansen RL, Phillips TM. Neuropeptides and neurotrophins in neonatal blood of children with autism, mental retardation, or cerebral palsy. *Neurology*, 48:597-606, 2000.

Neuringer M, Anderson GJ, Connor WE. The essentiality of n-3 fatty acids for the development and function of the retina and brain. *Ann Rev Nutr*, 18: 517–541, 1988.

Neuringer M, Reisbeck S, Janowsky J. The role of n-3 fatty acids in visual and cognitive development: current evidence and methods of assessment. *J Pediatr*, 125: 39–47, 1994.

Noterdaeme M, Amorosa H, Mildenberger K, Sitter S, Minow F. Evaluation of attention problems in children with autism and children with a specific language disorder. *Eur Child Adolesc Psychiatry*, 10:58-66, 2001.

Page T, Coleman M. Purine metabolism abnormalities in a hyperuricosuric subclass of autism. *Biochim Biophys Acta*, 1500:291-296, 2000

Posner MI and Petersen SE. The attention system of the human brain. *Annu. Rev. Neurosci*, 13:25-42, 1990.

Powers JM. Normal and defective neuronal membranes: structure and function: neuronal lesions in peroxisomal disorders. *Journal of Molecular Neuroscience*, 16:285-287, 2001.

Pudelkewicz C, Seufert J and Holman RT. Requirements of the female rat for linoleic and linolenic acids. *J. Nutr*, 94:138-46, 1968.

Richardson AJ, Cox IJ, Sargentoni J, Puri BK. Abnormal cerebral phospholipid metabolism in dyslexia indicated by phosphorous-31 magnetic resonance spectroscopy. *NMR Biomed*, 10: 309–314, 1997.

Richardson AJ, Ross MA. Fatty acid metabolism in neurodevelopmental disorder: a new perspective on associations between attention-deficit/hyperactivity disorder, dyslexia, dyspraxia and the autistic spectrum. *Prostaglandins Leukot Essent Fatty Acids*, 63:1-9, 2000

Rice-Evans CA. Formation of free radicals and mechanisms of action in normal biochemical processes and pathological states. In: Rice-Evans CA, Burdon RH, eds. Free radical damage and its control. Amsterdam: *Elsevier*; 131-53, 1994.

Richardson AJ and Puri BK. The potential role of fatty acids in attention-deficit/hyperactivity disorder. *Prostaglandins Leukot. Essent. Fatty Acids*, 63:79-87, 2000.

Richardson AJ and Puri BK. A randomized double-blind, placebo-controlled study of the effects of supplementation with highly unsaturated fatty acids on ADHD-related symptoms in children with specific learning difficulties. *Prog. Neuropsychopharmacol. Biol. Psychiatry*, 26:233-9, 2002.

Rosenberger-Debiesse J, Coleman M. Preliminary evidence for multiple etiologies in autism. *J Autism Dev Disord*, 16:385-392, 1986.

Scarnati R. An outline of hazardous side effects of Ritalin (methylphenidate). *Int J Addict*, 21:837-41, 1986.

Simeon JG and Wiggins DM. Pharmacotherapy of attention-deficit hyperactivity disorder. *Can. J. Psychiatry,* 38:443-8, 1993.

Smalheiser NR, Dissanayake S and Kapil A. Rapid regulation of neurite outgrowth and retraction by phospholipase A2-derived arachidonic acid and its metabolites. *Brain Res,* 721:39-48, 1996.

Sogut S, Zoroglu SS, Ozyurt H, Yilmaz HR, Ozugurlu F, Sivasli E, Yetkin O, Yanik M, Tutkun H, Savas HA, Tarakcioglu M, Akyol O. Changes in nitric oxide levels and antioxidant enzyme activities may have a role in the pathophysiological mechanisms involved in autism. *Clin Chim Acta,* 331:111-117, 2003

Stevens LJ, Zentall SS, Deck JL, Abate ML, Watkins BA, Lipp SR and Burgess JR. Essential fatty acid metabolism in boys with attention-deficit hyperactivity disorder. *Am. J. Clin. Nutr,* 62:761-8, 1995.

Swanson JM, Lerner M and Williams L. More frequent diagnosis of attention deficit-hyperactivity disorder. *N. Engl. J. Med.,* 333:944, 1995.

Swanson JM, McBurnett K, Wigal T and Pfiffner LJ. Effect of stimulant medication on children with attention deficit disorder: A "review of reviews." *Exceptional Children,* 60:154-61, 1993.

Szatmari P, Offord DR and Boyle MH. Correlates, associated impairments and patterns of service utilization of children with attention deficit disorder: findings from the Ontario Child Health Study. *J. Child Psychol. Psychiatry,* 30:205-17, 1989.

Stanley JA, Kipp H, Greisenegger K, Panchalingam K, Pettegrew JW, Keshavan MS and Bukstein OG. *In Vivo 31P Spectroscopy Evidence of Different Regional Neurodevelopmental Alterations in Children with ADHD.* Society for Neuroscience Abstracts, 2004.

Stordy BJ. Benefit of docosahexaenoic acid supplements to dark adaptation in dyslexia. *Lancet,* 346: 385, 1995.

Taylor KET, Richardson AJ. Visual function, fatty acids and dyslexia. *Prostaglandins Leukot Essent Fatty Acids,* 63: 89–93, 2000.

Terai K, Munesue T, Hiratani M. Excessive water drinking behavior in autism. *Brain Dev,* 21:103-106, 1999.

Todd RD. Genetics of childhood disorders: XXI. ADHD, part 5: a behavioral genetic perspective. *J. Am. Acad. Child Adolesc. Psychiatry,* 39:1571-3, 2000.

Vancassel S, Durand G, Barthelemy C, Lejeune B, Martineau J, Guilloteau D, Andres C, Chalon S. Plasma fatty acid levels in autistic children. *Prostaglandins Leukot Essent Fatty Acids.* 65:1-7, 2001.

Voigt RG, Llorente AM, Jensen CL, Fraley JK, Berretta MC and Heird WC. A randomized, double-blind, placebo-controlled trial of docosahexaenoic acid supplementation in children with attention-deficit/hyperactivity disorder. *J. Pediatr,* 139:189-96, 2001

Vreugdenhil M, Bruehl C, Voskuyl RA, Kang JX, Leaf A, Wadman WJ. Polyunsaturated fatty acids modulate sodium and calcium currents in CA1 neurons. *Proc Natl Acad Sci,* 93:12559-12563, 1996.

Ward PE. Potential diagnostic aids for abnormal fatty acid metabolism in a range of neurodevelopmental disorders. *Prostaglandins Leukot Essent Fatty Acids,* 63:65-68, 2000.

Willatts P, Forsyth JS. The role of long-chain polyunsaturated fatty acids in infant cognitive development. *Prostaglandins Leukot Essent Fatty Acids,* 63: 95–100, 2000.

Yao JK, Reddy RD, van Kammen DP. Oxidative damage and schizophrenia: an overview of the evidence and its therapeutic implications. *CNS Drugs*, 15:287-310, 2001.

Yehuda S, Rabinovitz S and Mostofsky DI. Essential fatty acids and sleep: mini-review and hypothesis. *Med. Hypotheses*, 50:139-45, 1998.

Young GS, Maharaj NJ and Conquer JA. Blood phospholipid fatty acid analysis of adults with and without attention deficit/hyperactivity disorder. *Lipids*, 39:117-23, 2004.

Yorbik O, Sayal A, Akay C, Akbiyik DI, Sohmen T. Investigation of antioxidant enzymes in children with autistic disorder. *Prostaglandins Leukot Essent Fatty Acids*, 67:341-343, 2002.

Yoshida S, Sato A, Okuyama H. Pathophysiological effects of dietary essential fatty acid balance on neural systems. *Jpn J Pharmacol*, 77: 11–22, 1998.

Zametkin AJ and Liotta W. The neurobiology of attention-deficit/hyperactivity disorder. *J. Clin. Psychiatry*, 59:17-23, 1998.

Zoroglu SS, Armutcu F, Ozen S, Gurel A, Sivasli E, Yetkin O, Meram I. Increased oxidative stress and altered activities of erythrocyte free radical scavenging enzymes in autism. *Eur Arch Psychiatry Clin Neurosci*, 254:143-147, 2004.

Section III: Therapeutic Prospects

In: Fatty Acids and Oxidative Stress...
Editors: R. Reddy and J.K. Yao, pp. 131-158
ISBN: 978-1-60021-677-0

Chapter IX

Fatty Acids and Oxidative Stress in Neurodevelopmental Disorders of Childhood: Treatment Implications

Alexandra J. Richardson and Marion A. Ross

Introduction

There is a high degree of comorbidity and familiality across a whole range of neurodevelopmental and psychiatric disorders, including conditions usually diagnosed in childhood - such as attention-deficit / hyperactivity disorder (ADHD), dyslexia or specific reading disabilities (SRD), dyspraxia or developmental coordination disorder (DCD) and autistic spectrum disorders (ASD) - and various disorders more often diagnosed in adults, such as depression, bipolar disorder and the schizophrenia spectrum of disorders. The associated features and traits that these conditions appear to have in common strongly suggest some common underlying risk factors at the level of biological predisposition. Increasing evidence – reviewed here - suggests that these biological risk factors may include functional deficiencies in certain highly unsaturated fatty acids and/or reduced antioxidant capacities (Peet et al, 2003; Richardson and Ross, 2000; Ross, 2000), either of which can impair brain development and function.

A major limitation of these 'diagnoses' of neurodevelopmental and psychiatric disorders is that they have little or no explanatory value, but are essentially descriptive labels for particular patterns of behaviour. Similarly, many of the techniques commonly used in neuroscientific research into these conditions (such as structural or functional brain imaging) are also rather limited in terms of what they can actually reveal about etiology. There is thus a clear need for new approaches to the identification and management of neurodevelopmental disorders, based on a better understanding of the complex etiological factors involved. The causes of these disorders are almost always multi-factorial, and these are also likely to vary between different individuals with the same diagnostic label. Genetic factors undoubtedly play a part, but the predisposing factors at this level are complex and in most cases these appear to involve the interaction of many different genes, at least some of which are widely distributed in the general population (Aherson and Curran, 2001; Fisher and DeFries, 2002; Muhle et al, 2004).

Only environmental factors can explain the apparent increasing prevalence of many of these conditions (Mandell et al, 2005), and it will be argued here that changes affecting nutrition are likely to be amongst these. In particular, there has been a particularly dramatic reduction in the dietary intake of omega-3 polyunsaturated fatty acids in most developed countries over the last 50 years, and this has already been linked with increasing rates of physical disorders such as cardiovascular disease, inflammatory disorders and certain cancers (Simopoulos, 2002). Low dietary omega-3 intake also has serious implications for many aspects of brain development and function, to which some individuals will be more vulnerable than others owing to individual differences in fatty acid metabolism.

Other environmental changes include the dramatic increase in the levels of pollutants to which humans are exposed as a result of industrialisation. Our air, water, and soil – and therefore also our food supply – now contain thousands of artificial chemicals in addition to heavy metals and other contaminants, and among their many potential effects on health is the depletion that these can cause in antioxidant reserves. Poor antioxidant status - from dietary deficiencies and/or increased exposure to environmental oxidants - is thought to contribute to a wide range of physical health problems, including heart disease and stroke, cancer and loss of vision. The brain is particularly vulnerable to oxidative stress; and again, some individuals will be more vulnerable than others for constitutional reasons.

Fatty acid and antioxidant metabolism actually lie at the interface of gene-environment interactions. Individual differences exist for many genes that influence the absorption, transport and utilisation of fatty acids, as well as others that influence the capacity of the brain and body to protect themselves from oxidative damage (to which polyunsaturated fats are particularly prone). Conversely, the expression of individual genetic differences of all kinds also depends heavily on the dietary intake of both fatty acids and other essential nutrients, both during development and throughout life. Dietary deficiencies in certain vitamins and minerals will impair antioxidant capacity, and in addition to other negative effects on physical and mental health, this will increase the probability of functional deficiencies in key fatty acids that are absolutely required for optimal brain function.

In view of the above, it is unfortunate that conventional approaches to the identification and management of neurodevelopmental disorders currently take no account of nutritional status. Further research in this area is clearly needed, but as this review will show, the study of fatty acids and oxidative stress in relation to these complex and inter-related conditions offers not only new insights into their etiology, but also new treatment possibilities, primarily in the form of nutritional interventions.

Childhood Neurodevelopmental Disorders

The commonest neurodevelopmental disorders of childhood include ADHD, dyslexia, dyspraxia and autistic spectrum disorders. Between them these conditions are now estimated to affect at least 20% of the school age population when mild-to-moderate forms are included, although more males than females are affected (Blaxill, 2004; Kadesjo and Gillberg, 1999; Rutter et al, 2004; Scahill and Schwab-Stone, 2000). All of these conditions usually persist into adulthood, with substantial costs not only to the individuals and families directly affected, but also to society as a whole.

Each of these disorders is defined by a different and relatively specific pattern of difficulties in behaviour and/or learning. Thus in ADHD, the core features involve attentional problems and/or hyperactive-impulsive behaviours; in dyslexia, specific difficulties with reading and writing; in dyspraxia or developmental coordination disorder (DCD), specific weaknesses in the planning and coordination of actions; and in the autistic spectrum, impaired social and communication skills. In practice, however, the overlaps between these conditions are high, and their boundaries with normal variation in behavioural and cognitive function are not clear-cut (Jones, 2000; Kadesjo and Gillberg, 1999, 2001; Willicut and Pennington, 2000). 'Pure' cases are the exception, not the rule, and within each category there is substantial heterogeneity.

These childhood neurodevelopmental conditions also show familial and other associations with certain psychiatric disorders including depression and bipolar disorder as well as schizophrenia spectrum and personality disorders (Biederman et al, 1987; Hellgren at al, 1994; Horrobin et al, 1995; Rasmussen and Gillberg, 2000). This and other evidence points to some common genetic factors in the predisposition to all of these conditions, and it has been proposed that these include constitutional anomalies of fatty acid metabolism (Bennett and Horrobin, 2000).

Omega-3 and Omega-6 Polyunsaturated Fatty Acids

Long-chain polyunsaturated fatty acids (LC-PUFA) from the omega-6 and omega-3 series are crucial for normal brain structure and function, but all of these must ultimately be derived from dietary sources. Among these, four highly unsaturated fatty acids (HUFA) are particularly important for the brain: DGLA and AA (omega-6), and EPA and DHA (omega-3). If not supplied directly by the diet, these can in theory be synthesised from the corresponding essential fatty acids (EFA), linoleic acid and alpha-linolenic acid respectively, as shown in Table 1.

Structurally, AA and DHA are key components of neuronal membranes, making up 15-20% of the brain's dry mass and more than 30% of the retina. In early life, both omega-3 and omega-6 LC-PUFA are critical in supporting brain growth and maturation. During prenatal development the placenta effectively doubles the levels circulating in maternal plasma (Crawford, 2000), and severe deficits may have permanent effects if they occur during critical periods of early development. AA is crucial to brain growth, and mild deficiencies are associated with low birth weight and reduced head circumference. It also plays a key role in the cellular processes underlying learning and memory. DHA is particularly concentrated in highly active membranes such as synapses and photoreceptors, and adequate supplies are essential for normal visual, motor and cognitive development (Carlson and Neuringer, 1999; Moriguchi et al, 2000; Uauy et al, 2001).

Table 1. Pathways for the Synthesis of Omega-6 and omega-3 fatty acids

OMEGA-6 series		Enzymes involved in HUFA synthesis	OMEGA-3 series	
Linoleic (LA)	18:2		Alpha-linolenic (ALA)	18:3
↓		*Delta 6- desaturase*	↓	
Gamma-linolenic (GLA)	18:3		Octadecatetraenoic	18:4
↓		*Elongase*	↓	
Dihomogamma-linolenic (DGLA)	**20:3**		Eicosatetraenoic	20:4
↓		*Delta 5-desaturase*	↓	
Arachidonic (AA)	**20:4**		**Eicosapentaenoic (EPA)**	**20:5**
↓		*Elongase*	↓	
Adrenic	22:4		Docosapentaenoic (DPA)	22:5
↓		*Elongase, Delta 6-desaturase, Beta-oxidation*	↓	
Docosapentaenoic (DPA)	22:5		**Docosahexaenoic (DHA)**	**22:6**

Throughout life, adequate supplies of LC-PUFA remain crucial for optimal brain function. They increase the fluidity of neuronal membranes, essential for efficient signal transduction, and some act as second messengers in chemical neurotransmitter systems as well as contributing to many other aspects of cell signalling (Bourre, 2004; Yehuda et al, 1999). Functionally, three LC-PUFA (DGLA, AA and EPA) are particularly important as substrates for the eicosanoids – highly bioactive hormone-like substances including prostaglandins, leukotrienes and thromboxanes. These derivatives play key roles in regulating blood flow, endocrine and immune functions and can modulate ion channels, neurotransmitter uptake, synaptic transmission, apoptosis and many other biological processes.

An adequate and appropriately balanced supply of LC-PUFA is thus required for normal brain function, both during early development and throughout life. Unfortunately, there are many possible reasons why their availability may be less than optimal, particularly in the case of omega-3 fatty acids.

First, these fatty acids are scarce in many modern diets, especially those in which highly processed foods predominate. This particularly applies to the omega-3 LC-PUFA most important for the brain (EPA and DHA), which are found in appreciable quantities only in fish and seafood. Their precursor, alpha-linolenic acid, is found in green vegetables and some nuts and seeds, but its conversion to EPA and DHA is limited, as discussed further below. Omega-6 fats are much more plentiful (vegetable oils are rich in linoleic acid, while meat and dairy produce contain AA pre-formed) and the last century has seen dramatic increases in the ratios of omega-6 to omega-3 in average Western-type diets, from approximately 3:1 to more than 20:1 in some cases.[8] Gene transfer studies have recently provided powerful evidence that the earlier ratios were much closer to the optimum for healthy human cells (Kang, 2003); and this relative disappearance of omega-3 from the diet has already been explicitly linked with increased rates of many disorders of both physical and mental health, including cardiovascular disease, immune disorders, depression and schizophrenia (Haag, 2003; Peet et al, 2003; Simopoulos, 1999). As discussed below, evidence is now mounting that it may also be a factor in the childhood neurodevelopmental disorders considered here.

Second, there may be difficulties in the synthesis of LC-PUFA from linoleic or alpha-linolenic acid. This conversion process is not very efficient in humans (Pawlosky et al, 2001; Salem et al, 1999) and can be adversely affected by diet and lifestyle as well as constitutional factors. Males appear particularly vulnerable to LC-PUFA deficiency (Huang and Horrobin, 1987; Marra and Alaniz, 1989), and show poorer EFA- LC-PUFA conversion than females (Burdge and Wootton, 2002; Burdge et al, 2002). These sex differences are of particular interest in view of the excess of males affected by most childhood neurodevelopmental disorders. LC-PUFA synthesis may also be impaired in atopic conditions such as eczema (Horrobin, 2000), which are commonly associated with ADHD, dyslexia, autism and related conditions. From these and other observations, constitutional inefficiencies in LC-PUFA synthesis were explicitly proposed in the first study implicating fatty acid abnormalities in ADHD (Colquhoun and Bunday, 1981).

Constitutional factors affecting other aspects of fatty acid metabolism may also reduce the availability of LC-PUFA. Fatty acids are constantly replaced and recycled during the normal turnover and remodelling of cell membranes and in the chemical cascades triggered by normal cell signalling processes. Phospholipase A2 (PLA2) enzymes cleave highly unsaturated fatty acids from the SN-2 position of membrane phospholipids, creating free fatty acids and lyso-phospholipids that require rapid recycling via least two further enzyme steps. A huge number of other enzymes are involved in the transport and utilization of fatty acids in the brain and body. Individual differences in their efficiency will have implications for optimal dietary requirements for LC-PUFA and genetic influences on them may contribute to the risks for various developmental and psychiatric disorders (Bennett and Horrobin, 2000).

The Potential Role of Fatty Acids in Childhood Neurodevelopmental Disorders

Many features associated with ADHD, dyslexia, dyspraxia and autistic spectrum disorders are consistent with deficiencies or imbalances in omega-3 and/or omega-6 fatty acids, as discussed in more detail elsewhere (Richardson and Puri, 2000; Richardson et al, 2003). These include the excess of males affected, slightly increased tendencies for pregnancy and birth complications and minor physical anomalies, and an increased frequency of atopic or auto-immune disorders in affected individuals and their relatives. Fatty acid abnormalities could also help to account for some of the key cognitive and behavioural features of these conditions, such as anomalous visual, motor, attentional or language processing, as well as associated difficulties with mood, digestion, temperature regulation and sleep.

Etiological factors in neurodevelopmental disorders such as ADHD, dyslexia, autism and related disorders are obviously both complex and multifactorial, and as the previous section makes clear, LC-PUFA have such profound and widespread influences on brain development and function that they could have numerous possible contributory roles both in the prenatal origins of these conditions and in their maintenance throughout childhood and later life. Brief consideration of some key themes in current theories of these neurodevelopmental disorders illustrates just a few possible mechanisms by which fatty acid deficiencies or imbalances could exert an influence.

Neurotransmitter Function

The limitations of simplistic neurotransmitter-imbalance models of psychiatric illness have long been apparent, but the many ways in which fatty acids and their metabolism can affect the functioning of conventional neurotransmitter systems – and the implications of this for many disorders of mental health and development - are only just starting to be fully appreciated (Haag, 2003; Peet et al, 2003).

ADHD is routinely treated with stimulant medications that increase the availability of dopamine, and this fact is reflected in all pharmacological theories of this condition. It is therefore relevant that chronic omega-3 deficiency has been shown to reduce dopamine (and its binding to D2 receptors) in frontal cortex, and is associated with comparable attentional and behavioural dysfunction in animal studies (Delion et al, 1996; Zimmer et al, 1998). Preliminary evidence suggests that some effects on dopaminergic systems of very early omega-3 deficiency may respond to subsequent dietary interventions better than others (Levant et al, 2004). In some children, controlled trials have shown that ADHD symptoms can be reduced by treatment with omega-3 fatty acids, as discussed below, but further exploration of this issue is still required.

Dietary supplementation with omega-3 can modulate the adrenal activation induced by mental stress (Delarue et al, 2003), and also affects serotonergic function, which could help to explain the benefits of omega-3 in controlled trials of depression (Nemets et al, 2002; Peet and Horrobin, 2002; Su et al, 2003) and in preventing or reducing hostility and aggression (Hamazaki et al, 1996; Zanarini and Frankenburg, 2003). These clinical features are commonly associated with the childhood disorders considered here, supporting the possibility that omega-3 fatty acids could also help in the management of these conditions.

Immune Function

Many aspects of normal brain development and plasticity are influenced by the immune system (Huh et al, 2000), and abnormal immune reactivity is increasingly implicated in a wide range of neurodevelopmental and psychiatric disorders (Anderson, 1996; Jyonouchi et al, 2002; Krause etal, 2002; Muller et al, 2000; O'Brien et al, 2004). Pro-inflammatory cytokines have been shown to alter the function of monoamine neurotransmitters implicated in stress, sleep disorders and depression (Leonard and Song, 1999; Song et al, 1998). The relative effects of omega-3 and omega-6 fatty acids on immune function are complex (Harbige, 2003), but eicosanoid derivatives of AA and EPA generally have opposing actions such that an excess of omega-6 relative to omega-3 (as found in most modern diets in developed countries) will tend to increase tendencies towards inflammation and autoimmune reactivity via these and other mechanisms (Calder, 2003).

Abnormal immune responses in autism reported to date include both enhanced autoimmunity and reduced immune function (Ashwood and Van de Water, 2004), and a recent post-mortem study has found evidence of active neuroinflammatory processes in the brain tissue of autistic subjects.[60] It has also been shown that maternal neuronal antibodies from the serum of mothers of children with neurodevelopmental disorders can induce motor and other learning difficulties in the offspring when injected into pregnant mice (Dalton et al, 2003; Vincent et al, 2002).. Similarly, the minor neuroanatomical anomalies reported in dyslexic brains post-mortem are consistent with inflammatory processes during prenatal development (Taylor et al, 2001, and vulnerability to such immunological processes could very plausibly be increased by relative omega-3 deficiencies. It is almost 40 years since an early report showed that susceptibility of the rat brain to allergic encephalomyelitis was increased by dietary insufficiency of polyunsaturated fatty acids (Clausen and Moller, 1967), and abundant evidence since then has supported the critical importance of omega-3 for brain structure and function, both during early development and throughout life (Bourre, 2004).

Some of the visual and other sensory processing deficits associated with dyslexia are thought to originate from autoimmune influences in utero (Stein and Walsh, 1967). Whether dietary supplementation with omega-3 in later life may improve these aspects of visual function in dyslexia is currently under investigation, but DHA is known to play a unique role in visual signal transduction (Litman et al, 2001). Of particular interest is the fact that EPA rather than DHA has generally been thought to be responsible for the immuno-modulatory effects of fish oils, via its eicosanoid derivatives (and this might perhaps help to explain why preliminary evidence suggests that EPA may be more effective than DHA in the treatment of neurodevelopmental and psychiatric disorders, as discussed below). However, recent studies have shown that DHA also gives rise to derivatives that have anti-inflammatory properties (Hong at al, 2003; Serhan et al, 2004). Further exploration of these findings is clearly needed, but meanwhile they only serve to emphasise that omega-3 fatty acids have a wide range of effects on immune system functioning.

Gut-Brain Interactions

The gut is a key interface in the highly complex interactions between the immune system and the brain; and many substances regarded primarily as neurotransmitters also play crucial signalling roles in the gut. Digestive disorders appear to be unusually prevalent in children with neurodevelopmental disorders, especially within the autistic spectrum (Horvath et al, 1999). In early life, dietary polyunsaturated fatty acids are important to the establishment and maintenance of healthy gut flora and can influence Th1 and Th2 programming by gut associated lymphoid tissues, with consequences for autoimmune or allergic responses (Das, 2002; Wallace et al, 2001). Relative omega-3 deficiencies may predispose to gut inflammation and associated increases in membrane permeability in a wide variety of different ways (Teitelbaum and Allan Walker, 2001), with detrimental effects on both nutrient absorption and detoxification processes. Immune-gut disturbances of this kind have many potential implications. The so-called 'leaky gut' theories of autism propose that if the integrity of the gut wall is compromised, for which there appears to be evidence in some cases (D'Eufemia et al, 1996), peptides or other substances could enter the bloodstream and exert deleterious effects on brain function. While some aspects of this theory remain controversial, there is already enough evidence to support further investigation.

Experimental Evidence for Fatty Acid Abnormalities in Childhood Developmental and Psychiatric Disorders

Certain physical signs have been reliably associated with polyunsaturated fatty acid deficiency in animal studies, including excessive thirst, frequent urination, dull, dry or rough-textured hair and skin, and soft or brittle nails. A greater prevalence or severity of these signs has now been reported in connection with ADHD, dyslexia and autistic spectrum disorders (Baker, 1985; Bell et al, 2000; Colquhoun and Bunday, 1981; Richardson et al, 2000; Stevens et al, 1995, 1996; Taylor et al, 2000). Although these kinds of physical symptoms can obviously have other causes, their association with these conditions is consistent with essential fatty acid deficiencies.

There are also several reports of reduced blood LC-PUFA concentrations in children with ADHD-type difficulties relative to matched controls (Arnold et al, 1994; Bekaroglu et al, 1996; Chen et al, 2004; Colquhoun and Bunday, 1981; Mitchell et al, 1987; Stevens et al, 1995, 1996), and one with similar findings in adults (Young et al, 2004). The precise pattern of results varies between studies, but reductions of AA as well as DHA have usually been apparent, and plasma concentrations appear to show the most consistent findings. One recent study still found reduced plasma LC-PUFA in ADHD children despite significantly *elevated* concentrations of the same fatty acids in red cell membranes (Burgess and Stevens, 2003). Some studies of children from the USA showed reductions of both omega-3 and omega-6 LC-PUFA in plasma phospholipids despite no lack of the precursor EFA, a pattern that could reflect difficulties in EFA- LC-PUFA conversion and/or an unusually rapid rate of LC-PUFA breakdown and loss (Stevens et al, 1995, 1996). In a Taiwanese study, however, reductions of red blood cell (RBC) linoleic acid as well as AA and DHA were found in ADHD children relative to controls, despite their similar dietary fatty acid intake (Chen et al, 2004), making difficulties with *in vivo* conversion a less likely explanation.

Some studies have also explored associations between biochemical and behavioural measures. In a combined sample of 96 ADHD and control boys, low omega-6 status was related only to some physical health measures (such as dry skin and hair, frequency of colds, and antibiotic use), but not to parental ratings of behaviour or learning (Stevens et al, 1996). By contrast, low omega-3 status was associated not only with physical signs of fatty acid deficiency (particularly excessive thirst, frequent urination and dry skin) but also with both behavioural problems (including conduct disorder, hyperactivity-impulsivity, anxiety, temper tantrums and sleep problems) and learning difficulties in these children, irrespective of the ADHD diagnosis.

In autistic spectrum subjects LC-PUFA abnormalities have been reported in both plasma (Vancassel et al, 2001) and RBC membranes (Bell et al, 2000; 2004). Findings include particular reductions in omega-3 LC-PUFA, an elevated AA:EPA ratio and an apparent increased susceptibility to breakdown of membrane fatty acids, possibly reflecting increased oxidative stress, as discussed further below.

In dyslexia, blood biochemical studies to date are few, but include one early case report of fatty acid deficiency in a dyslexic child (Baker, 1985) and the finding of elevated levels of a Type IV PLA2 enzyme in RBC membranes of dyslexic adults relative to controls (MacDonell et al, 2000). PLA2 enzymes release highly unsaturated fatty acids from membranes, so this finding would be consistent with an increased tendency towards deficiency in these fatty acids. Preliminary results from new studies have also shown direct correlations between omega-3 concentrations in RBC polar lipids and reading ability in both dyslexic and non-dyslexic adults (Bell et al, 2004), while poor working memory performance - a hallmark of dyslexia - was associated with low omega-3 status only in the dyslexic group (Ross et al, 2004).

These studies broadly support the proposal that fatty acid abnormalities play a part in these childhood neurodevelopmental disorders, but at present the interpretation of blood biochemical measures remains difficult, even leaving aside the different methodologies used in different studies. Blood biochemical measures of fatty acid status reflect complex interactions between influences of diet and other environmental factors with constitutional aspects of fatty acid metabolism. Not enough is yet known about the normal variability of omega-3 and omega-6 fatty acids in compartments such as RBC versus plasma, or triglyceride versus phospholipid fractions. Furthermore, there is also a real question about the significance of "normative values" drawn from populations consuming diets that contain less than optimal proportions of omega-3 in particular, as the epidemiological evidence suggests (Simopoulos, 1999). Further investigations of the clinical significance of these kinds of measures are therefore needed, ideally using samples drawn from the general community.

Randomised Controlled Trials of Fatty Acid Treatment in Childhood Neurodevelopmental Disorders

Intervention studies provide the most direct evidence that fatty acids can play a role in childhood neurodevelopmental disorders. Anecdotal evidence, case reports and open studies have suggested possible benefits from fatty acid treatment in these conditions (Baker 1985; Stordy et al, 1995; 2000), but randomised controlled trials (RCT) are essential to demonstrate causality. Published studies of this kind remain few, and only an overview summary can be provided here, but further details of most of these studies can be found in a recent review (Richardson, 2004).

Two early RCTs assessed the effects of evening primrose oil (providing the omega-6 fatty acid GLA) in hyperactive or ADHD children (Aman et al, 1987; Arnold et al, 1989). Unfortunately the study designs were not optimal in either case, involving treatment periods of only one month as well as potential confounds from the use of full treatment crossover. Few benefits from such supplementation were apparent from these trials, although some positive trends were noted.

Emphasis then shifted towards treatment with omega-3 fatty acids, for which there is a clearer rationale in terms of the relative lack of these fatty acids from modern diets in developed countries. However, two studies have found no benefits for children with ADHD-type difficulties from supplementation with the omega-3 fatty acid DHA. The first involved 63 children aged 6-12 years with a DSM-IV diagnosis of ADHD, treated for four months with either 345mg/day of pure algal-source DHA or placebo in addition to maintenance stimulant medication (Voigt et al, 2001). Compared with placebo, active treatment significantly increased DHA in plasma phospholipids, but no group differences were found for any behavioural outcomes, which included computerized tests of inattention and impulsivity as well as clinical and parent-ratings of ADHD symptoms. The second such RCT involved 40 Japanese children aged 6-12 years attending a summer camp for their ADHD-type difficulties (Hirayama et al, 2004). They received either foods fortified with DHA (at a dosage of approximately 0.5g / day) or indistinguishable control foods for 2 months. A wide range of measures was used pre-and post-treatment, but the only significant difference between treatment groups involved greater improvements over time on visual and auditory memory tests for those on placebo.

In contrast to these negative results for DHA, three other studies have shown some reduction in ADHD symptoms in children treated with combined omega-3 / omega-6 supplements. The first of these involved 50 ADHD-type children, pre-selected for physical signs consistent with fatty acid deficiency and treated for four months with a supplement containing fish oil and evening primrose oil in a 4:1 ratio (providing 480 mg DHA, 80 mg EPA, 96 mg GLA, 40 mg AA, and 24 mg alpha-tocopheryl acetate daily) or an olive oil placebo (Stevens et al, 2003). Both groups showed improvements, but significant benefits for active treatment over placebo were found for two of sixteen outcomes – parent-rated conduct problems and teacher-rated attentional difficulties. In addition, oppositional defiant behaviour dropped from clinical to sub-clinical levels in significantly more children receiving active treatment.

The second positive study used very similar treatments but involved children with a primary diagnosis of dyslexia who also showed ADHD-related symptoms (Richardson and Puri, 2002). After 12 weeks of supplementation (providing 480 mg DHA, 186 mg EPA, 96mg GLA, 42mg AA and 60 IU vitamin E as dl-α tocopherol per day), scores for anxiety, attentional difficulties and general behaviour problems were significantly lower for the 15 children on active treatment than 14 on the olive oil placebo. In a 12-week follow-up stage involving one-way treatment crossover, similar improvements in ADHD-related symptoms were then seen in the placebo group when they switched to active treatment (Richardson, 2003).

The third positive study involved 117 children with developmental coordination disorder (DCD) / dyspraxia who were aged between 5 and 12 years, attending mainstream schools, and receiving no other treatment for their behavioural or learning difficulties (Richardson and Montgomery, 2005). This is therefore the largest RCT to date of fatty acid treatment in children with neurodevelopmental disorders. Active treatment was a supplement containing 80% fish oil and 20% evening primrose oil (supplying 558 mg EPA, 174 mg DHA and 60 mg GLA plus 9.6mg Vitamin E (as alpha-tocopherol per day). The placebo contained olive oil, carefully matched with the active treatment for both colour and flavour. In the main phase of the study, children were followed for 3 months of treatment in parallel groups; there was then a one-way treatment crossover (placebo to active treatment) for a further 3 months. In view of the substantial comorbidity between DCD/dyspraxia and both ADHD and dyslexia, primary outcome measures were chosen to assess functioning in all three of these domains, involving changes in age-standardised measures of motor skills, reading and spelling, and ADHD-related symptoms as assessed by teacher ratings. Results showed highly significant improvements in reading and spelling for active treatment versus placebo, as well as highly significant reductions in ADHD-related symptoms. No differences were observed, however, on objective measures of motor skills. Over the 3-month follow-up phase, clear crossover effects were then apparent in the group switching from placebo to active treatment, while those continuing with active treatment maintained or increased their earlier improvements.

These findings support the idea that fatty acid supplementation may be of benefit in childhood neurodevelopmental disorders, although further studies are clearly needed both to replicate the existing findings and to determine optimal dosage and composition of fatty acid treatment, as well as to identify those children most likely to benefit. The similarity in the treatment effect sizes for ADHD-related symptoms between both DCD children and those selected with standard criteria for dyslexia (Richardson and Montgomery, 2005; Richardson and Puri, 2002) suggests that these results may generalise more widely, but the applicability of these results to children with DSM-IV ADHD still needs further investigation.

At the time of writing, no controlled trials of fatty acid treatment in autistic spectrum disorders have yet been reported, but there are many anecdotal reports of benefits from omega-3 supplementation, and some indications that this can help to normalize certain biochemical measures (Bell et al, 2004), as discussed further below.

The Potential Role of Oxidative Stress in Childhood Neurodevelopmental Disorders

Oxidative stress refers to a state of imbalance in which the burden of oxygen free radicals (from either internal metabolic processes or external sources) exceeds the body's antioxidant capacities for neutralising these. Oxygen free radicals (OFR) are produced within the body as a normal part of numerous metabolic processes, and at low levels they are essential for cell signalling. In physiological health, there is a balance between the levels of OFR needed for normal cellular function and the levels of antioxidants that protect against undue oxidative damage. When OFR exceed current antioxidant capacity, the resulting oxidative stress can interfere with normal metabolism and otherwise disrupt cellular processes by damaging lipids, proteins and DNA.

Increasing susceptibility to oxidative stress is an important feature of normal ageing, and it is also heavily implicated in various neurological and neurodegenerative disorders (Metodiewa and Koska, 2000; Rao and Balachandran, 2002; Smythies, 1999; Sofic et al, 2002). A substantial research literature now indicates that oxidative stress may be a key feature of schizophrenia (as amply discussed elsewhere in this volume) (Prabhakaran et al, 2004; Reddy et al, 2003; Yao et al, 2001), but very few studies have yet focused on the childhood neurodevelopmental disorders considered here, in which this may also be a factor (Johnson, 2001). Only with respect to autistic spectrum disorders has this idea been explicitly investigated. Studies have generally been small, and their methodologies and findings quite diverse, but a review of the current evidence does appear to support the idea that autism is associated with increased oxidative stress (McGinnis, 2004), as discussed further below. Evidence that oxidative stress may play a role in other neurodevelopmental disorders is largely indirect, but clearly deserves further investigation. Nutrition has a very significant impact on antioxidant capacity, which adds to the case for a serious consideration of dietary factors in the etiology and management of these disorders.

Dietary Influences on Antioxidant Capacity

Around half of the plasma antioxidant capacity comes from albumin and urate. The remainder is generally provided by the activities of antioxidants that are obtained directly from the diet – such as Vitamins C and E, or the flavonoids, carotenoids and other substances found in fruit and vegetables. Other antioxidants are synthesised within the body, such as the enzymes glutathione peroxidase (GPx) and superoxide dismutase (SOD), depending on the availability of the component substances needed, which include the key minerals zinc and selenium as well as copper and manganese.

Unfortunately, general population surveys show that many children in developed countries are lacking one or more of the vitamins and minerals needed for optimal antioxidant protection. Thus Vitamin C deficiency was found in around 5% of UK children from all age groups (Gregory, 2000); and low dietary intakes of Vitamin E affect a significant proportion of children both in the USA (Bendich, 1992) and in Europe (Gregory, 2000; Malvy et al, 1989; Serra Majen et al, 2001). Zinc deficiency is even more common, affecting up to 40% of UK children in some age groups (Doyle et al, 1994; Gregory, 2000). Selenium status varies according to geographical region, reflecting differences in soil content (generally high in the USA but low in many regions within Europe) as well as diet. Selenium deficiency was found in only 5% of German school children (Piechotowski et al, 2002), but a UK study found that 35-60% of young male prisoners had intakes below the LRNI (Eves and Gesch, 2003). Less attention has been paid to copper and manganese (although the former usually varies reciprocally with zinc status), but both have important influences on brain function, as discussed below.

Dietary Antioxidants and Brain Function

Low Vitamin C status has been associated with cognitive decline and dementia in the elderly (Charlton et al, 2004; Martin et al, 2002), although a causal relationship has not yet been demonstrated. Vitamin E deficiency can produce overt neurological symptoms, and particularly neuromuscular disorders (Sokol, 1990); and in adults it is associated with neurodegenerative disorders such as Alzheimer's disease (Rinaldi et al, 2003), Parkinson's disease (Paraskevas et al, 2003) and the motor dysfunction syndrome of tardive dyskinesia often found in schizophrenia. In children with cystic fibrosis, cognitive impairments were found to be linked with low Vitamin E status in particular (Koscik et al, 2004).

Zinc is important for numerous physiological functions including antioxidant defences, immune functioning and cell division and replication. It is particularly crucial in the development and maintenance of the brain and nervous system, and is an essential cofactor for the synthesis of highly unsaturated fatty acids. Zinc is also necessary for the synthesis of serotonin and melatonin; and it also plays a key role in blood glucose regulation, so zinc deficiency can have negative effects on mood and arousal via these mechanisms. Zinc deficiency also impairs the sense of taste and smell, which is likely to exacerbate poor food choices. Reduced zinc levels have been related to both immune system dysfunction and treatment-resistance in depressed patients (Maes et al, 1994; 1997), and zinc is one of many trace elements that is often reduced in schizophrenia and other psychiatric disorders including anorexia (Katz et al, 1987; Yanik et al, 2004).

Selenium is a key component of the glutathione peroxidase, deiodinase, and thioredoxine reductase families of enzymes. It is essential for normal thyroid hormone metabolism (along with iodine, iron and zinc), and also plays a key role (with zinc and other micronutrients) in the functioning of both the immune system (Fenencik and Ebringer, 2003) and the brain (Chen and Berry, 2003). Low selenium status has been linked with low mood, and although there is currently no evidence that dietary supplementation can improve mood states, this possibility remains under investigation (Benton, 2002; Hawkes and Hornbostel, 1996).

Copper is an essential trace element required for a number of oxidative enzymes, including cytochrome c oxidase in the mitochondria, Cu/Zn superoxide dismutase and caeruloplasmin (important in controlling free radicals, defences against infection and the intracellular oxidation of $Fe2^{+}$ to $Fe3^{+}$). Significantly increased levels of copper have been found in the plasma of patients with schizophrenia compared to healthy controls (Yanik et al, 2004), and elevated copper is usually found with zinc deficiency, as there is a reciprocal balance between these minerals. An elevated Cu/Zn ratio has also been associated with aggressive and antisocial behaviour in both general and psychiatric populations (Tokdemir et al, 2003).

Manganese is an essential trace element that acts as a co-factor for enzymes such as hexokinase, superoxide dismutase, and xanthine oxidase. Manganese is required for efficient sugar metabolism, for the production of cartilage, and for the manufacture of cyclic AMP - a cellular second messenger. Neurological symptoms can follow from either a deficiency or an excess of manganese. Industrial exposure to excessive levels of manganese has long been known to lead to brain toxicity, involving psychiatric symptoms and/or a Parkinson-like syndrome. Manganese deficient animals suffer impaired growth, reproductive problems, and a shortened life span, and low levels of manganese (along with other trace elements) have been reported in schizophrenia and bipolar disorder (Yanik et al, 2004). Early clinical reports suggested that treatment with manganese may alleviate tardive dyskinesia in this condition (Hoffer, 1977), and recent genetic studies suggest that interactions between polymorphisms for dopamine and manganese superoxide dismutase may confer vulnerability to schizophrenia (Zhang et al, 2003).

Superoxide dismutase helps protect against the damaging effects of the superoxide free radical (O^{2-}), and studies of genetically altered mice suggest that the regulation of brain extracellular SOD (EC-SOD) may be important in cognitive function (Fattman et al, 2003). Mice expressing either higher or lower than normal levels of EC-SOD showed impaired memory and learning abilities, although under conditions of high motivation these mice were not incapable of learning (Levin et al, 1998; 2000). This finding might have implications for other cognitive impairments that are related to motivational states, such as some of those associated with ADHD. The selenium dependent antioxidant enzyme, GSH-Px is vital in defending cells against the toxic effects of hydrogen peroxide and other peroxides including lipid peroxides. Increased erythrocyte SOD and GSH-Px has been reported in children with Down syndrome, in whom GSH-Px was also significantly correlated with memory function (Brugge et al, 1999).

Experimental Evidence for Oxidative Stress in Childhood Neurodevelopmental Disorders

Antioxidant Enzymes

One early study of antioxidant enzymes in autism reported abnormalities of GSH-Px in erythrocytes and SOD I in platelets (Golse et al, 1978). Although this work was not followed up at the time, a recent study found significantly lower activity levels of plasma GSH-Px, erythrocyte SOD and GSH-Px in 45 children with autism relative to 41 controls (Yorbik et al, 2002), and the authors noted that the low levels of these antioxidants in autistic subjects could render the brain unusually vulnerable to the damaging effects of free radicals.

Another study reported significantly increased activity of erythrocyte SOD in 27 subjects with autism compared to 26 age- and sex-matched controls, together with significantly increased levels of thiobarbituric acid-reactive substances (TBARS) – an index of lipid peroxidation – in the autism group (Zoroglu et al, 2004). Yet another study found no difference in either TBARS or SOD activity in plasma between a group of 27 subjects with autism and 30 matched controls, but the autistic subjects showed significantly increased plasma GSH-Px and RBC nitric oxide levels (Sogut et al, 2003).

Abnormalities in the metabolism of methionine and homocysteine have been associated with various other neurological disorders,[144] so these pathways were assessed in a recent study of 20 children with autism and 33 controls (James et al, 2004). The metabolic profile of the autistic children was consistent with both an impaired capacity for methylation (shown by a significantly lower ratio of S-adenosylmethionene to S-adenosylhomocysteine) and increased oxidative stress (as indexed by a significantly lower redox ratio of reduced glutathione to oxidized glutathione).

Although results from these studies are not entirely consistent, and none of these studies controlled for dietary intake or nutritional status, the overall pattern does suggest abnormalities of antioxidant enzyme activity in autism. This issue is further complicated by the methods used in these studies for measuring lipid peroxidation, some of which may give erroneous results, as discussed further below.

Vitamins and Minerals

Several reports indicate low zinc status in children with ADHD (Baker, 1985; Colquhoun and Bunday, 1981; Starobrat-Hermelin, 1998), and potential benefits from supplementation. In one randomised controlled trial, zinc status was lowered by the consumption of the food colouring tartrazine in ADHD children but not controls (Ward et al, 1990). In another, zinc status appeared to moderate the response to treatment with either omega-6 fatty acids or stimulant medication (Arnold et al, 2000). Two randomised controlled trials have since found benefits from zinc supplementation in ADHD children, either as monotherapy (Bilici et al, 2004) or as an adjunct to stimulant medication (Akhondzadeh et al, 2004).

In one of these experimental studies, the ADHD children were found to be deficient in a number of other blood minerals in addition to zinc, including copper, iron, magnesium and calcium (Starobrat-Hermelin, 1998). A recent study from France found abnormally low ferritin levels in ADHD children, and suggested that this might reflect a primary abnormality of iron metabolism (Kanofal et al, 2004). By contrast, a Taiwanese study primarily investigating fatty acid status found significantly higher levels of iron – both in blood biochemical measures and from dietary intake records - in ADHD children versus controls (Chen et al, 2004). Further research is clearly needed to elucidate these findings.

Deficiency of zinc (and other minerals) has also been reported in dyslexia (Grant et al, 1988; 2004). Few studies have investigated vitamin status in childhood developmental disorders; but a preliminary supplementation study in children with autism found some improvement in the severity of their symptoms when given very high doses of vitamin C over a 30 week period (Dolske et al, 1993).

Lipid Peroxidation

Increased levels of lipid peroxidation (measured by levels of malonyldialdehyde) have been reported in the plasma of children with autism compared with their developmentally normal siblings (Chauhan et al, 2004). Levels of the proteins ceruloplasmin and transferrin were also assessed in this study, and lower levels of both were found in the autistic group, suggesting a lack of carrier protein for both copper and iron. Unfortunately, however, as with some of the studies already discussed, the method used in this study to measure lipid peroxidation gives ambiguous results.

There are many different lipid peroxidation pathways, but one of the most common end products (in both enzymatic and non-enzymatic peroxidation pathways) is malondialdehyde. A simple colorimetric assay to measure thiobarbituric-reactive-substances (TBARS) was originally developed for use in the food industry, and later became a tool for assessing lipid peroxidation in human tissue samples. Subsequent evidence, however, has shown that many other substances (such as sugars and proteins) can react with thiobarbituric acid to produce the pink colour. Modified versions of this assay (e.g. using HPLC to separate the malondialdehyde complex from the other contaminants (Young and Trimble, 1991), or extraction of the malondialdehyde complex and fluorimeter quantification of derived butane (Yagi, 1984) have improved its accuracy but these modified TBARS assays are accessible to far fewer researchers. Commercial colorimetric assay kits have now been developed to measure malondialdehyde using reactants other than thiobarbituric acid, so these now tend to be the method of choice.

Breath Hydrocarbons

There are thought to be at least 3500 volatile compounds in breath (Phillips, 1992), some originating from the environment and others reflecting metabolic processes within the body. Altered levels of some of these volatiles are a well-known feature of certain disease states, such as the increase in ketones associated with diabetes. Analysis of breath can also provide an index of lipid peroxidation by quantification of either ethane or pentane. Ethane originates from the peroxidation of n-3 fatty acids and pentane from n-6 fatty acids, and in theory it should be possible to measure these compounds in breath using gas chromatography (GC) with either mass spectrometry (MS) or flame ionisation detection (FID) (Aghdassi and Allard, 2000; Phillips, 1999). To date, however, there has been little agreement on normal breath levels of these hydrocarbons. Different studies have not only used different methods of quantifying ethane and pentane, but have also used a variety of units to express the results (e.g. relating breath hydrocarbon output to factors such as the weight of the subject, the time period or the volume of breath). Before collection, some researchers have opted for a 'washout period' whereby the subject breathes in hydrocarbon-free air for varying periods of time, while others have not. Methods of breath sample collection also vary from very elaborate methods to much simpler ones, and identification and quantification can be carried out using GC/MS or FID with suitable standard gases. Automatic thermal desorption (ATD) is the most commonly used method of breath capture since volatiles from breath are concentrated on to a suitable adsorbent in a stainless steel tube. The ATD tube can be sealed and then transported to the laboratory for analysis.

Hydrocarbons appear to be metabolised by the p450 system in the liver and animal studies indicate that their half-life in the body is inversely proportional to the number of carbons in the molecule (Wade and van Rij, 1985). Therefore pentane has a much shorter half-life in the body than ethane. However it is still unclear whether breath pentane is endogenous or whether it originates from ambient air since it is highly soluble in lipids (Springfield and Levitt, 1994). Others have reported a negative gradient of breath pentane with ambient air (Phillips et al, 1994). Breath ethane measures may be a better index of oxidative stress than pentane, since ethane has a longer metabolism in humans and is also less prone to misidentification than pentane (Caileux and Allain, 1993).

To date there is only one published study of breath hydrocarbon levels in neurodevelopmental disorders, in which higher levels of ethane were reported in the breath of ADHD children relative to controls (Ross et al, 2003). Unfortunately, however, these researchers omitted to measure the environmental, background levels of ethane, which show significant variability both throughout the day and according to location (Ross, personal communication). Given that some of the controls provided breath samples in a completely different geographical region from those with ADHD, whose samples were themselves collected in different locations, it is impossible to draw any meaningful conclusions from this particular study. Nonetheless, the measurement of breath ethane could be a promising research tool if properly conducted.

Urinary Isoprostanes

Some of the products of oxidative stress are eliminated from the body in the urine. The isoprostanes for example are a product of the non-enzymatic product of lipid peroxidation and are thought by some to be one of the best markers of lipid peroxidation. The isoprostanes are structurally similar to prostaglandins originating from fatty acids such as arachidonic acid, docosahexaenoic acid and eicosapentaenoic acid. These isoprostanes are often referred to as the F_2-isoprostanes, since they are structurally similar to prostaglandin $F_{2\alpha}$ ($PGF_{2\alpha}$).

The F_2-isoprostanes are found in plasma, usually esterified to phospholipids although some can be found in the unbound 'free' state. F_2-isoprostanes have also been measured in breath condensate and in one study they were found to be elevated in those suffering with acute lung injury or acute respiratory distress syndrome (Carpenter et al, 1998).

F_2- isoprostanes are also excreted in urine and these can be easily measured. Urinary concentrations of F_2-isoprostanes are increased in smokers (Obata et al, 2000), and there is evidence that increasing the dietary intakes of n-3 fatty acids of non-insulin dependent diabetics decreases urinary F_2-isprostanes (Mori et al, 1999). This suggests that n-3 fatty acids may have some antioxidant activity, which will be discussed later in this chapter. Furthermore rats supplemented with vitamin E have also shown decreased F_2-isoprostanes in urine (Sodergren et al, 2000).

One study using this method has reported increased isoprostanes in autism (McGinnis, 2005). The non-invasive procedure makes urinary F_2- isoprostanes an ideal measure of oxidative stress in children, so further studies of this kind are clearly warranted.

Oxidative Stress and Fatty Acids

Oxidative stress can cause the breakdown of membrane lipids via the process of lipid peroxidation, and polyunsaturated fatty acids are particularly susceptible to free radical attack due to the large number of double bonds they contain. For this reason, the findings of apparent deficiencies of LC-PUFA in neurodevelopmental disorders could well be the result of increased oxidative stress, although they could also reflect other aspects of either fatty acid metabolism or diet.

Phospholipase A2 Enzymes

During periods of oxidative stress, the activity of phospholipase A2 enzymes is increased. These enzymes release fatty acids from the sn2 position of membrane phospholipids (which usually hold highly unsaturated fatty acids, particularly the 20 carbon fatty acids used for eicosanoid synthesis).

Elevations of a cytosolic type IV PLA2 enzyme in red blood cell membranes have been reported in subjects with autistic spectrum disorder, in addition to other fatty acid abnormalities (Bell et al, 2000; 2004). As noted earlier, significantly increased levels of the same PLA2 in red cell membranes have also been reported in dyslexic adults compared with controls (MacDonell et al, 2000), a finding that would be consistent with increased oxidative stress. Elevated PLA2 was associated with both regressive and classical forms of autism, as well as Asperger's syndrome, although initial observations of a greater vulnerability to breakdown of highly unsaturated fatty acids in membranes (consistent with elevated PLA2 activity) appeared to be specific to those with regressive forms of autism. Subjects who supplemented their diets with EPA were found to have significantly reduced PLA2 levels compared to those not taking supplementation.

Omega-3 vs α-Tocopherol

Vitamin E, particularly α-tocopherol, is important for the antioxidant protection of LC-PUFA, both in the cell membrane and in food, including fatty acid supplements. Because polyunsaturated fatty acids are particularly vulnerable to lipid peroxidation, it has generally been assumed that a high dietary intake of LC-PUFA could increase oxidative stress *in vivo* unless matched by an increased amount of antioxidants such as vitamin E for protection. (This has been the rationale for adding antioxidants such as Vitamin E - and/or sometimes herbal extracts such as oil of thyme or rosemary - to some specialist fish oil supplements)

While this may be the case for omega-6 fatty acids, evidence has been emerging to suggest that omega-3 fatty acids may not necessarily require additional antioxidant protection, and instead can actually reduce oxidative stress (Barbosa et al, 2003; Erdogan et al, 2004; Mori et al, 2004). In one recent study addressing this issue, high supplementation of natural tocopherols was carried out in preterm infants fed with LC-PUFA supplemented formula, standard formula or breast milk (Kaempf Rotzoll et al, 2003). The additional LC-PUFA not only failed to cause tocopherol depletion in cell membranes, but actually led to *higher* levels of alpha- and gamma-tocopherol in plasma, erythrocytes, platelets, monocytes, and polymorphonuclear leukocytes, and higher gamma-tocopherol uptake in all cell fractions studied, suggesting that LC-PUFA supplementation may improve tocopherol solubility and stability in biological membranes.

In this context, it is notable that in one controlled trial of fatty acid treatment in ADHD-type children, improvements in blood Vitamin E status following treatment were significantly correlated with several measures of improved behaviour and learning (as were improvements in omega-3 status) (Stevens et al, 2003). To date, the clinical trials of fatty acid supplementation in childhood neurodevelopmental disorders with positive findings have all used supplements to which at least some Vitamin E had been added (Richardson and Montgomery, 2005; Richardson and Puri, 2002; Stevens et al, 2003), hence this finding clearly merits further investigation.

Conclusion

The evidence reviewed here provides a compelling case for considering fatty acid and antioxidant status as potential factors in the etiology of neurodevelopmental disorders. While the direct evidence from experimental studies and treatment trials in children with these conditions is still fairly limited, it is nonetheless more than suggestive; and the critical importance of both fatty acids and antioxidant protection to brain development and function is beyond question.

Fatty acid metabolism and antioxidant status are inevitably very closely linked, and their interaction could help to account for some of the known risk factors for neurodevelopmental disorders. For example, sex differences in fatty acid metabolism have been well-documented, and these arise partly from the female steroid hormone estrogen that also appears to have some antioxidant capacity in membranes in addition to other neuroprotective roles (Sawada and Shimohama, 2000). This could be one factor that may help to explain the increased vulnerability of males to childhood neurodevelopmental disorders, although others are also likely to be relevant. Many environmental factors are also important, and ADHD in particular has repeatedly been associated with maternal smoking and alcohol consumption. Experimental evidence of the long-term neurotoxicity associated with prenatal tobacco and alcohol exposure indicates that this is caused by increased formation of free radicals and oxidative stress (Li and Wang, 2004). From current evidence, it seems more than likely that modern diets relatively lacking in essential nutrients are another factor contributing to the apparent increase in childhood behavioural and learning difficulties, but the potential role of nutrition in both the prevention and the treatment of these disorders requires further investigation with properly designed studies.

Meanwhile, there is already a strong case for new approaches to childhood neurodevelopmental disorders that include nutritional and biochemical assessments – currently neglected in the diagnosis and management of these conditions. Given the heterogeneity within each of the main diagnostic categories, the descriptive nature of these, and the high comorbidity between them, we would also argue that a symptom-based or trait-based approach to these conditions may well be more fruitful than an exclusive focus on current diagnostic categories.

References

Aghdassi, E. and Allard, J.P., Breath alkanes as a marker of oxidative stress in different clinical conditions. *Free radical biology and medicine,* 2000. 28(6): p. 880-6.

Akhondzadeh, S., Mohammadi, M.-R., and Khademi, M., Zinc sulphate as an adjunct to methylphenidate for the treatment of attention deficit hyperactivity disorder in children: A double blind and randomized trial. *BMC Psychiatry,* 2004. 4: p. 9.

Aman, M.G., Mitchell, E.A., and Turbott, S.H., The effects of essential fatty acid supplementation by Efamol in hyperactive children. *J Abnorm Child Psychol,* 1987. 15(1): p. 75-90.

Anderson, J.L., The immune system and major depression. *Adv Neuroimmunol,* 1996. 6(2): p. 119-29.

Arnold, L.E., *et al.*, Gamma-linolenic acid for attention-deficit hyperactivity disorder: placebo-controlled comparison to D-amphetamine. *Biol Psychiatry,* 1989. 25(2): p.222-8.

Arnold, L.E., *et al.*, Potential link between dietary intake of fatty acids and behavior: Pilot exploration of serum lipids in attention-deficit hyperactivity disorder. *Journal of Child and Adolescent Psychopharmacology,* 1994. 4(3): p. 171-182.

Arnold, L.E., Pinkham, S.M., and Votolato, N., Does zinc moderate essential fatty acid and amphetamine treatment of attention-deficit/hyperactivity disorder? *J Child Adolesc Psychopharmacol,* 2000. 10(2)): p. 111-7.

Asherson, P.J. and Curran, S., Approaches to gene mapping in complex disorders and their application in child psychiatry and psychology. *Br J Psychiatry,* 2001. 179: p. 122-8.

Ashwood, P. and Van de Water, J., Is autism an autoimmune disease? *Autoimmunity reviews,* 2004. 3(7-8): p. 557-62.

Baker, S.M., A biochemical approach to the problem of dyslexia. *J Learn Disabil,* 1985. 18(10): p. 581-4.

Barbosa, D.S., *et al.*, Decreased oxidative stress in patients with ulcerative colitis supplemented with fish oil omega-3 fatty acids. *Nutrition Burbank, Los Angeles County, Calif.,* 2003. 19(10): p. 837-42.

Bekaroglu, M., *et al.*, Relationships between serum free fatty acids and zinc, and attention deficit hyperactivity disorder: a research note. *J Child Psychol Psychiatry,* 1996. 37(2): p. 225-7.

Bell, J.G., *et al.* Membrane fatty acids, reading and spelling in dyslexic and non-dyslexic adults. in *ISSFAL 6th International Congress*. 2004. Brighton, UK.

Bell, J.G., *et al.*, Apparent fatty acid deficiency in autistic spectrum disorders. *Prostaglandins Leukot Essent Fatty Acids,* 2004. In press.

Bell, J.G., *et al.*, Essential fatty acids and phospholipase A2 in autistic spectrum disorders. *Prostaglandins, Leukotrienes and Essential Fatty Acids,* 2004. 71(4): p. 201-4.

Bell, J.G., *et al.*, Red blood cell fatty acid compositions in a patient with autistic spectrum disorder: a characteristic abnormality in neurodevelopmental disorders? *Prostaglandins Leukot Essent Fatty Acids,* 2000. 63(1-2)): p. 21-5.

Bendich, A., Vitamin E status of US children. *Journal of the American College of Nutrition,* 1992. 11(4): p. 441-4.

Bennett, C.N. and Horrobin, D.F., Gene targets related to phospholipid and fatty acid metabolism in schizophrenia and other psychiatric disorders: an update. *Prostaglandins Leukot Essent Fatty Acids,* 2000. 63(1-2): p. 47-59.

Benton, D., Selenium intake, mood and other aspects of psychological functioning. *Nutr Neurosci,* 2002. 5(6): p. 363-74.

Biederman, J., *et al.*, High rate of affective disorders in probands with attention deficit disorder and in their relatives: a controlled family study. *Am J Psychiatry,* 1987. 144(3): p. 330-3.

Bilici, M., *et al.*, Double-blind, placebo-controlled study of zinc sulfate in the treatment of attention deficit hyperactivity disorder. *Prog Neuropsychopharmacol Biol Psychiatry,* 2004. 28(1): p. 181-90.

Blaxill, M.F., What's going on? The question of time trends in autism. *Public health reports Washington, D.C. 1974,* 2004. 119(6): p. 536-51.

Bourre, J.M., Roles of unsaturated fatty acids (especially omega-3 fatty acids) in the brain at various ages and during ageing. *Journal of nutrition, health and aging,* 2004. 8(3): p. 163-74.

Brugge, K., *et al.*, Correlations of glutathione peroxidase activity with memory impairment in adults with Down syndrome. *Biological psychiatry,* 1999. 46(12): p. 1682-9.

Burdge, G.C. and Wootton, S.A., Conversion of alpha-linolenic acid to eicosapentaenoic, docosapentaenoic and docosahexaenoic acids in young women. *Br J Nutr,* 2002. 88(4): p. 411-20.

Burdge, G.C., Jones, A.E., and Wootton, S.A., Eicosapentaenoic and docosapentaenoic acids are the principal products of alpha-linolenic acid metabolism in young men. *Br J Nutr,* 2002. 88(4): p. 355-63.

Burgess, J.R. and Stevens, L.J., Essential fatty acids in relation to attention deficit hyperactivity disorder, in *Phospholipid Spectrum Disorders in Psychiatry and Neurology,* Peet, M., Glen, I., Horrobin, D.F. (eds.), Editor. 2003, Marius Press: Carnforth. p. 511-519.

Cailleux, A. and Allain, P., Is pentane a normal constituent of human breath? *Free radical research communications,* 1993. 18(6): p. 323-7.

Calder, P.C., N-3 polyunsaturated fatty acids and inflammation: from molecular biology to the clinic. *Lipids,* 2003. 38(4): p. 343-52.

Carlson, S.E. and Neuringer, M., Polyunsaturated fatty acid status and neurodevelopment: a summary and critical analysis of the literature. *Lipids,* 1999. 34(2):. p. 171-8.

Carpenter, C.T., Price, P.V., and Christman, B.W., Exhaled breath condensate isoprostanes are elevated in patients with acute lung injury or ARDS. *Chest,* 1998. 114(6): p. 1653-9.

Charlton, K.E., *et al.*, Lowered plasma vitamin C, but not vitamin E, concentrations in dementia patients. *journal of nutrition, health and aging,* 2004. 8(2): p. 99-107.

Chauhan, A., *et al.*, Oxidative stress in autism: increased lipid peroxidation and reduced serum levels of ceruloplasmin and transferrin--the antioxidant proteins. *Life sciences,* 2004. 75(21): p. 2539-49.

Chen, J. and Berry, M.J., Selenium and selenoproteins in the brain and brain diseases. *Journal of neurochemistry,* 2003. 86(1): p. 1-12.

Chen, J.R., *et al.*, Dietary patterns and blood fatty acid composition in children with attention-deficit hyperactivity disorder in Taiwan. *Journal of Nutritional Biochemistry,* 2004. 15(8): p. 467-72.

Clausen, J. and Moller, J., Experimental allergic encephalomyelitis provoked in rats after developmental lack of polyunsaturated tally acids. *Acta Neurol Scand,* 1967. 43(Suppl): p. 74.

Colquhoun, I. and Bunday, S., A lack of essential fatty acids as a possible cause of hyperactivity in children. *Med Hypotheses,* 1981. 7(5): p. 673-9.

Crawford, M., Placental delivery of arachidonic and docosahexaenoic acids: implications for the lipid nutrition of preterm infants. *Am J Clin Nutr,* 2000. 71(1 Suppl): p. 275S-84S.

Dalton, P., *et al.*, Maternal neuronal antibodies associated with autism and a language disorder. *Annals of neurology,* 2003. 53(4): p. 533-7.

Das, U.N., Essential fatty acids as possible enhancers of the beneficial actions of probiotics. *Nutrition,* 2002. 18(9): p. 786.

Delarue, J., *et al.*, Fish oil prevents the adrenal activation elicited by mental stress in healthy men. *Diabetes Metab,* 2003. 29(3): p. 289-95.

Delion, S., *et al.*, alpha-Linolenic acid dietary deficiency alters age-related changes of dopaminergic and serotoninergic neurotransmission in the rat frontal cortex. *J Neurochem,* 1996. 66(4): p. 1582-91.

D'Eufemia, P., *et al.*, Abnormal intestinal permeability in children with autism. *Acta Paediatrica,* 1996. 85(9): p. 1076-1079.

Dolske, M.C., *et al.*, A preliminary trial of ascorbic acid as supplemental therapy for autism. *Progress in neuro psychopharmacology and biological psychiatry,* 1993. 17(5): p. 765-74.

Doyle, W., *et al.*, Nutritional status of schoolchildren in an inner city area. *Archives of disease in childhood,* 1994. 70(5): p. 376-81.

Erdogan, H., *et al.*, Effect of fish oil supplementation on plasma oxidant/antioxidant status in rats. *Prostaglandins, leukotrienes, and essential fatty acids,* 2004. 71(3): p. 149-52.

Eves, A. and Gesch, B., Food provision and the nutritional implications of food choices made by young adult males, in a young offenders' institution. *Journal of human nutrition and dietetics the official journal of the British Dietetic Association,* 2003. 16(3): p. 167-79.

Fattman, C.L., Schaefer, L.M., and Oury, T.D., Extracellular superoxide dismutase in biology and medicine. *Free radical biology and medicine,* 2003. 35(3): p. 236-56.

Ferencik, M. and Ebringer, L., Modulatory effects of selenium and zinc on the immune system. *Folia microbiologica,* 2003. 48(3): p. 417-26.

Fisher, S.E. and DeFries, J.C., Developmental dyslexia: Genetic dissection of a complex cognitive trait. *Nat Rev Neurosci,* 2002. 3(10): p. 767-780.

Golse, B., *et al.*, Perturbation de deux enzymes: la superoxyde-dismutase I et la glutathion-peroxydase dans la psychose infantile de developpement (autisme infantile). *Revue neurologique,* 1978. 134(11): p. 699-705.

Grant, E.C., Developmental dyslexia and zinc deficiency. *Lancet,* 2004. 364(9430): p. 247-8.

Grant, E.C., *et al.*, Zinc deficiency in children with dyslexia: concentrations of zinc and other minerals in sweat and hair. *Br Med J (Clin Res Ed)*, 1988. 296(6622)): p. 607-9.

Gregory, J.e.a., *UK National Diet and Nutrition Survey: young people aged 4 to 18 years. Volume 1: Findings*. 2000, London: The Stationery Office.

Haag, M., Essential fatty acids and the brain. *Canadian Journal of Psychiatry*, 2003. 48: p. 195-203.

Hamazaki, T., *et al.*, The effect of docosahexaenoic acid on aggression in young adults. A placebo-controlled double-blind study. *J Clin Invest*, 1996. 97(4): p. 1129-33.

Harbige, L.S., Fatty acids, the immune response, and autoimmunity: a question of n-6 essentiality and the balance between n-6 and n-3. *Lipids*, 2003. 38(4): p. 323-41.

Hawkes, W.C. and Hornbostel, L., Effects of dietary selenium on mood in healthy men living in a metabolic research unit. *Biological psychiatry*, 1996. 39(2): p. 121-8.

Hellgren, L., *et al.*, Children with deficits in attention, motor control and perception (DAMP) almost grown up: psychiatric and personality disorders at age 16 years. *J Child Psychol Psychiatry*, 1994. 35(7): p. 1255-71.

Hirayama, S., Hamazaki, T., and Terasawa, K., Effect of docosahexaenoic acid-containing food administration on symptoms of attention-deficit/hyperactivity disorder - a placebo-controlled double-blind study. *European journal of clinical nutrition*, 2004. 58(3): p. 467-73.

Ho, P.I., *et al.*, Multiple aspects of homocysteine neurotoxicity: Glutamate excitotoxicity, kinase hyperactivation and DNA damage. *J NEUROSCI RES*, 2002. 70(5): p. 694-702.

Hoffer, A., Tardive dyskinesia treated with manganese. *Canadian Medical Association journal*, 1977. 117(8): p. 859.

Hong, S., *et al.*, Novel docosatrienes and 17S-resolvins generated from docosahexaenoic acid in murine brain, human blood, and glial cells. Autacoids in anti-inflammation. *J Biol Chem*, 2003. 278(17): p. 14677-87.

Horrobin, D.F., Essential fatty acid metabolism and its modification in atopic eczema. *Am J Clin Nutr*, 2000. 71(1 Suppl): p. 367S-72S.

Horrobin, D.F., Glen, A.I., and Hudson, C.J., Possible relevance of phospholipid abnormalities and genetic interactions in psychiatric disorders: the relationship between dyslexia and schizophrenia. *Med Hypotheses*, 1995. 45(6)): p. 605-13.

Horvath, K., *et al.*, Gastrointestinal abnormalities in children with autistic disorder. *Journal of Pediatrics*, 1999. 135(5): p. 559-563.

Huang, Y.S. and Horrobin, D.F., Sex differences in n-3 and n-6 fatty acid metabolism in EFA-depleted rats. *Proc Soc Exp Biol Med*, 1987. 185(3): p. 291-6.

Huh, G.S., *et al.*, Functional requirement for class I MHC in CNS development and plasticity. *Science*, 2000. 290(5499): p. 2155-9.

James, S.J., *et al.*, Metabolic biomarkers of increased oxidative stress and impaired methylation capacity in children with autism. *American journal of clinical nutrition*, 2004. 80(6): p. 1611-7.

Johnson, S., Micronutrient accumulation and depletion in schizophrenia, epilepsy, autism and Parkinson's disease? *Medical hypotheses*, 2001. 56(5): p. 641-5.

Jones, G.S., Autistic spectrum disorder: diagnostic difficulties. *Prostaglandins Leukot Essent Fatty Acids*, 2000. 63(1-2): p. 33-6.

Jyonouchi, H., Sun, S., and Itokazu, N., Innate immunity associated with inflammatory responses and cytokine production against common dietary proteins in patients with autism spectrum disorder. *Neuropsychobiology,* 2002. 46(2): p. 76-84.

Kadesjo, B. and Gillberg, C., Developmental coordination disorder in Swedish 7-year-old children. *J Am Acad Child Adolesc Psychiatry,* 1999. 38(7): p. 820-8.

Kadesjo, B. and Gillberg, C., The comorbidity of ADHD in the general population of Swedish school-age children. *J Child Psychol Psychiatry,* 2001. 42(4): p. 487-92.

Kaempf Rotzoll, D.E., Hellstern, G., and Linderkamp, O., Influence of long-chain polyunsaturated fatty acid formula feeds on vitamin E status in preterm infants. *International journal for vitamin and nutrition research; Internationale Zeitschrift fur Vitamin und Ernahrungsforschung; Journal international de vitaminologie et de nutrition,* 2003. 73(5): p. 377-87.

Kang, J.X., The importance of omega-6/omega-3 fatty acid ratio in cell function: the gene transfer of omega-3 fatty acid desaturase. *World Rev Nutr Diet,* 2003. 92: p. 23-36.

Katz, R.L., *et al.*, Zinc deficiency in anorexia nervosa. *Journal of adolescent health care official publication of the Society for Adolescent Medicine,* 1987. 8(5): p. 400-6.

Konofal, E., *et al.*, Iron deficiency in children with attention-deficit/hyperactivity disorder. *Archives of Pediatrics and Adolescent Medicine,* 2004. 158(12): p. 1113-5.

Koscik, R.L., *et al.*, Cognitive function of children with cystic fibrosis: deleterious effect of early malnutrition. *Pediatrics,* 2004. 113(6): p. 1549-58.

Krause, I., *et al.*, Immune factors in autism: a critical review. *J Autism Dev Disord,* 2002. 32(4): p. 337-45.

Leonard, B.E. and Song, C., Stress, depression, and the role of cytokines. *Adv Exp Med Biol,* 1999. 461: p. 251-65.

Levant, B., Radel, J.D., and Carlson, S.E., Decreased brain docosahexaenoic acid during development alters dopamine-related behaviors in adult rats that are differentially affected by dietary remediation. *Behav Brain Res,* 2004. 152(1): p. 49-57.

Levin, E.D., Brucato, F.H., and Crapo, J.D., Molecular overexpression of extracellular superoxide dismutase increases the dependency of learning and memory performance on motivational state. *Behavior genetics,* 2000. 30(2): p. 95-100.

Levin, E.D., *et al.*, Molecular manipulations of extracellular superoxide dismutase: functional importance for learning. *Behavior genetics,* 1998. 28(5): p. 381-90.

Li, Y. and Wang, H., In utero exposure to tobacco and alcohol modifies neurobehavioral development in mice offspring: consideration a role of oxidative stress. *Pharmacological research the official journal of the Italian Pharmacological Society,* 2004. 49(5): p. 467-73

Litman, B.J., *et al.*, The role of docosahexaenoic acid containing phospholipids in modulating G protein-coupled signaling pathways: visual transduction. *J Mol Neurosci,* 2001. 16(2-3): p. 237-42; discussion 279-84.

MacDonell, L.E., *et al.*, Increased levels of cytosolic phospholipase A2 in dyslexics. *Prostaglandins Leukot Essent Fatty Acids,* 2000. 63(1-2): p. 37-9.

Maes, M., *et al.*, Hypozincemia in depression. *Journal of affective disorders,* 1994. 31(2): p. 135-40.

Maes, M., *et al.*, Lower serum zinc in major depression is a sensitive marker of treatment resistance and of the immune/inflammatory response in that illness. *Biological psychiatry,* 1997. 42(5): p. 349-58.

Malvy, J.M., *et al.*, Retinol, beta-carotene and alpha-tocopherol status in a French population of healthy children. *International journal for vitamin and nutrition research,* 1989. 59(1): p. 29-34.

Mandell, D.S., *et al.*, Trends in diagnosis rates for autism and ADHD at hospital discharge in the context of other psychiatric diagnoses. *Psychiatric services Washington, D.C.,* 2005. 56(1): p. 56-62.

Marra, C.A. and de Alaniz, M.J., Influence of testosterone administration on the biosynthesis of unsaturated fatty acids in male and female rats. *Lipids,* 1989. 24(12): p. 1014-9.

Martin, A., *et al.*, Effects of fruits and vegetables on levels of vitamins E and C in the brain and their association with cognitive performance. *journal of nutrition, health and aging,* 2002. 6(6): p. 392-404.

McGinnis, W.R., Oxidative stress in autism. *Alternative therapies in health and medicine,* 2004. 10(6): p. 22-36.

McGinnis, W.R., Oxidative stress in autism. *Alternative therapies in health and medicine,* 2005. 11(1): p. 19.

Metodiewa, D. and Koska, C., Reactive oxygen species and reactive nitrogen species: relevance to cyto(neuro)toxic events and neurologic disorders. An overview. *Neurotoxicity research,* 2000. 1(3): p. 197-233.

Mitchell, E.A., *et al.*, Clinical characteristics and serum essential fatty acid levels in hyperactive children. *Clin-Pediatr-Phila,* 1987. 26(8): p. 406-11.

Mori, A., *et al.*, Natural antioxidants may prevent posttraumatic epilepsy: a proposal based on experimental animal studies. *Acta medica Okayama,* 2004. 58(3): p. 111-8.

Mori, T.A., *et al.*, Effect of dietary fish and exercise training on urinary F2-isoprostane excretion in non-insulin-dependent diabetic patients. *Metabolism clinical and experimental,* 1999. 48(11): p. 1402-8.

Moriguchi, T., Greiner, R.S., and Salem, N., Jr., Behavioral deficits associated with dietary induction of decreased brain docosahexaenoic acid concentration. *J Neurochem,* 2000. 75(6): p. 2563-73.

Muhle, R., Trentacoste, S.V., and Rapin, I., The genetics of autism. *Pediatrics,* 2004. 113(5): p. e472-86.

Muller, N., *et al.*, The immune system and schizophrenia. An integrative view. *Ann N Y Acad Sci.,* 2000. 917: p. 456-67.

Nemets, B., Stahl, Z., and Belmaker, R.H., Addition of omega-3 fatty acid to maintenance medication treatment for recurrent unipolar depressive disorder. *Am J Psychiatry,* 2002. 159(3): p. 477-9.

Obata, T., *et al.*, Smoking and oxidant stress: assay of isoprostane in human urine by gas chromatography-mass spectrometry. *Journal of chromatography. B, Biomedical sciences and applications,* 2000. 746(1): p. 11-5.

O'Brien, S.M., Scott, L.V., and Dinan, T.G., Cytokines: abnormalities in major depression and implications for pharmacological treatment. *Hum Psychopharmacol,* 2004. 19(6): p. 397-403.

Paraskevas, G.P., *et al.*, Plasma levels of antioxidant vitamins C and E are decreased in vascular parkinsonism. *Journal of the neurological sciences,* 2003. 215(1-2): p. 51-5.

Pawlosky, R.J., *et al.*, Physiological compartmental analysis of alpha-linolenic acid metabolism in adult humans. *J Lipid Res,* 2001. 42(8): p. 1257-65.

Peet, M. and Horrobin, D.F., A dose-ranging study of the effects of ethyl-eicosapentaenoate in patients with ongoing depression despite apparently adequate treatment with standard drugs. *Arch Gen Psychiatry,* 2002. 59(10): p. 913-9.

Peet, M., Glen, I., and Horrobin, D.F., eds. *Phospholipid spectrum disorders in psychiatry and neurology*. 2nd ed. 2003, Marius Press: Carnforth.

Phillips, M., Breath tests in medicine. *Scientific American,* 1992. 267(1): p. 74-9.

Phillips, M., *et al.*, Variation in volatile organic compounds in the breath of normal humans. *Journal of chromatography. B, Biomedical sciences and applications,* 1999. 729(1-2): p. 75-88.

Phillips, M., Greenberg, J., and Awad, J., Metabolic and environmental origins of volatile organic compounds in breath. *Journal of clinical pathology,* 1994. 47(11): p. 1052-3.

Piechotowski, I., *et al.*, Serumselenkonzentrationen bei Schulkindern in Baden-Wurttemberg und ihre gesundheitliche Bewertung. *Gesundheitswesen Bundesverband der Arzte des Offentlichen Gesundheitsdienstes Germany,* 2002. 64(11): p. 602-7.

Prabakaran, S., *et al.*, Mitochondrial dysfunction in schizophrenia: evidence for compromised brain metabolism and oxidative stress. *Molecular psychiatry,* 2004. 9(7): p. 684-97, 643.

Rao, A.V. and Balachandran, B., Role of oxidative stress and antioxidants in neurodegenerative diseases. *Nutritional neuroscience,* 2002. 5(5): p. 291-309.

Rasmussen, P. and Gillberg, C., Natural outcome of ADHD with developmental coordination disorder at age 22 years: a controlled, longitudinal, community-based study. *J Am Acad Child Adolesc Psychiatry,* 2000. 39(11): p. 1424-31.

Reddy, R., Keshavan, M., and Yao, J.K., Reduced plasma antioxidants in first-episode patients with schizophrenia. *Schizophrenia research,* 2003. 62(3): p. 205-12.

Richardson, A.J. and Montgomery, P., The Oxford-Durham study: a randomised controlled trial of fatty acid supplementation in children with developmental coordination disorder. *Pediatrics,* 2005. 115(5): p. 1360-1366.

Richardson, A.J. and Puri, B.K., A randomized double-blind, placebo-controlled study of the effects of supplementation with highly unsaturated fatty acids on ADHD-related symptoms in children with specific learning difficulties. *Prog Neuropsychopharmacol Biol Psychiatry,* 2002. 26(2): p. 233-9.

Richardson, A.J. and Puri, B.K., The potential role of fatty acids in attention-deficit/hyperactivity disorder. *Prostaglandins Leukot Essent Fatty Acids,* 2000. 63(1-2): p. 79-87.

Richardson, A.J. and Ross, M.A., Fatty acid metabolism in neurodevelopmental disorder: a new perspective on associations between attention-deficit/hyperactivity disorder, dyslexia, dyspraxia and the autistic spectrum. *Prostaglandins Leukot Essent Fatty Acids,* 2000. 63(1-2): p. 1-9.

Richardson, A.J., Clinical trials of fatty acid supplementation in ADHD, in *Phospholipid Spectrum Disorders in Psychiatry and Neurology,* Peet M, G.I., Horrobin DF. (eds), Editor. 2003, Marius Press: Carnforth. p. 529-41.

Richardson, A.J., Clinical trials of fatty acid treatment in ADHD, dyslexia, dyspraxia and the autistic spectrum. *Prostaglandins Leukot Essent Fatty Acids,* 2004. 70(4): p. 383-390.

Richardson, A.J., Cyhlarova, E., and Puri, B.K., Clinical and biochemical fatty acid abnormalities in dyslexia, dyspraxia, and schizotypy: an overview., in *Phospholipid Spectrum Disorders in Psychiatry and Neurology,* Peet M, G.I., Horrobin DF. (eds), Editor. 2003, Marius Press: Carnforth. p. 477-90.

Richardson, A.J., *et al.*, Fatty acid deficiency signs predict the severity of reading and related difficulties in dyslexic children. *Prostaglandins Leukot Essent Fatty Acids,* 2000. 63(1-2): p. 69-74.

Rinaldi, P., *et al.*, Plasma antioxidants are similarly depleted in mild cognitive impairment and in Alzheimer's disease. *Neurobiology of aging,* 2003. 24(7): p. 915-9.

Ross, B.M., *et al.*, Increased levels of ethane, a non-invasive marker of n-3 fatty acid oxidation, in breath of children with attention deficit hyperactivity disorder. *Nutritional neuroscience,* 2003. 6(5): p. 277-81.

Ross, M.A., Comment on: Increased levels of ethane, a non-invasive marker of n-3 fatty acid oxidation, in children with attention-deficit / hyperactivity disorder. *Nutritional Neuroscience - submitted,* 2005.

Ross, M.A., Could oxidative stress be a factor in neurodevelopmental disorders? *Prostaglandins Leukot Essent Fatty Acids,* 2000. 63(1-2): p. 61-3.

Ross, M.A., *et al.* Working memory and schizotypal traits in relation to fatty acid status in dyslexia. in *ISSFAL 6th International Conference.* 2004. Brighton, UK.

Rutter, M., *et al.*, Sex differences in developmental reading disability: new findings from 4 epidemiological studies. *JAMA the journal of the American Medical Association,* 2004. 291(16): p. 2007-12.

Salem, N., Jr., *et al.*, In vivo conversion of linoleic acid to arachidonic acid in human adults. *Prostaglandins Leukot Essent Fatty Acids,* 1999. 60(5-6): p. 407-10.

Sawada, H. and Shimohama, S., Neuroprotective effects of estradiol in mesencephalic dopaminergic neurons. *Neuroscience and biobehavioral reviews,* 2000. 24(1): p. 143-7.

Scahill, L. and Schwab Stone, M., Epidemiology of ADHD in school-age children. *Child and adolescent psychiatric clinics of North America,* 2000. 9(3): p. 541-55, vii.

Serhan, C.N., *et al.*, Resolvins, docosatrienes, and neuroprotectins, novel omega-3-derived mediators, and their aspirin-triggered endogenous epimers: an overview of their protective roles in catabasis. *Prostaglandins Other Lipid Mediat,* 2004. 73(3-4): p. 155-72.

Serra Majem, L., *et al.*, Risk of inadequate intakes of vitamins A, B1, B6, C, E, folate, iron and calcium in the Spanish population aged 4 to 18. *International journal for vitamin and nutrition research,* 2001. 71(6): p. 325-31.

Simopoulos, A.P., Essential fatty acids in health and chronic disease. *American journal of clinical nutrition,* 1999. 70(3 Suppl): p. 560S-569S.

Simopoulos, A.P., The importance of the ratio of omega-6/omega-3 essential fatty acids. *Biomed Pharmacotherapy,* 2002. 56(8): p. 365-379.

Smythies, J., The neurotoxicity of glutamate, dopamine, iron and reactive oxygen species: functional interrelationships in health and disease: a review-discussion. *Neurotoxicity research,* 1999. 1(1): p. 27-39.

Sodergren, E., *et al.*, Vitamin E supplementation decreases basal levels of F(2)-isoprostanes and prostaglandin f(2alpha) in rats. *Journal of nutrition,* 2000. 130(1): p. 10-4.

Sofic, E., *et al.*, Serum antioxidant capacity in neurological, psychiatric, renal diseases and cardiomyopathy. *Journal of neural transmission Vienna, Austria 1996,* 2002. 109(5-6): p. 711-9.

Sogut, S., *et al.*, Changes in nitric oxide levels and antioxidant enzyme activities may have a role in the pathophysiological mechanisms involved in autism. *Clinica chimica acta; International journal of clinical chemistry,* 2003. 331(1-2): p. 111-7.

Sokol, R.J., Vitamin E and neurologic deficits. *Advances in pediatrics,* 1990. 37: p. 119-48.

Song, C., *et al.*, The inflammatory response system and the availability of plasma tryptophan in patients with primary sleep disorders and major depression. *J Affect Disord,* 1998. 49(3): p. 211-9.

Springfield, J.R. and Levitt, M.D., Pitfalls in the use of breath pentane measurements to assess lipid peroxidation. *Journal of lipid research,* 1994. 35(8): p. 1497-504.

Starobrat-Hermelin, B., [The effect of deficiency of selected bioelements on hyperactivity in children with certain specified mental disorders]. *Ann Acad Med Stetin,* 1998. 44: p. 297-314.

Stein, J. and Walsh, V., To see but not to read; the magnocellular theory of dyslexia. *Trends Neurosci,* 1997. 20(4): p. 147-52.

Stevens, L., *et al.*, EFA supplementation in children with inattention, hyperactivity and other disruptive behaviours. *Lipids,* 2003. 38(10): p. 1007-1021.

Stevens, L.J., *et al.*, Essential fatty acid metabolism in boys with attention-deficit hyperactivity disorder. *Am J Clin Nutr,* 1995. 62(4)): p. 761-8.

Stevens, L.J., *et al.*, Omega-3 fatty acids in boys with behavior, learning, and health problems. *Physiol-Behav,* 1996. 59(4-5): p. 915-20.

Stordy, B.J., Benefit of docosahexaenoic acid supplements to dark adaptation in dyslexics. *Lancet,* 1995. 346(8971): p. 385.

Stordy, B.J., Dark adaptation, motor skills, docosahexaenoic acid, and dyslexia. *Am J Clin Nutr,* 2000. 71(1 Suppl): p. 323S-6S.

Su, K.P., *et al.*, Omega-3 fatty acids in major depressive disorder. A preliminary double-blind, placebo-controlled trial. *Eur Neuropsychopharmacol,* 2003. 13(4): p. 267-71.

Taylor, K.E., *et al.*, Dyslexia in adults is associated with clinical signs of fatty acid deficiency. *Prostaglandins Leukot Essent Fatty Acids,* 2000. 63(1-2)): p. 75-8.

Taylor, K.E., Richardson, A.J., and Stein, J.F., Could platelet activating factor play a role in developmental dyslexia? *Prostaglandins Leukot Essent Fatty Acids,* 2001. 64(3): p. 173-80.

Teitelbaum, J.E. and Allan Walker, W., Review: the role of omega 3 fatty acids in intestinal inflammation. *J Nutr Biochem,* 2001. 12(1): p. 21-32.

Tokdemir, M., *et al.*, Blood zinc and copper concentrations in criminal and noncriminal schizophrenic men. *Archives of andrology,* 2003. 49(5): p. 365-8.

Uauy, R., *et al.*, Essential fatty acids in visual and brain development. *Lipids,* 2001. 36(9): p. 885-95.

Vancassel, S., *et al.*, Plasma fatty acid levels in autistic children. *Prostaglandins Leukot Essent Fatty Acids,* 2001. 65(1): p. 1-7.

Vargas, D.L., *et al.*, Neuroglial activation and neuroinflammation in the brain of patients with autism. *Annals of neurology,* 2005. 57(1): p. 67-81.

Vincent, A., *et al.*, Maternal antibody-mediated dyslexia? Evidence for a pathogenic serum factor in a mother of two dyslexic children shown by transfer to mice using behavioural studies and magnetic resonance spectroscopy. *J NEUROIMMUNOL,* 2002. 130(1-2): p. 243-247.

Voigt, R.G., *et al.*, A randomized, double-blind, placebo-controlled trial of docosahexaenoic acid supplementation in children with attention-deficit/hyperactivity disorder. *J Pediatr,* 2001. 139(2)): p. 189-96.

Wade, C.R. and van Rij, A.M., In vivo lipid peroxidation in man as measured by the respiratory excretion of ethane, pentane, and other low-molecular-weight hydrocarbons. *Analytical biochemistry,* 1985. 150(1): p. 1-7.

Wallace, F.A., *et al.*, Dietary fatty acids influence the production of Th1- but not Th2-type cytokines. *Journal of leukocyte biology,* 2001. 69(3): p. 449-57.

Ward, N.I., *et al.*, The influence of the chemical additive tartrazine on the zinc status of hyperactive children: a double-blind placebo-controlled Study. *Journal of Nutritional Medicine,* 1990. 1(1): p. 51-58.

Willcutt, E.G. and Pennington, B.F., Psychiatric comorbidity in children and adolescents with reading disability. *J Child Psychol Psychiatry,* 2000. 41(8): p. 1039-48.

Yagi, K., Assay for blood plasma or serum. *Methods in enzymology,* 1984. 105: p. 328-31.

Yanik, M., *et al.*, Plasma manganese, selenium, zinc, copper, and iron concentrations in patients with schizophrenia. *Biological trace element research,* 2004. 98(2): p. 109-17.

Yao, J.K., Reddy, R.D., and van Kammen, D.P., Oxidative damage and schizophrenia: an overview of the evidence and its therapeutic implications. *CNS drugs,* 2001. 15(4): p. 287-310.

Yehuda, S., Rabinovitz, S., and Mostofsky, D.I., Essential fatty acids are mediators of brain biochemistry and cognitive functions. *Journal of Neuroscience Research,* 1999. 56(6): p. 565-570.

Yorbik, O., *et al.*, Investigation of antioxidant enzymes in children with autistic disorder. *Prostaglandins Leukot Essent Fatty Acids,* 2002. 67(5): p. 341-343.

Young, G.S., Maharaj, N.J., and Conquer, J.A., Blood phospholipid fatty acid analysis of adults with and without attention deficit/hyperactivity disorder. *Lipids,* 2004. 39(2): p. 117-23.

Young, I.S. and Trimble, E.R., Measurement of malondialdehyde in plasma by high performance liquid chromatography with fluorimetric detection. *Annals of clinical biochemistry,* 1991. 28(Pt 5): p. 504-8.

Zanarini, M.C. and Frankenburg, F.R., Omega-3 Fatty acid treatment of women with borderline personality disorder: a double-blind, placebo-controlled pilot study. *Am J Psychiatry,* 2003. 160(1): p. 167-9.

Zhang, Z.J., *et al.*, Interaction between polymorphisms of the dopamine D3 receptor and manganese superoxide dismutase genes in susceptibility to tardive dyskinesia. *Psychiatric genetics,* 2003. 13(3): p. 187-92.

Zimmer, L., *et al.*, Chronic n-3 polyunsaturated fatty acid diet-deficiency acts on dopamine metabolism in the rat frontal cortex: a microdialysis study. *Neurosci Lett,* 1998. 240(3): p. 177-81.

Zoroglu, S.S., *et al.*, Increased oxidative stress and altered activities of erythrocyte free radical scavenging enzymes in autism. *European archives of psychiatry and clinical neuroscience,* 2004. 254(3): p. 143-7.

In: Fatty Acids and Oxidative Stress...
Editors: R. Reddy and J.K. Yao, pp. 159-170
ISBN: 978-1-60021-677-0

Chapter X

Essential Fatty Acid Supplementation in Schizophrenia and Mood Disorders

Robin Emsley, Piet Oosthuizen and Susan J van Rensburg

Introduction

This chapter reviews the published case reports and clinical trials in the literature in which the essential fatty acids (EFAs) have been tested as therapeutic agents in the mood disorders and schizophrenia. The rationale and supportive evidence for the use of these compounds as therapeutic agents has been dealt with in preceding chapters.

Mood Disorders

The mood disorders are a highly prevalent group of disorders, with lifetime rates according to the United States of America National Comorbidity Study of 17.1% for major depression, 1.6% for mania and 6.4% for dysthymia (Kessler et al, 1994). They are the cause of much suffering and distress to individuals and society, and both unipolar and bipolar mood disorder rank high on the list of *the* most disabling disorders (Murray and Lopez, 1997).

Major advances have been made in the field of psychopharmacology, and the introduction of a new generation of antidepressants, mood stabilizers and antipsychotics has changed the way we treat patients with unipolar and bipolar mood disorders. However, in spite of this, for many patients there remain significant unmet treatment needs. For example, currently available treatments have a delayed onset of action, do not always treat the full spectrum of symptoms, are not effective in all patients (i.e. some patients are treatment refractory), and although less so than with previous medications, there are still significant side-effects. There is, therefore a great need for additional therapeutic options that can effectively address these shortcomings, either as stand alone treatment or as supplementation to current treatment.

EFAs as Supplemental Therapy in Major Depression

In a report of a single case by Puri et al (2001), the omega-3 essential fatty acid ethyl-eicosapentaenoic acid (E-EPA) was given as supplemental treatment to conventional antidepressants for a severely depressed, treatment-refractory and suicidal male patient with a seven-year history of unremitting depressive symptoms. The addition of E-EPA resulted in a dramatic and sustained clinical improvement in all of the symptoms of depression, including the suicidal ideation. The response was noted within a month of initiation of the treatment (Puri et al, 2001).

In an open label, pilot study in a small sample (n=7), omega-3 fatty acid supplementation given to euthymic pregnant females with a past history of postpartum depression between the 34th and 36th week of pregnancy failed to have any significant effect. Four subjects developed a depressive episode (Marangell et al, 2004).

In a study by Su et al (2003), the investigators conducted an 8-week, double-blind, placebo-controlled trial comparing omega-3 EFAs (6.6 g/day) with placebo as supplemental treatment, in 28 subjects with major depressive disorder. Participants in the omega-3 PUFA group had a significantly greater reductions in Hamilton Rating Scale for Depression scores compared to those in the placebo group ($P < 0.001$), and the treatment was well tolerated.

In a 4-week parallel group double blind add-on design, the effects of the ethyl ester of eicosapentaenoic acid (E-EPA), as an adjunct to maintenance antidepressant treatment for DSM-IV recurrent unipolar depressive disorder were investigated (Nemets et al, 2002). With one exception, the sample included only patients who had experienced episodes of major depression while receiving antidepressant treatment for at least the preceding 3 months. Twenty patients with a current diagnosis of major depressive disorder participated, and had at least 3 weeks of antidepressant therapy at the current dose and a Hamilton Depression Rating Scale Score of at least 18. E-EPA was significantly superior to placebo at week 2 ($p<0.001$), week 3 ($p<0.001$), and week 4 ($p<0.001$). The mean reduction of the Hamilton depression scale score in patients receiving E-EPA was 12.4 points, compared with 1.6 in patients receiving placebo. Six of 10 patients receiving E-EPA achieved a 50% reduction in Hamilton depression score, compared to only one of 10 receiving placebo. The placebo group was remarkable for its low rate of improvement. No clinically relevant side effects were reported. The authors pointed out that it is not possible in this study to distinguish whether E-EPA augments antidepressant action in a manner similar to the mood stabilizers, or whether it has antidepressant properties of its own.

To further test the antidepressant effects of E-EPA a dose-ranging study was conducted in 70 patients with persistent major depressive disorder (Hamilton depression score of ≥ 15) despite ongoing treatment with an adequate dose of a recognised antidepressant (Peet and Horrobin, 2002). Subjects were randomized, on a double-blind basis, to either placebo or to E-EPA at dosages of 1, 2, or 4 g/d as supplement to their standard medication for the 12 weeks duration of the trial. Subjects were assessed by means of the 17-item Hamilton Depression Rating Scale, the Montgomery-Asberg Depression Rating Scale, and the Beck Depression Inventory. Forty-six (88%) of 52 patients receiving E-EPA and 14 (78%) of 18 patients receiving placebo completed the study. No serious adverse events were reported in either group, and the treatment was well tolerated. The 1 g/d E-EPA group showed a significantly better outcome than the placebo group on all 3 rating scales. In the intention-to-treat analysis, 5 (29%) of 17 subjects receiving placebo and 9 (53%) of 17 subjects receiving 1 g/d of E-EPA achieved a 50% reduction on the Hamilton Depression Rating Scale score. In the per-protocol analysis, the corresponding figures were 3 (25%) of 12 subjects for placebo and 9 (69%) of 13 subjects for the 1-g/d group. The 2-g/d group showed no indication of efficacy, and the 4-g/d group showed only a trend level effect toward improvement. In order to investigate whether the therapeutic effect was restricted to certain symptoms only, individual items and component scores of the assessment scales were compared between groups. It was found that all of the individual items on all of the rating scales improved with the 1-g/d dosage of E-EPA vs. placebo, with particularly strong beneficial effects on items rating depression, anxiety, sleep, lassitude, libido, and suicidality. The authors concluded that the trial was too small to draw firm conclusions about the doses other than 1g/day, and that larger studies would be required to investigate possible beneficial effects of higher doses. However, treatment with E-EPA at a dosage of 1 g/d was effective in treating depression in subjects who remained depressed despite adequate standard therapy (Peet and Horrobin, 2002).

Monotherapy in Major Depression

In a double-blind, placebo-controlled study, 36 subjects with major depressive disorder were randomly assigned to receive the omega-3 fatty acid docosahexaenoic acid (DHA) 2 g/day, or placebo for 6 weeks. Response was defined a priori as ≥ 50% reduction in the score on the Montgomery-Asberg Depression Rating Scale. Response rates were 27.8% in the DHA group and 23.5% in the placebo group. The difference in response rates between groups was not statistically significant., indicating that DHA monotherapy failed to show a significant effect in subjects with major depression (Marangell et al, 2003;2004).

Bipolar Disorder

Stoll et al (1999) conducted a double-blind, placebo-controlled study in 30 subjects with bipolar disorder, comparing omega-3 fatty acids (total dosage 9.6 g/d, comprising 6.2g EPA and 3.4g DHA) with placebo, either as adjunctive medication or as monotherapy for at least one month and up to 4 months. The patients met DSM-IV (APA, 2004) diagnostic criteria for bipolar disorder, and had experienced a manic or hypomanic episode within the preceding 12 months. Patients were randomly assigned to parallel groups after stratifying for sex, recent lithium use, and a history of rapid cycling. The primary outcome measure was the duration of time to exit the blinded phase of the study for treatment of an emergent mood episode. Overall, the omega-3 fatty acid patient group had a significantly longer period of remission than the placebo group ($p = .002$), and for nearly every other outcome measure, the omega-3 fatty acid group fared better than the placebo group. The exception was the Young Mania Rating Scale scores, possibly because subjects were not very manic at entry to the study. A subset of eight subjects who were not receiving mood stabilizers at the start of the trial was also examined. The four who received omega-3 fatty acids as monotherapy remained in remission significantly longer than the four who received placebo only. The omega-3 fatty

Table 1. Published clinical trials of EFAs in mood disorders

Authors	Essential fatty acid	Subjects	Nature of study	Sample size	Treatment duration (wks)	Outcome
Puri et al, 2001 (3)d}	E-EPA	Severe, refractory depression	Supplemental, unblinded case report	1	ongoing	Dramatic and sustained improvement
Marangell et al, 2004 (4)2)	Omega-3 FAs	Pregnant subjects with previous post-partum depression	Double blind, placebo controlled, preventive	7	8	Did not prevent post-partum depressive episodes in 4 of the 7 subjects.
Su et al, 2003 (5)3)	6.6g/day omega-3 PUFAs	Major depressive disorder	Supplemental, double blind, placebo controlled	28	8	Greater improvements for omega-3 PUFA group
Nemets et al, 2002 (6)4)	2 g/d E-EPA	Recurrent, unipolar, major depressive disorder	Supplemental, double blind, placebo controlled	20	4	Greater improvements for E-EPA at weeks 2, 3 and 4
Peet & Horrobin, 2002 (7)5)	1, 2 and 4 g/d E-EPA	Persistent major depressive disorder	Supplemental, double blind, placebo controlled	70	12	Significantly better outcome than placebo on all 3 rating scales for 1 g/d E-EPA only
Marangell et al, 2003 (8)d}	2 g/d DHA	Major depressive disorder	Monotherapy, double blind, placebo controlled	36	6	Response rates did not differ significantly
Stoll et al, 1999 (10)d}	Omega-3 FAs (6.2 g/d EPA + 3.4 g/d DHA)	Bipolar disorder	Adjunctive or monotherapy	30	4 to 16	Omega-3 group had a significantly longer period of remission and fared better on most outcome measures

acids were also well tolerated, with the most commonly reported adverse effect being loose stools (10). In an accompanying editorial commentary it was noted that certain design aspects of the study make the data of this study difficult to interpret. Patients were not sufficiently ill for this to be regarded as an acute phase study, and it could also not be regarded as a typical maintenance study because subjects were not sufficiently stable prior to randomization. Furthermore, the study duration of 120 days was unusually short. Concerns were also expressed regarding some of the secondary analyses. Nevertheless, this study was described as representing a landmark attempt in drug development for bipolar disorder in which a naturally occurring compound was evaluated, and a series of double-blind, placebo-controlled studies evaluating the efficacy and dose-response relationships of the omega-3 fatty acids in the various phases of bipolar disorder was recommended.

Schizophrenia

Mental health problems are an increasing part of the global health burden, and now rate as one of the largest causes of lost years of quality of life (Sadik, 1992). Schizophrenia obviously forms a significant part of this burden. It imposes a disproportionately large economic burden on patients and their families, health care systems and society, because of its early onset, devastating effects, and usually lifelong course (Glazer and Johnstone, 1997). While the introduction of conventional antipsychotics 50 years ago heralded the first major breakthrough in the treatment of schizophrenia and other psychotic disorders, these compounds have serious limitations in terms of both efficacy and tolerability. Patients treated with these agents often have varying degrees of treatment refractoriness, suffer frequent relapses, develop prominent functional impairment and experience distressing and disabling side-effects (Emsley, 1999). The most important side-effects are extrapyramidal symptoms (EPS), particularly tardive dyskinesia (TD).

The introduction of the second generation antipsychotics (SGAs) has effectively addressed some of these problems. There is accumulating evidence of significant efficacy and tolerability advantages over their predecessors. In particular, it has been shown that the newer drugs are less likely to induce acute EPS (Thomas and Lewis, 1998). But even the SGAs have troublesome side effects, particularly weight gain, dyslipidaemias and the development of insulin resistance and diabetes mellitus (Allison et al, 1999). Furthermore, the high acquisition costs of the SGAs has put them beyond the reach of many patients worldwide (Emsley et al, 1999). As is the case with the mood disorders therefore, there is a great need for additional treatment options that are effective, safe, well tolerated and affordable. Considerable preclinical evidence, reviewed in the preceding chapters, suggests that the omega-3 fatty acids might be effective in the treatment of schizophrenia, and this has led to the conduction of a number of clinical trials.

Supplemental Therapy

Case Reports

A 30-year-old male with severe, treatment refractory schizophrenia was treated with a fatty acid supplement comprising 30 ml/day of emulsion rich in EPA over a period of 6 months. For the 2 years prior to the study his clinical profile had remained unchanged, and antipsychotic medication had been discontinued because of lack of efficacy. The patient was assessed at monthly intervals. A marked improvement in his condition was noted at 2 months. He continued to improve thereafter, and at 6 months very few active symptoms were observed. These findings provided early clinical support for a possible beneficial effect for certain fatty acids in the management of schizophrenia (Richardson et al, 1999). Because of the potential risks associated with antipsychotic drugs during pregnancy, it was decided to treat a 30-year-old pregnant married woman with chronic schizophrenia, who experienced an episode of acute exacerbation of psychotic symptoms with an open trial of omega-3 PUFAs monotherapy. She was reported to show a dramatic improvement in both positive and negative symptoms, and at the same time a significant increase of omega-3 PUFA composition in erythrocyte membranes was observed. There were no adverse effects reported with the treatment. On the basis of their findings, the authors proposed that omega-3 PUFAs could be a safe treatment option to pregnant schizophrenic women (Su et al, 2001).

Open label Clinical Trials

In an extension to a study in which substantial depletions of fatty acids were found in red cell membranes (Peet et al, 1995), dietary analysis revealed no deficiency of fatty acid intake in the patients with schizophrenia, although greater intake of omega-3 fatty acids appeared to be associated with less prominent symptomatology. Patients were then given dietary supplementation with 10 g per day of concentrated fish oil over a period of six weeks. Significant improvement was observed in their symptoms of psychosis, and this was associated with an increased level of omega-3 fatty acids in the red cell membranes (Laugharne et al, 1996).

In another open label design, a mixture of EPA/DHA (180:120 mg) and antioxidants (vitamin E/C, 400 IU:500 mg) was given orally twice daily for 4 months to 33 subjects with schizophrenia. Post-treatment levels of RBC EPUFAs were found to be significantly higher than the pre-treatment levels as well as levels in normal controls without any significant increase in plasma peroxides. Psychiatric symptoms were observed to improve insofar as Brief Psychiatric Rating Scale, Positive and Negative Syndrome Scale (PANSS) and a quality of life scale score all showed reductions (Arvindakshan et al, 2003).

Randomised Controlled Trials

To date, four randomised, placebo controlled EPA supplementary studies in schizophrenia have been reported in the English literature. In a study designed to assess the effects of two different omega-3 fatty acids, EPA and DHA, 45 outpatients with schizophrenia with persistent symptoms while on stable antipsychotic medication were assessed. They were randomised to receive treatment either with EPA, DHA or placebo for 12 weeks, as a supplement to their usual antipsychotic medication. Symptoms were assessed by means of the PANSS. The EPA group showed significantly greater PANSS total score reductions compared with both the DHA and placebo groups. A greater response rate, defined as >25% PANSS total score reduction, was also found in the EPA group. There was a significant association between the change in positive PANSS scores and baseline omega-3 fatty acid levels within the EPA group. Baseline EPA emerged as a significant predictor of improvement in clinical scores (Peet et al, 2001).

A study involving a relatively large sample which was also conducted over the longest period reported negative (Fenton et al, 2001). This was a North American, multi-site, randomised, placebo-controlled trial of ethyl-EPA supplementation for residual symptoms and cognitive impairment. The sample comprised 87 outpatients meeting DSM-IV criteria for schizophrenia or schizoaffective disorder, with persistent symptoms despite antipsychotic treatment. Subjects had had been on unchanged medication in the 30 days prior to the trial, and significant residual symptoms were judged to be present (defined as either one or more positive and/or negative symptom scores > 4 or total PANSS scores > 45 with a score of 3 or more on at least 3 positive or negative items on the PANSS scale). Participants were randomly assigned to receive either ethyl-EPA 3 g/day (N=43) or placebo (N=44) in a 16-week, double-blind design, as supplementation to their usual antipsychotic treatment. Assessments took place at baseline and at weeks 1, 2, 4, 8, 12, and 16, and cognitive testing was performed at baseline and at week 16. No significant differences were found between the groups in positive or negative symptoms, mood, cognitive function, or global impression ratings. Results were similar for the intention-to-treat (N=87) and completer (N=75) analyses. The mean reduction in PANSS total scores from baseline to endpoint was 5 for the ethyl-EPA group and 6 for the placebo group. The AA/EPA ratio change (used to index pre- and post-treatment fatty acid composition in red blood cells) from baseline to endpoint was not significantly associated with any efficacy variable.

An attempted dose-ranging multi-centre study investigating the effects of E-EPA as supplemental medication in patients with persistent schizophrenic symptoms was conducted in the United Kingdom (Peet and Horrobin, 2002). The sample comprised 115 patients with DSM-IV-defined schizophrenia. Thirty-one of the subjects were receiving clozapine, 48 other second generation antipsychotics and 36 conventional antipsychotics. Patients were randomised to receive, in addition to their other antipsychotic medication, either E-EPA 1g/d, E-EPA 2g/day, E-EPA 4g/day or placebo. The study period was 12 weeks. The a priori primary efficacy measure was change from baseline to 12 weeks on the PANSS total score and its sub-scales. All groups improved significantly from baseline but there were no significant differences between groups. Specifically, no difference was found between EPA and placebo in terms of the primary efficacy outcome measure. There was a large mean reduction in total PANSS scores in the placebo group (-16.6). Although his was a negative study in terms of the primary outcome measure, an interesting finding was reported in the clozapine subset of patients. In patients on EPA 2 g/day there were improvements on the PANSS and its sub-scales, but there was no difference between active treatment and placebo. However, patients on clozapine showed little placebo response, but a statistically significant reduction of symptoms if they received EPA. This effect was greatest at 2 g/day. A positive relationship was reported between improvement on rating scales and rise in red blood cell AA concentration. No treatment-related side effects or adverse biochemical or haematological effects were reported. Subjects on 2 and 4 g/day EPA showed significant reductions in triglyceride levels, which had been elevated by clozapine. Limiting factors in this study include the small sample sizes considering the multiple comparison groups, and the large placebo effect size.

In a randomized, placebo-controlled study conducted by ourselves (Emsley et al, 2002) we investigated the effects of E-EPA as supplemental treatment in 40 subjects with persistent symptoms of schizophrenia. The subjects had chronic schizophrenia, with a degree of refractoriness insofar as they experienced persistent positive symptoms despite ongoing antipsychotic therapy. The latter comprised either conventional (usually depot) antipsychotics, or clozapine. The participants were randomised 1:1 to receive either EPA 3 g/day or placebo, in addition to their existing treatment. This was a fixed-dose, double-blinded study over 12 weeks. For the primary efficacy measure, the E-EPA group were reported to have a significantly greater reduction in the Positive and Negative Syndrome Scale (PANSS) total scores (p=0.03) at 12 weeks. The difference was significant at 3 weeks already, suggesting a possible early onset of action. A further finding of significantly greater reduction in Extrapyramidal Symptom Rating Scale (ESRS) dyskinesia subscale scores for the E-EPA group was also reported at 12 weeks (p=0.008). We concluded that these results provided further evidence that E-EPA may be an effective, safe and well tolerated supplemental treatment in schizophrenia, with an additional benefit of improving tardive dyskinesia. The prospect of an effective, safe, well-tolerated and affordable treatment for schizophrenia is of great importance, particularly to countries where the high acquisition costs of the second generation antipsychotics have put them beyond the reach of most patients (Emsley et al, 1999).

Monotherapy

Case Report

The administration of EPA to a subject with schizophrenia who had never been exposed to treatment with conventional antipsychotic medication was described in the literature (Puri et al, 2001). Treatment led to a clear-cut and sustained improvement in both positive and negative symptoms, which was reported to be accompanied by a correction in erythrocyte membrane abnormalities in both omega-3 and omega-6 poly-unsaturated fatty acids and with reduced neuronal membrane phospholipid turnover. MRI scans done a year before EPA treatment reported cerebral atrophy, but subsequent scans reported reversal of atrophy after six months of EPA treatment. These results suggest that EPA can reverse both the phospholipid abnormalities previously described in schizophrenia and cerebral atrophy. The authors proposed that their findings provide further evidence in support of the membrane phospholipid model of schizophrenia (Arvindakshan et al, 2003).

Randomised Controlled Trial

Only one clinical trial has investigated and EPA as monotherapy in schizophrenia (Peet et al, 2001). The study population comprised 30 participants with DSM-IV diagnosed schizophrenia who where not receiving medication at the time of presentation with acute exacerbation or as first-episode of psychosis to a clinic in India. Participants were randomly allocated in a double blind treatment arm to receive either EPA 2g/day (in the form of EPA-enriched oil (Kirunal), or identical placebo, for a 12 week treatment period. The aim was to use trial medication as a sole treatment whenever possible, but for ethical reasons, to introduce conventional antipsychotic when considered clinically imperative. Outcome measures included the need for, and duration of, conventional antipsychotic treatment, and changes in PANSS scores. By the end of the study, all 12 subjects on placebo, but only eight of 14 subjects on EPA, were taking antipsychotic drugs. The EPA subjects also had significantly lower scores on the PANSS compared to the placebo group.

Table 2. Published clinical trials of EFAs in schizophrenia

Authors	Essential fatty acid and dose	Study design	Sample size	Duration (wks)	Outcome
Laugharne et al, 1996 (20)	10g/day concentrated fish oil	Open label, supplemental	23	6 weeks	Significant improvement
Arvindakshan et al, 2003 (21)	EPA/DHA(180:120mg) (vit E/C, 400IU:500 mg)	Open label, supplemental	33	4 months	Reduction in symptoms
Peet et al, 2001 (22)	EPA 2g/day EPA-enriched oil (Kirunal)	Randomised, double blinded, supplemental	45 (including DHA arm)	12 weeks	EPA significantly better than placebo and DHA
Fenton et al, 2001 (23)	3g/day E-EPA	Randomised, double blind, supplemental	87	116 weeks	No difference
Peet & Horrobin, 2002 (7)	1, 2 and 4g/day E-EPA	Randomised, double blind, Dose-ranging supplemental	115	12 weeks	No difference on primary outcome
Emsley et al, 2002 (24)	3g/day E-EPA	Randomised, double blind, supplemental	40	12 weeks	EPA significantly better
Peet et al, 2001 (22)	EPA 2g/day EPA-enriched oil (Kirunal)	Randomised, double blind, monotherapy	30	12 weeks	Fewer EPA treated subjects required additional antipsychotic medication

EFAs and Schizophrenia: Conclusions

With regard to the published randomised controlled trials comparing supplemental EPA to placebo in schizophrenia, results have been mixed. Two studies reported positive findings, and two negative findings on the primary outcome measure. Reasons for discrepant results are not immediately clear given the similarities in patient samples, omega-3 fatty acid used, trial design and duration, assessment instruments etc. However, there were possible differences that could have explained some of the discrepancies. For example, background fatty acid intake may have been substantially different between the populations studied. Fatty acids are known to compete with one another for uptake, so that a diet rich in fatty acids may result in less uptake of EPA. Dietary fatty acid content is likely to differ substantially, for example, between North America, the UK, South Africa and India. The placebo response rates in the various studies also differed considerably, making it difficult to draw comparisons across samples. Finally, most studies involved chronic, treatment refractory samples – subjects least likely to respond to any form of treatment. When keeping this in mind, the 50% rate of positive studies can be regarded as very encouraging considering the fact that 25% of studies comparing novel antipsychotics as monotherapy with placebo failed to differentiate between the active compound and placebo (Laughren, 2001).

Overall Conclusions

Much of the clinical research conducted in the field of essential fatty acids can be criticised for methodological limitations. Samples are generally small, and study designs sometimes lack the rigour often seen with multinational clinical trials supported by the private sector. However, taken together, there is a strengthening signal suggesting important roles specifically for EPA in mood disorders and schizophrenia. There is preliminary evidence that 1 to 2 g/d EPA is an effective supplementary treatment in major depressive disorder and 2 to 3 g/d EPA in schizophrenia. There may be additional benefits in bipolar mood disorder and TD. Many more well-designed clinical trials are urgently required to further elucidate the role of EFAs in mood disorders and schizophrenia. It would be of great interest to investigate any antidepressant, antipsychotic or mood stabilising properties of the EFAs. Longer term trials investigating safety aspects of these agents are also urgently required. In future studies more attention should be paid to aspects such as background diet, nutritional status and substance abuse. Laboratory analyses should include lipograms, glucose, vitamins B12, B6 and folic acid, homocysteine and the hormones cortisol, dehydroepiandrosterone-sulphate, testosterone and prolactin. In order to provide a true double blind, "Fish flavour" should also be added to the placebo capsules.

References

Allison DB, Mentore JL, Heo M, Chandler LP, Cappelleri JC, Infante MC et al. Antipsychotic-induced weight gain: a comprehensive research synthesis. *Am J Psychiatry* 1999; 156(11):1686-1696.

American Psychiatric Association. *Diagnostic and Statistical Manual of Mental Disorders*. Fourth ed. Washington,D.C.: American Psychiatric Association, 1994.

Arvindakshan M, Sitasawad S, Debsikdar V, Ghate M, Evans D, Horrobin DF et al. Essential polyunsaturated fatty acid and lipid peroxide levels in never-medicated and medicated schizophrenia patients. *Biol Psychiatry* 2003; 53(1):56-64.

Calabrese JR. Fish oils and bipolar disorder: A promising but untested treatment. *Arch Gen Psychiatry* 1999; 56:413-414.

Emsley R, Myburgh C, Oosthuizen P, van Rensburg SJ. Randomized, placebo-controlled study of ethyl-eicosapentaenoic acid as supplemental treatment in schizophrenia. *Am J Psychiatry* 2002; 159(9):1596-1598.

Emsley RA, Oosthuizen PP, Joubert AF, Hawkridge SM, Stein DJ. Treatment of schizophrenia in low-income countries. *Int J Neuropsychopharmcol* 1999; 2(4):321-325.

Emsley RA. Partial response to antipsychotic treatment: the patient with enduring symptoms. *J Clin Psychiatry* 1999; 60 Suppl 23:10-13.

Fenton WS, Dickerson F, Boronow J, Hibbeln JR, Knable M. A placebo-controlled trial of omega-3 fatty acid (ethyl eicosapentaenoic acid) supplementation for residual symptoms and cognitive impairment in schizophrenia. *Am J Psychiatry* 2001; 158(12):2071-2074.

Glazer WM, Johnstone BM. Pharmacoeconomic evaluation of antipsychotic therapy for schizophrenia. *J Clin Psychiatry* 1997; 58 Suppl 10:50-54.

Kessler RC, McGonagle KA, Zhao S, Nelson CB, Hughes M, Eshleman S et al. Lifetime and 12-month prevalence of DSM-III-R psychiatric disorders in the United States. Results from the National Comorbidity Survey. *Arch Gen Psychiatry* 1994; 51(1):8-19.

Laugharne J, Mellor JE, Peet M. Fatty acids and schizophrenia. *Lipids* 1996; 31(Suppl): S163-S165.

Laughren TP. The scientific and ethical basis for placebo-controlled trials in depression and schizophrenia: an FDA perspective. *Eur Psychiatry* 2001; 16:418-423.

Marangell LB, Martinez JM, Zboyan HA, Chong H, Puryear LJ. Omega-3 fatty acids for the prevention of postpartum depression: negative data from a preliminary, open-label pilot study. *Depress Anxiety* 2004; 19(1):20-23.

Marangell LB, Martinez JM, Zboyan HA, Kertz B, Kim HF, Puryear LJ. A double-blind, placebo-controlled study of the omega-3 fatty acid docosahexaenoic acid in the treatment of major depression. *Am J Psychiatry* 2003; 160(5):996-998.

Murray CJ, Lopez AD. Alternative projections of mortality and disability by cause 1990-2020: Global Burden of Disease Study. *Lancet* 1997; 349(9064):1498-1504.

Nemets B, Stahl Z, Belmaker RH. Addition of omega-3 fatty acid to maintenance medication treatment for recurrent unipolar depressive disorder. *Am J Psychiatry* 2002; 159(3): 477-479.

Peet M, Brind J, Ramchand CN, Shah S, Vankar GK. Two double-blind placebo-controlled pilot studies of eicosapentaenoic acid in the treatment of schizophrenia. *Schizophr Res* 2001; 49(3):243-251.

Peet M, Horrobin D. A dose-ranging study of the effects of ethyl-eicosapentaenoate in patients with ongoing depression despite apparently adequate treatment with standard drugs. *Arch Gen Psychiatry* 2002; 59(10):913-919.

Peet M, Laugharne J, Rangarajan N, Horrobin D, Reynolds G. Depleted red cell membrane essential fatty acids in drug-treated schizophrenic patients. *J Psychiatr Res* 1995; 29(3):227-232.

Puri BK, Counsell SJ, Hamilton G, Richardson AJ, Horrobin DF. Eicosapentaenoic acid in treatment-resistant depression associated with symptom remission, structural brain changes and reduced neuronal phospholipid turnover. *Int J Clin Pract* 2001; 55(8): 560-563.

Richardson AJ, Easton T, Gruzelier JH, Puri BK. Laterality changes accompanying symptom remission in schizophrenia following treatment with eicosapentaenoic acid. *Int J Psychophysiol* 1999; 34(3):333-339.

Sadik N. Public policy and private decisions: world population and world health in the 21st century. *Journal of public health policy* 1992; 13:133-9

Stoll AL, Severus WE, Freeman MP, Reuter S, Zboyan HA, Diamond E et al. Omega 3 fatty acids in bipolar disorder: a preliminary double-blind, placebo-controlled trial. *Arch Gen Psychiatry* 1999; 56(5):407-412.

Su KP, Huang SY, Chiu CC, Shen WW. Omega-3 fatty acids in major depressive disorder. A preliminary double-blind, placebo-controlled trial. *Eur Neuropsychopharmacol* 2003; 13(4):267-271.

Su KP, Shen WW, Huang SY. Omega-3 fatty acids as a psychotherapeutic agent for a pregnant schizophrenic patient. *Eur Neuropsychopharmacol* 2001; 11(4):295-299.

Thomas CS, Lewis S. Which atypical antipsychotic. *Br J Psychiatry* 1998; 172:106-109.

In: Fatty Acids and Oxidative Stress...
Editors: R. Reddy and J.K. Yao, pp. 171-184
ISBN: 978-1-60021-677-0

Chapter XI

Bioactive Lipids, Niacin Sensitivity and Cognition in Schizophrenia

Gregor E. Berger, Tina Proffitt, Mirabel Mcconchie, Stephen J. Wood and Patrick D. Mcgorry

Introduction

Schizophrenia and related disorders are often associated with cognitive dysfunction, in particular impairment in (working) memory and attention (Goldberg et al, 1993; Goldman-Rakic, 1994; Hoff et al, 1992; Weinberger et al, 1992). Most of the research efforts to explain such memory dysfunctions have focused on neurotransmitter systems, such as the dopaminergic and serotonergic (Weinberger and Gallhofer, 1997), and more recently GABAergic (Lewis et al, 1999) and muscarinic (Crook et al, 2000) systems. However, memory not only depends upon synaptic transmission and associated signaling events but also synaptic plasticity (Goldman-Rakic, 1996; 1999) which is strongly associated with the release of bioactive lipids such as arachidonic acid (AA, 20:4 [n-6]), and docosahexanoic acid (DHA) (Piomelli, 1994). Bioactive lipids are abundant in the central nervous system, in particular in synapses and the retina, and play a crucial role in neurodevelopment Crawford, 1993) and memory formation (Wainright et al, 1994).

More recent evidence suggests that bioactive lipids, such as AA, are reduced in peripheral and central cell membranes of patients with schizophrenia (for a systematic review see Chapter 8), and preclinical studies suggest that the relationship between peripheral and central lipid composition is probably genetically determined. A recent clinical study in drug-naïve first episode psychosis patients measuring arachidonic acid and the in vivo break down of membrane lipids using phosphorus spectroscopy further supports such a link (Yao et al, 2002). In addition to reviewing the evidence that niacin sensitivity defines such an intermediate phenotype within the schizophrenia spectrum and discuss its potential clinical and research implications, we present preliminary findings of an ongoing study investigating the relationship between niacin sensitivity as a marker of bioactive lipid availability and cognition.

Bioactive Lipids in Schizophrenia

Feldberg (1976) and Horrobin (1977) independently proposed that schizophrenia might be related to a disturbance of prostaglandin metabolism. Horrobin argued that a reduced prevalence of inflammatory disorders, in particular rheumatoid arthritis (Oken and Schulzer, 1999), reduced pain sensitivity (Dworkin, 1994; Kudoh et al, 2000), altered thermo-sensitivity Hermesh et al, 2000), and (anecdotal) case reports of remission of psychotic symptoms during inter-current physical illness and febrile episode (Fuller-Torrey et al, 2000; Lipper and Werman, 1977) point towards a deficiency of arachidonic acid and its derivates (e.g. prostaglandins) in schizophrenia and related disorders.

Niacin Sensitivity and Schizophrenia

An elegant way to assess the availability of AA is to measure niacin sensitivity. The observation that patients ingesting large doses of niacin respond with a skin flush reaction of the upper body goes back to the middle of the last century. Morrow et al (1992) were the first to investigate the underlying biochemical mechanism of the reaction. His group observed that topical administration of methyl-nicotinate to the forearm induced a flush reaction and cutaneous erythema that were both closely associated with a dramatic release of prostaglandin D_2 (PGD_2). Not only was this release equivalent to a 58- to 122-fold increase in PGD_2, no increase was found in the contralateral arm indicating that the PGD_2 was released locally from the site of methyl-nicotinate exposure. The release of PGD_2 occurs in a dose-dependent manner over the concentration range of 10^{-3} to 10^{-1} M and is not accompanied by a release of histamine, suggesting that it is not from mast cells. Morrow et al. concluded that the skin has to be the major site of release, and that the cutaneous vasodilation occurs primarily due to the release of PGD_2 from niacin responsive cells that reside there. Therefore, niacin sensitivity gives indirect information about the availability of PGD_2, its precursor AA, and the capacity of cells to release these substrates (e.g. via activation of phospholipases and cyclo-oxygenases).

Early descriptions of reduced upper body flush response after intake of large doses of niacin in some patients with schizophrenia (Hoffer, 1969) and its restoration after remission of psychotic symptoms, led to the suggestion that niacin response could be used to monitor treatment response (Rybakowski and Weterle, 1991). Two controlled studies trying to quantify the flush reaction using methodologies such as repeated malar temperature or blood flow measures presented mixed results, probably due to different approaches and low nicotinic acid doses (Rybakowski and Weterle, 1991; Wilson and Douglass, 1986). A more sensitive methodology introduced by Wilkin (1982), using temperature differences between skin and the body core, was successfully used in a clinical population by Hudson at al (1997), who demonstrated that after oral intake of 200mg niacin 42.9% of patients with schizophrenia had no change in temperature, whereas 94% of patients with bipolar disorder and 100% of normal controls did. The authors suggested that an impaired niacin skin flush response to an oral niacin challenge dose selects a subgroup of patients within the schizophrenia spectrum with a underlying disturbance of the phospholipid - AA - eicosanoid messenger system (Hudson et al, 1997). However, the oral form of the niacin challenge test never achieved substantial resonance in clinical or research communities, probably because of inconsistent results, but also due to the subjective inconvenience associated with the niacin skin flush reaction of the upper body.

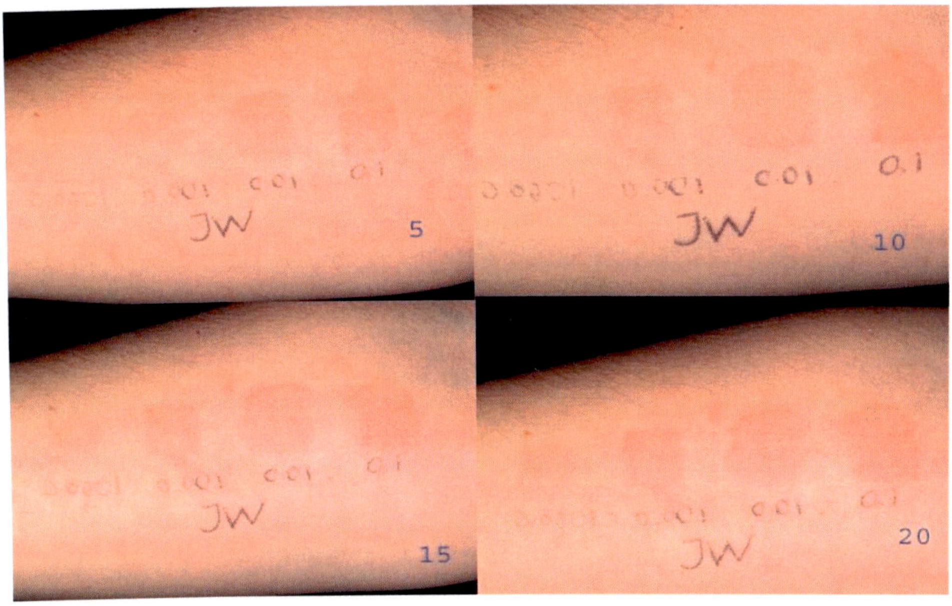

Figure 1. Photograph of normal niacin flush reaction (6100 Pixel, Fuji S1, ring flush).

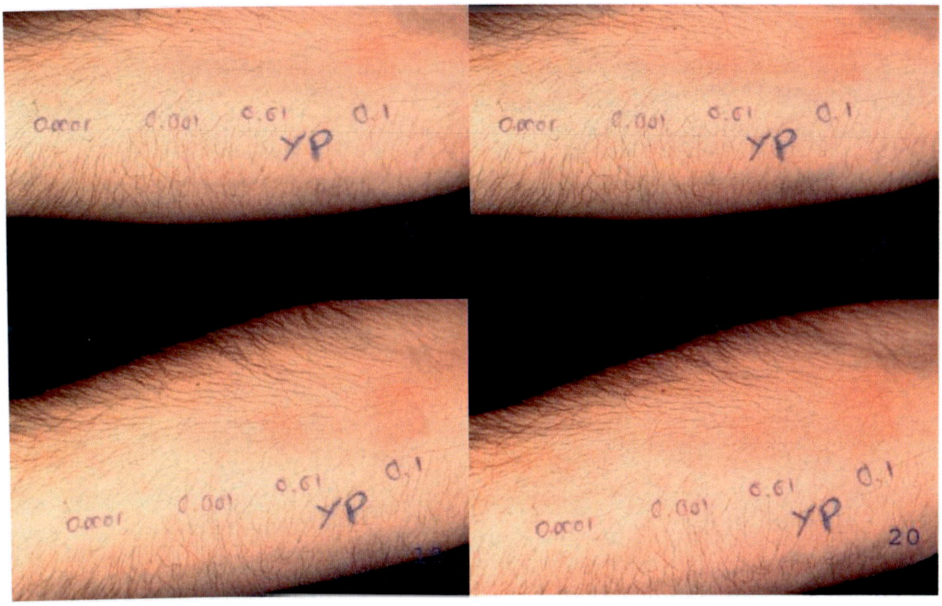

Figure 2. Photograph of impaired niacin flush reaction (6100 Pixel, Fuji S1, ring flush).

The topical variant of the niacin challenge test Morrow et al, 1992) is more promising and provides an easy way to select a biochemically-defined intermediate phenotype of potential relevance for genetic, neuroimaging, pharmacological and/or neuropsychological studies (Ward, 2000).

In patients with chronic schizophrenia Ward et al (Ward et al, 1998) demonstrated that over 70% of patients had either no or a markedly reduced erythema compared to normal controls after topical application of four different concentrations of niacin (0.1M, 0.01M, 0.001M and 0.0001M nicotinic acid solution). These findings were later replicated in both medicated (Puri et al, 2001) and unmedicated (Shah et al, 2000) patients with chronic schizophrenia. All studies demonstrated that a substantial proportion (between 40-50%) of patients with schizophrenia had a clearly reduced flush reaction. However a proportion of the patients also showed a niacin flush reaction indistinguishable from normal controls.

There are a few studies that investigated the niacin skin flush test in other disorders then schizophrenia. Patients with alcohol abuse or dependency (Fiedler et al, 1986), social phobia (Bouwer and Stein, 1998) or during manic exacerbation of bipolar disorder (Hudson et al, 1997) seem to be niacin sensitive or in some cases even hyper-responsive, whereas patients with Huntington's disease (Puri, 2001) demonstrate an impaired reaction similar to those with schizophrenia. Future research needs to address the question of specificity and state-dependency of the topical niacin skin flush reaction more systematically.

The quantification of the topical niacin flush reaction has shown promise (Smesny et al, 2000). However, the emerging edema at high nicotinic acid concentrations in niacin sensitive probands seems to jeopardize the use of quantitative methodologies such as optical reflection spectroscopy (ORS) (Smesny et al, 2003). A recent study comparing a semi-quantitative, descriptive rating scale with ORS could demonstrate that ORS was able to separate early psychosis patients from controls in a similar way as rating scales, especially at lower niacin concentrations. As soon as edema developed (what normally occurs at higher concentrations in niacin sensitive cases), the positive correlation between descriptive and quantitative measures disappeared. Therefore, the influence of edema has to be taken into account if using a quantitative technique such as ORS to measure niacin response. On the other hand the high sensitivity of ORS to detect skin colour changes may offer the possibility to reduce the number of necessary test solutions from four to two (0.001 M and 0.01 M). Furthermore, a combined use of spectroscopic measures at lower niacin concentrations and assessment of the time of "earliest" skin response is highly effective in distinguishing subjects with normal or disturbed skin response (sensitivity 92 %, specificity 84 %) (Smesny et al, 2003). However, sensitivity of a simplified test protocol has to be further studied yet. In contrast to ORS the qualitative rating scale was able to separate patients from controls at all concentrations over all four time points (Smesny et al, 2003).

Our group started to use the topical variant of the niacin skin flush test in the context of a double-blind, randomized, placebo-controlled eicosapentaenoic acid (EPA) supplementation trial in first-episode psychosis patients. We developed and evaluated our own descriptive assessment scale, and defined a subgroup of about 40-50% of patients with a clearly impaired niacin skin reaction (unpublished data, see figure 3).

Table 1. Summary of published studies using niacin sensitivity in schizophrenia

Reference	Oral Niacin Dose (mg)	Subjects	Methodology	Results
Hoffer A (1962) Horrobin, D (1980)	up to 3000	NA	Visual Inspection	Case reports of diminished flush response after oral intake of niacin in patients with SCZ; restitutio of flush reaction after remission of symptoms
Wilson et al (1986)	100	16	Photoplethysmography	No quantitative significant difference between SCZ and CTL flushing
Fiedler et al (1986)	25 (iv)	12	Malar skin temperature	No significant difference between SCZ and CTL flushing
Rybakowski et al (1991)	200	33	Visual inspection + Handheld thermometer to measure skin temp	Blunted flush in 24% of SCZ group vs. 0% of CTL group ($p<0.05$)
Hudson et al (1997)	200	28	Visual inspection + Thermocouple sensors to measure skin temp to calculate 'thermal index'	Blunted flush in 43% of SCZ group vs. 0% of CTL group ($p<0.05$) and 0% of Bipolar group, no correlation between antipsychotic dose and flush response
Glen et al (1996)	200	128	Visual Inspection	Blunted flush in 60% of SCZ group (predominantly negative symptom patients)
Hudson et al (1999)	200	23	Visual Inspection + Skin Temperature	Blunted flush in 43% of SCZ group vs. 0% of CTL group ($p<0.05$)
	Topical niacin conc.			
Ward et al (1998)	0.001	38	Visual Inspection	Blunted flush in 83% of SCZ group vs. 23% of CTL group ($p<0.05$)
Waldo et al		10	Visual Inspection	No overlap between pedigrees with impaired niacin flush reaction compared with those who have delayed evoked potential
Das at al (1999)	0.001	19	Visual Inspection	Blunted flush in 50% of SCZ group
Shah et al (2000)	0.1, 0.01, 0.001, 0.0001	56 (32medicated/ 24unmedicated)	Visual Inspection	Blunted flush in SCZ group vs CTL at three higher concentrations, no significant effect of medication status on flush response

Table 1. Continued

Reference	Oral Niacin Dose (mg)	Subjects	Methodology	Results
Puri et al (2001/2002)	0.001 (all four conc.)	21 (27)	Visual Inspection (Volumetic index)	Sensitivity of niacin skin test of 90% and a specificity of 75% (the volumetric index resulted in a sensitivity of 78% and a specificity of 65%)
Smesny et al (2003)	0.1, 0.01, 0.001, 0.0001	25	Optical Reflection Spectroscopy (ORS), compared to visual inspection using descriptive assessment scale (see table 1)	Descriptive scale separated patients from controls at all concentrations, quantification with ORS demonstrated positive correlation at lower concentrations (0.0001, 0.001), but inverse correlation at higher concentrations (0.01, 0.1) due to edema

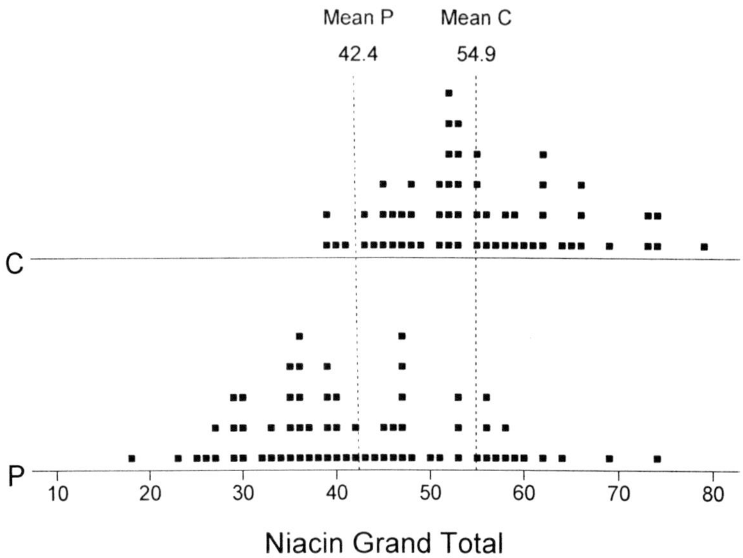

Figure 3. Distribution of composite score Niacin Grand Total (integrating all four time points and concentrations). C Normal controls, P First episode Patients (unpublished data, presented at the 11th Biennial Winter Workshop on Schizophrenia).

These data are in keeping with the findings of Glen et al (1994), which suggested a bimodal distribution of AA and DHA levels in cell membranes of patients with schizophrenia and related disorders. Further, our work supports the idea of using niacin sensitivity (as a indirect measure of AA availability) to stratify schizophrenia patients into two groups (Hudson et al, 1999).

Niacin sensitivity is dependant on the release of AA from membrane phospholipids and the generation of prostaglandins. Interesting is the counterintuitive finding for Tavaras et al

(2003), finding an inverse relationship between PLA_2 activity and niacin sensitivity resulting in a increase of 30% iPLA2 activity in niacin insensitive compared to sensitive patients. It was unclear what type of PLA_2 activity was measured. Genetic association studies have found a link between the number of poly-adenosine repeats of the $cPLA_2$ (Type IVA) gene on chromosome 1 and schizophrenia, with the presence of more than 50 repeats accounting for 40% of enrolled subjects (Hudson et al, 1996; Wei et al, 1998). In addition, polymorphisms of the $cPLA_2$ promoter region (Ban I) showed a significant allelic frequencies (A2/A2) differences between patients with schizophrenia compared to normal controls (Peet et al, 1998; Ramchand et al, 1999). However, not all groups replicated these findings (Frieboes et al, 2001; Price et al, 1997), most likely because few studies have used niacin sensitivity to stratify the sample.

Bioactive Lipids and Cognition

Pre-clinical studies link cognitive function to bioactive lipid metabolism in EFA-deprived animals (Wainwright et al., 1992). Of particular interest was a study demonstrating that EFA-deprived rats needed up to three times longer in the water maze test compared to normally fed rats. The water maze test is a measure of visuospatial memory that probes hippocampal

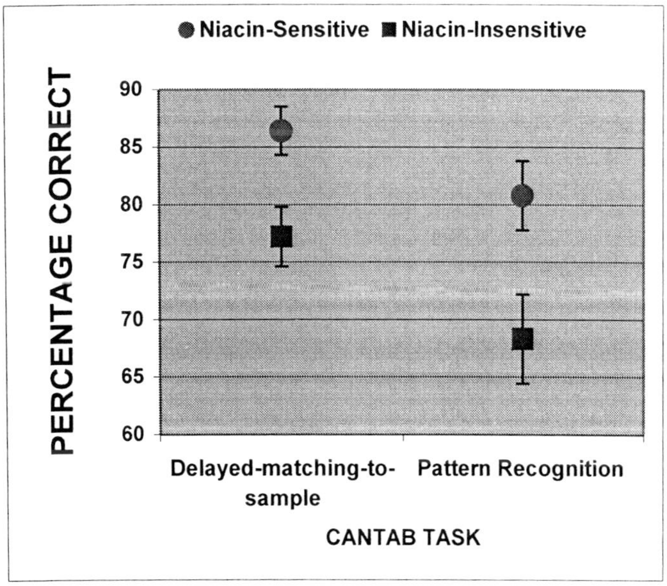

NB. Explanation of tasks as follows:

1. Delayed-Matching-to-Sample - participants were shown a complex abstract pattern, simultaneous with, or followed 0, 4, & 12 seconds later by presentation of 4 'choice' patterns, only one of which exactly matched the initial stimulus. The summary score represents 'Percent Correct' matches across all conditions.
2. Pattern Recognition - participants were shown two sets of 12 abstract patterns followed by a two-alternative forced-choice recognition test. 'Percent Correct' scores were calculated.

Figure 4. CANTAB Recognition memory task performance in the Niacin-Insensitive and Niacin-Sensitive groups [mean (s.e.)].

function. Our group has demonstrated that a similar measure is impaired in established illness (Wood et al, 2002) and may decline over the early phase of the disorder (unpublished data). We present here some preliminary results of an ongoing study that has investigated this relationship in 46 first-episode (FE) psychosis patients aged between 15 and 29 years. As predicted from the animal evidence, we found that recognition memory was the only cognitive domain significantly associated with niacin insensitivity (see Figure 4). This impairment does not appear to be a result of poor attention or perceptual matching, lower premorbid IQ, age or gender differences or greater levels of psychopathology. Further, the results do not suggest that the niacin-insensitive group show a particular problem maintaining information over a delay. The problem appears to be greater difficulty with encoding of complex stimuli and/or discrimination of targets from multiple, highly similar distracters, which may specifically implicate perirhinal and entorhinal cortex (Baxter & Murray, 2001; Murray & Bussey, 2001). Volumetric changes have been found in these regions in FE psychosis and FE schizophrenia (Pantelis and Maruff, 2002), but specific links with disturbed phospholipid metabolism are yet to be demonstrated. Our data also provide support for our earlier finding of impaired recognition memory in early psychosis (Wood et al, 2002).

EFA-Supplementation and Cognition

The importance of bioactive lipids and EFA supplementation has been studied extensively in animals (reviewed by Wainwright, 1997; 1999). Animals (mice, rats, cats, primates) fed an EFA deficient or depleted diet showed significantly aberrant sleeping patterns, adipsia, polydipsia, anorexia, hyperalgesia, reduced self-grooming activity, decreased exploration and social interaction, as well as a disruption of established food-motivated lever-pressing responses. EFA supplementation had a positive impact on 18 out of 24 altered behavioural variables in EFA deficient animals (reviewed by Wainwright, 1997). Improvements in behavioural tasks after EFA supplementation were correlated with alterations in brain lipid composition and were specific for different brain areas such as the temporal lobe (Goustard-Langelier et al, 1999: Marteinsdottir et al, 1998). The n-3/n-6 ratio in the diet was of particular importance in determining the availability of EFA for the incorporation of EFA into animal brains with a preference for n-3 EFA, especially EPA and DHA (Lin et al, 1990; Wainright, 1997). There is now growing evidence that EPA competes with AA by inhibiting AA metabolism via inhibition of 5-lipoxygenase activity at low concentrations, and PLA_2 activity at high concentrations (Saku et al, 1999). The structural similarity between AA and EPA could explain the enhanced effectiveness of EPA if it behaves as an analogue of AA (Whelan, 1999). In addition, preclinical studies demonstrating an inhibitory effect of EPA on protein kinase C (Padma and Das, 1999; Seung Kim et al, 2001) suggests it may be neuroprotective (Manji and Lenox, 1999).

We are only aware of one positive placebo-controlled study (using a cross-over design) reporting improved memory function in patients with chronic schizophrenia with tardive dyskinesia randomized to evening primrose oil supplementation as add on treatment to antipsychotics (Vaddadi et al., 1989). We recently completed a double-blind randomized, placebo-controlled study using 2g EPA in 80 drug-naïve or early treated FE psychosis patients. Forty-nine patients had extensive neuropsychological assessments at baseline and after twelve weeks of add-on EPA or placebo treatment. Comparison of the effect size of overall cognitive change (unweighted composite score) marginally favored EPA over placebo (however, not statistically significant). Further preliminary analysis indicated that spatial working memory 'strategy' scores improved in the EPA treatment group compared to the placebo group with a four-fold greater effect size of change, however, again without reaching statistical significance (unpublished results).

In conclusion, preclinical data support the role of bioactive lipids, especially EFA, in brain development and behavioral neurobiology, which can be influenced by dietary intake of such lipids. Future research has to address the question if EFA supplementation can influence EFA deficiencies and GPL alterations described in schizophrenia and whether normalization in EFA levels is associated with improvement in cognition and better outcome in schizophrenia and related disorders.

Conclusions

We conclude that the niacin flush test is a tool to select a subgroup of patients (intermediate phenotype) with a disturbed membrane phospholipid – arachidonic acid – eicosanoid pathway. We have replicated previous findings of niacin insensitivity in about 40-50% of first-episode psychosis patients compared to normal controls, which is associated with greater dysfunction in medial temporal lobe regions involved in recognition memory. Future research has to investigate if EFA supplementation may improve some of these dysfunctions.

Acknowledgements

Dr Berger was supported by the Swiss National Science Foundation, the M. & W. Lichtenstein Foundation (University of Basel), the National Health and Medical Council of Australia (NH&MRC ID 209062), the Stanley Medical Research Centre and the Colonial Foundation (ORYGEN Research Centre). Dr Stephen Wood was supported by NH&MRC, Welcome Trust, and the Woods Family Trust. We are grateful to Leonie Simpson for conducting some of the cognitive assessments and Deidre Smith for her help in some of the psychopathology assessments.

References

Agranoff, B.W., A.B. Joyce, and A.K. Hajra, *Lipids*, in Basic neurochemistry: molecular, cellular and medical aspects, G.J. Siegel and B.W. Agranoff, Editors. 1998, Lippincott-Raven Publishers: Philadelphia. p. 47-68.

Baxter MG, Murray EA. Opposite relationship of hippocampal and rhinal cortex damage to delayed nonmatching-to-sample deficits in monkeys. *Hippocampus* 2001;11:61-71.

Bazan, N.G. and M.J. Serou, Second messengers, long-term potentiation, gene expression and epileptogenesis. *Adv Neurol*, 1999. 79: p. 659-64.

Bazan, N.G., et al., Bioactive lipids in excitatory neurotransmission and neuronal plasticity. *Neurochemistry International*, 1997. 30(2): p. 225-31.

Berger, G.E., et al., Implications of lipid biology for the pathogenesis of schizophrenia. *Aust N Z J Psychiatry*, 2002. 36(3): p. 355-66.

Bix, G.J. and G.D. Clark, Platelet-activating factor receptor stimulation disrupts neuronal migration In vitro. *J Neurosci*, 1998. 18(1): p. 307-18.

Bouwer, C. and D. Stein, Hyperresponsivity to nicotinic acid challenge in generalized social phobia: a pilot study. *Eur Neuropsychopharmacol*, 1998. 8(4): p. 311-3.

Calabrese, G., et al., 31Phosphorus magnetic resonance spectroscopy of the temporal lobes in schizophrenia. *Biological Psychiatry*, 1992. 32(1): p. 26-32.

Chen, C., et al., Attenuated LTP in hippocampal dentate gyrus neurons of mice deficient in the PAF receptor. *J Neurophysiol*, 2001. 85(1): p. 384-90.

Connor, W.E., M. Neuringer, and D.S. Lin, Dietary effects on brain fatty acid composition: the reversibility of n-3 fatty acid deficiency and turnover of docosahexaenoic acid in the brain, erythrocytes, and plasma of rhesus monkeys. *Journal of Lipid Research,* 1990. 31(2): p. 237-47.

Crawford, M., The role of essential fatty acids in neural development: implications for perinatal nutrition. *Am J Clin Nutr*, 1993. 57(5 Suppl): p. 703S-709S; discussion 709S-710S.

Crook, J.M., et al., Decreased muscarinic receptor binding in subjects with schizophrenia: a study of the human hippocampal formation. *Biol Psychiatry*, 2000. 48(5): p. 381-8.

Das, U.N., Essential fatty acids, lipid peroxidation and apoptosis. *Prostaglandins Leukot Essent Fatty Acids*, 1999. 61(3): p. 157-63.

Dworkin, R.H., Pain insensitivity in schizophrenia: a neglected phenomenon and some implications. *Schizophrenia Bulletin*, 1994. 20(2): p. 235-48.

Feldberg, W., Possible association of schizophrenia with a disturbance in prostaglandin metabolism: a physiological hypothesis. *Psychol Med*, 1976. 6(3): p. 359-69.

Fenton, W.S., J. Hibbeln, and M. Knable, Essential fatty acids, lipid membrane abnormalities, and the diagnosis and treatment of schizophrenia. *Biological Psychiatry*, 2000. 47(1): p. 8-21.

Fenton, W.S., J. Hibbeln, and M.B. Knable, Essential fatty acids, lipid membrane abnormalities, and the diagnosis and treatment of schizophrenia. *Biological Psychiatry*, 2000. 47: p. 8-21.

Fiedler, P., A. Wolkin, and J. Rotrosen, Niacin-induced flush as a measure of prostaglandin activity in alcoholics and schizophrenics. *Biol Psychiatry*, 1986. 21(13): p. 1347-50.

Frieboes, R.M., et al., Lack of association between schizophrenia and the phospholipase-A(2) genes cPLA2 and sPLA2. *American Journal of Medical Genetics*, 2001. 105(3): p. 246-9.

Fukuzako, H., et al., Changes in levels of phosphorus metabolites in temporal lobes of drug-naive schizophrenic patients. *American Journal of Psychiatry*, 1999. 156(8): p. 1205-8.

Fuller Torrey, E., R. Rawlings, and R.H. Yolken, The antecedents of psychoses: a case-control study of selected risk factors. *Schizophr Res*, 2000. 46(1): p. 17-23.

Gattaz, W.F., et al., Increased plasma phospholipase-A2 activity in schizophrenic patients: reduction after neuroleptic therapy. *Biological Psychiatry*, 1987. 22(4): p. 421-6.

Glen, A.I., et al., A red cell membrane abnormality in a subgroup of schizophrenic patients: evidence for two diseases. *Schizophrenia Research*, 1994. 12(1): p. 53-61.

Glen, I., *Family studies in Schizophrenia*, in Phospholipid Spectrum Disorder in Psychiatry, M. Peet, I. Glen, and D. Horrobin, Editors. 1999, Marius Press: Carnforth, Lancashire, UK. p. 145-156.

Goldberg, T.E., et al., Learning and memory in monozygotic twins discordant for schizophrenia. *Psychol Med*, 1993. 23(1): p. 71-85.

Goldman-Rakic, P.S., Regional and cellular fractionation of working memory. *Proc Natl Acad Sci U S A*, 1996. 93(24): p. 13473-80.

Goldman-Rakic, P.S., The physiological approach: functional architecture of working memory and disordered cognition in schizophrenia. *Biol Psychiatry*, 1999. 46(5): p. 650-61.

Goldman-Rakic, P.S., Working memory dysfunction in schizophrenia. *J Neuropsychiatry Clin Neurosci*, 1994. 6(4): p. 348-57.

Goustard-Langelier, B., et al., n-3 and n-6 fatty acid enrichment by dietary fish oil and phospholipid sources in brain cortical areas and nonneural tissues of formula-fed piglets. *Lipids*, 1999. 34(1): p. 5-16.

Grassi, S., et al., Role of platelet-activating factor in long-term potentiation of the rat medial vestibular nuclei. *J Neurophysiol*, 1998. 79(6): p. 3266-71.

Hermesh, H., et al., Heat intolerance in patients with chronic schizophrenia maintained with antipsychotic drugs. *American Journal of Psychiatry*, 2000. 157(8): p. 1327-9.

Hoff, A.L., et al., Neuropsychological functioning of first-episode schizophreniform patients. *Am J Psychiatry*, 1992. 149(7): p. 898-903.

Hoffer, A., Adverse effects of niacin in emergent psychosis. *Journal of the American Medical Association (JAMA)*, 1969. 207(7): p. 1355.

Horrobin, D.F., A.I. Glen, and K. Vaddadi, The membrane hypothesis of schizophrenia. *Schizophr Res*, 1994. 13(3): p. 195-207.

Horrobin, D.F., et al., Fatty acid levels in the brains of schizophrenics and normal controls. *biological psychiatry*, 1991. 30(8): p. 795-805.

Horrobin, D.F., Niacin flushing, Prostaglandin E3 and evening pimrose oil. A possible objective test monitoring therapy in schizophrenia. *Journal of Orthomolecular Psychiatry*, 1980. 9: p. 33-34.

Horrobin, D.F., Schizophrenia as a prostaglandin deficiency disease. *Lancet*, 1977. 1(8018): p. 936-7.

Horrobin, D.F., The membrane phospholipid hypothesis as a biochemical basis for the neurodevelopmental concept of schizophrenia. *Schizophrenia Research*, 1998. 30(3): p. 193-208.

Hudson, C., et al., Clinical subtyping reveals significant differences in calcium-dependent phospholipase A2 activity in schizophrenia. *Biol Psychiatry*, 1999. 46(3): p. 401-5.

Hudson, C.J., et al., Genetic variant near cytosolic phospholipase A2 associated with schizophrenia. *Schizophr Res*, 1996. 21(2): p. 111-6.

Hudson, C.J., et al., The niacin challenge test: clinical manifestation of altered transmembrane signal transduction in schizophrenia? *biological psychiatry*, 1997. 41(5): p. 507-13.

Kudoh, A., H. Ishihara, and A. Matsuki, Current perception thresholds and postoperative pain in schizophrenic patients. *Reg Anesth Pain Med*, 2000. 25(5): p. 475-9.

Lewis, D.A., et al., Altered GABA neurotransmission and prefrontal cortical dysfunction in schizophrenia. *Biol Psychiatry*, 1999. 46(5): p. 616-26.

Lim, S.Y. and H. Suzuki, Effect of dietary docosahexaenoic acid and phosphatidylcholine on maze behavior and fatty acid composition of plasma and brain lipids in mice [In Process Citation]. *Int J Vitam Nutr Res*, 2000. 70(5): p. 251-9.

Lin, D.S., et al., Effects of dietary n-3 fatty acids on the phospholipid molecular species of monkey brain. *Journal of Neurochemistry*, 1990. 55(4): p. 1200-7.

Lipper, S. and D.S. Werman, Schizophrenia and intercurrent physical illness: a critical review of the literature. *Comprehensive Psychiatry*, 1977. 18(1): p. 11-22.

Manji, H.K. and R.H. Lenox, Ziskind-Somerfeld Research Award. Protein kinase C signaling in the brain: molecular transduction of mood stabilization in the treatment of manic-depressive illness. *Biological Psychiatry*, 1999. 46(10): p. 1328-51.

Marteinsdottir, I., et al., Changes in dietary fatty acids alter phospholipid fatty acid composition in selected regions of rat brain. *Progresses in Neuropsychopharmacolology and Biological Psychiatry*, 1998. 22(6): p. 1007-21.

Morrow, J.D., et al., Identification of skin as a major site of prostaglandin D2 release following oral administration of niacin in humans. *Journal of Investigational Dermatology,* 1992. 98(5): p. 812-5.

Murray EA, Bussey TJ. Consolidation and the medial temporal lobe revisited: methodological considerations. *Hippocampus* 2001;11:1-7.

Oken, R.J. and M. Schulzer, At issue: schizophrenia and rheumatoid arthritis: the negative association revisited. *Schizophrenia Bulletin*, 1999. 25(4): p. 625-38.

Padma, M. and U.N. Das, Effect of cis-unsaturated fatty acids on the activity of protein kinases and protein phosphorylation in macrophage tumor (AK-5) cells in vitro. Prostaglandins, *Leukotriens and Essential Fatty Acids*, 1999. 60(1): p. 55-63.

Pantelis, C. and P. Maruff, The cognitive neuropsychiatric approach to investigating the neurobiology of schizophrenia and other disorders. *J Psychosom Res*, 2002. 53(2): p. 655-64.

Peet, M., et al., Association of the Ban I dimorphic site at the human cytosolic phospholipase A2 gene with schizophrenia. *Psychiatric Genetics*, 1998. 8(3): p. 191-2.

Pettegrew, J.W., et al., Alterations in brain high-energy phosphate and membrane phospholipid metabolism in first-episode, drug-naive schizophrenics. A pilot study of the dorsal prefrontal cortex by in vivo phosphorus 31 nuclear magnetic resonance spectroscopy [see comments]. *Archives of General Psychiatry*, 1991. 48(6): p. 563-8.

Piomelli, D., Eicosanoids in synaptic transmission. *Crit Rev Neurobiol*, 1994. 8(1-2): p.65-83.

Price, S.A., et al., Lack of association between schizophrenia and a polymorphism close to the cytosolic phospholipase A2 gene. *Psychiatr Genet*, 1997. 7(3): p. 111-4.

Puri, B.K., et al., The niacin skin flush test in schizophrenia: a replication study. *Int J Clin Pract*, 2001. 55(6): p. 368-70.

Puri, B.K., Impaired phospholipid-related signal transduction in advanced Huntington's disease. *Exp Physiol,* 2001. 86(5): p. 683-5.

Ramchand, C.N., et al., Phospholipase A2 Gene Polymorphism and associated Biochemical Alterations in Schizophrenia, in *Phospholipid Spectrum Disorder in Psychiatry*. 1999, Marius Press: Carnforth, Lancashire, UK. p. 31-39.

Ross, B.M., et al., Differential alteration of phospholipase A2 activities in brain of patients with schizophrenia. *Brain Research*, 1999. 821(2): p. 407-13.

Rotrosen, J. and A. Wolkin, Phopholipid and Prostaglandin Hypotheses of Schizophrenia, in *Psychopharmacology. The Third Generation of Progress*., H. Meltzer, Editor. 1987, Raven Press: New York. p. 759-764.

Rybakowski, J. and R. Weterle, Niacin test in schizophrenia and affective illness. *Biological Psychiatry*, 1991. 29(8): p. 834-6.

Saku, N., J. Kobayashi, and S. Kitamura, Eicosapentaenoic acid modulates arachidonic acid metabolism in rat alveolar macrophages activated by silica. Prostaglandins, *Leukotriens, and Essential Fatty Acids*, 1999. 61(1): p. 51-4.

Seung Kim, H.F., et al., Inhibitory effects of omega-3 fatty acids on protein kinase C activity in vitro. *Molecular Psychiatry*, 2001. 6(2): p. 246-8.

Shah, S.H., et al., Unmedicated schizophrenic patients have a reduced skin flush in response to topical niacin [letter]. *Schizophrenia Research*, 2000. 43(2-3): p. 163-4.

Smesny, S., et al., [Diseases of phospholipid metabolism as possible pathogenetic factors in schizophrenia. Current findings and critical evaluation]. *Fortschr Neurol Psychiatr*, 2000. 68(7): p. 301-12.

Smesny, S., et al., Potential use of the topical niacin skin test in early psychosis-a combined approach using optical reflection spectroscopy and a descriptive rating scale. *J Psychiatr Res*, 2003. 37(3): p. 237-47.

Spector, A.A., Plasma free fatty acid and lipoproteins as sources of polyunsaturated fatty acid for the brain. *J Mol Neurosci*, 2001. 16(2-3): p. 159-165.

Stanley, J.A., et al., Membrane phospholipid metabolism and schizophrenia: an in vivo 31P-MR spectroscopy study. *Schizophrenia Research*, 1994. 13(3): p. 209-15.

Stanley, J.A., In vivo magnetic resonance spectroscopy and its application to neuropsychiatric disorders. *Can J Psychiatry*, 2002. 47(4): p. 315-26.

Tavares, H., et al., Increased phospholipase A2 activity in schizophrenia with absent response to niacin. *Schizophr Res*, 2003. 61(1): p. 1-6.

Thomas, K.L. and M. Huttunen, Neural migration, pro-inflammatory interleukins and periventricular leukomalacia: involvement in schizophrenia. *Trends Neurosci*, 1999. 22(9): p. 389-90.

Vaddadi KS, Courtney P, Gilleard CJ, Manku MS, Horrobin DF. A double-blind trial of essential fatty acid supplementation in patients with tardive dyskinesia. *Psychiatry Res* 1989; 27:313-23.

Volz, H.P., et al., 31Phosphorus magnetic resonance spectroscopy of the dorsolateral prefrontal region in schizophrenics--a study including 50 patients and 36 controls. *Biological Psychiatry*, 1998. 44(6): p. 399-404.

Wainwright, P., Essential Fatty Acids and behaviour. Is there a role for the Eicosanoids?, in Handbook of Essential Fatty Acid Biology. *Biochemistry, Physiology, and Behavioral Neurobiology*, S. Yehuda and D. Mostofky, Editors. 1997, Human Press,: Totwa, NJ. p. 299-341.

Wainwright, P.E., et al., An imbalance of dietary essential fatty acids retards behavioral development in mice. *Physiology and Behaviour*, 1999. 66(5): p. 833-9.

Wainwright, P.E., et al., Brain and behavioral effects of dietary n-3 deficiency in mice: a three generational study. *Dev Psychobiol*, 1994. 27(7): p. 467-87.

Ward, P.E., et al., Niacin skin flush in schizophrenia: a preliminary report. *Schizophrenia Research*, 1998. 29(3): p. 269-74.

Ward, P.E., Potential diagnostic aids for abnormal fatty acid metabolism in a range of neurodevelopmental disorders. *Prostaglandins Leukot Essent Fatty Acids*, 2000. 63(1-2): p. 65-8.

Wei, J., K.H. Lee, and G.P. Hemmings, Is the cPLA2 gene associated with schizophrenia? [letter]. *Molecular Psychiatry*, 1998. 3(6): p. 480-1.

Weinberger, D.R. and B. Gallhofer, Cognitive function in schizophrenia. *Int Clin Psychopharmacol*, 1997. 12 Suppl 4: p. S29-36.

Weinberger, D.R., et al., Evidence of dysfunction of a prefrontal-limbic network in schizophrenia: a magnetic resonance imaging and regional cerebral blood flow study of discordant monozygotic twins. *Am J Psychiatry*, 1992. 149(7): p. 890-7.

Whelan, J., Antagonistic effects of dietary arachidonic acid and n-3 polyunsaturated fatty acids. *The Journal of Nutrition*, 1996. 126(4 Suppl): p. 1086S-91S.

Wieraszko, A., et al., Long-term potentiation in the hippocampus induced by platelet-activating factor. *Neuron*, 1993. 10(3): p. 553-7.

Williams, J.H., et al., Arachidonic acid induces a long-term activity-dependent enhancement of synaptic transmission in the hippocampus. *Nature*, 1989. 341(6244): p. 739-42.

Wilson, D.W. and A.B. Douglass, Niacin skin flush is not diagnostic of schizophrenia. *Biological Psychiatry*, 1986. 21(10): p. 974-7.

Wood, S.J., et al., Visuospatial memory and learning in first-episode schizophreniform psychosis and established schizophrenia: a functional correlate of hippocampal pathology? *Psychol Med*, 2002. 32(3): p. 429-38.

Yacubian, J., et al., 31P-spectroscopy of frontal lobe in schizophrenia: alterations in phospholipid and high-energy phosphate metabolism. *Schizophr Res*, 2002. 58(2-3): p. 117-22.

Yao, J., et al., Correlations between peripheral polyunsaturated fatty acid content and in vivo membrane phospholipid metabolites. *Biol Psychiatry*, 2002. 52(8): p. 823-30.

Yao, J.K. and D.P. van Kammen, Red blood cell membrane dynamics in schizophrenia. I. Membrane fluidity. *Schizophrenia Research*, 1994. 11(3): p. 209-16.

Yao, J.K., et al., Correlation between RBC fatty acids and ^{31}P MRS brain measures in schizophrenia. *Society for Biological Psychiatry*, 2000. 47: p. 41.

Yao, J.K., S. Leonard, and R.D. Reddy, Membrane phospholipid abnormalities in postmortem brains from schizophrenic patients. *Schizophrenia Research*, 2000. 42(1): p. 7-17.

In: Fatty Acids and Oxidative Stress...
Editors: R. Reddy and J.K. Yao, pp. 185-192

ISBN: 978-1-60021-677-0

Chapter XII

Schizophrenia, CAD and Omega-3 Fatty Acids

Ali F. Sonel, Ravinder Reddy and Jeffrey K. Yao

Introduction

Although cardiovascular disease was a relatively rare cause of mortality earlier in this century, it is the single most common cause of death in the developed world and accounts for almost 1 million fatalities each year in the United States alone. Of these cardiovascular deaths, nearly half result directly from coronary artery disease and another 20 percent from stroke.

There has been a growing appreciation of increased mortality from CAD in psychiatric populations (Harris and Barraclough, 1998), and notably in patients with depression (Lett et al, 2004) and schizophrenia (Goff et al, 2005; Hennekens et al, 2005). While antidepressant and antipsychotic drugs have been associated with cardiac morbidity and mortality (Glassman et al, 2003; Strauss et al, 2004), there might be lifestyle factors (smoking, obesity, lack of exercise) that contribute to increased CAD risk.

Coronary Artery Disease: Risk Factors

Certain risk factors are associated with the development of coronary artery disease. In order to be considered a risk factor, a specific feature must predate the onset of coronary artery disease and have biological plausibility. Risk factors may be related to acquired behavior (such as smoking), an inherited trait (such as familial hyperlipidemia), or a laboratory measure (such as cholesterol or C-reactive protein).

Cigarette consumption remains the single most important modifiable risk factor for coronary artery disease and the leading preventable cause of death in the United States, where it accounts for more than 400,000 deaths annually (Anon, 2002) with 35-40 percent being due to ischemic heart disease. On the other hand, smoking cessation can reduce coronary heart disease mortality by 36 percent (Critchley and Capewell, 2003).

In contrast to smoking, hypertension is often a silent cardiovascular risk factor and its prevalence is steadily increasing. In the most recent National Health and Nutrition Examination Survey, 28.7 percent of subjects evaluated had either elevated blood pressure or reported use of antihypertensive medications (Hajjar and Kotchen, 2003).

Varying degrees of glucose intolerance, which increases risk of type II diabetes mellitus and premature atherothrombosis is prevalent in nearly 35 million Americans (Grundy et al, 2002). Patients with diabetes have a two- to eight-fold higher rate of future cardiovascular events as compared with age- and ethnically matched non-diabetic individuals (Howard et al, 2002). The risk of cardiovascular disease increases long before the onset of clinical diabetes. Although hyperglycemia is associated with microvascular disease, insulin resistance itself promotes atherosclerosis even before it produces frank diabetes, and available data corroborate the role of insulin resistance as an independent risk factor for atherothrombosis. This finding has prompted recommendations for increased surveillance for the metabolic syndrome, a cluster of glucose intolerance and hyperinsulinemia accompanied by hypertriglyceridemia, low HDL levels, hypofibrinolysis, hypertension, microalbuminuria, a predominance of small, dense LDL particles, and central obesity.

The relationship between cholesterol and atherosclerosis is widely accepted. Reductions in LDL cholesterol can reduce coronary events in broad populations. First shown in survivors of myocardial infarction with relatively high cholesterol levels, recent clinical trials have extended the benefits of hydroxymethylglutaryl coenzyme A (HMG-CoA) reductase inhibitors to individuals without atherosclerotic disease and with average and even at goal cholesterol levels. The HMG-CoA reductase inhibitors lowered cholesterol levels 20 to 60 percent and reduced coronary events by up to one-third over a 5-year period. The risk of hypercholesterolemia begins early in life. In the Bogalusa Heart Study, the burden of risk factors for atherosclerosis including hypercholesterolemia correlate with autopsy-proven fatty streak and raised lesion formation in the arterial tree. A recent compilation of three major observational studies underscored the importance of cholesterol levels in young adulthood to long-term cardiovascular risk (Stamler et al, 2002). Several studies have found that coronary heart disease risk correlates inversely with the high-density lipoprotein (HDL) cholesterol levels. In fact, patients with angiographically proven coronary artery disease more often have low levels of HDL than high levels of LDL, as defined by current criteria. The process of reverse cholesterol transport may explain in part the apparent protective role of HDL against coronary death. HDL can also carry antioxidant enzymes that may reduce the levels of oxidized phospholipids in atheromatous lesions that may enhance atherogenesis.

In contrast to the compelling evidence favoring a causal role for LDL in atherogenesis, the role of triglycerides still engenders controversy. First, triglyceride levels tend to vary inversely with HDL levels and demonstration of a clear effect of triglycerides in development of atherothrombosis and cardiac mortality independent of HDL levels has been difficult. Second, the level of triglycerides in the blood depends extensively on diet and may be different than the actual exposure of the artery wall to triglyceride rich lipoprotein particles such as very-low density lipoproteins. However, elevated triglyceride levels should enter into the overall risk assessment for an individual and stimulate consideration of the reason for triglyceride elevation.

Coronary Artery Disease Risk in Schizophrenia

There is excess and premature mortality in schizophrenia, from both natural and unnatural causes, when compared to the general population (Dalmau et al, 1997; Harris and Barraclough, 1998; Hennekens et al, 2005). While suicide and other violent causes of death significantly contribute to increased mortality rates in schizophrenia, deaths from other causes such as cardiovascular diseases are also more frequent (Brown, 1997). In fact, cardiopulmonary disorders accounted for 50% of deaths in a longitudinal follow up study of patients with schizophrenia (Goff et al, 2005).

The high mortality from CAD in patients with schizophrenia is likely due to excess of the same risk factors associated with CAD in the general population, such as obesity, cigarette smoking, high blood levels of cholesterol and triglycerides, and unhealthy diet (Hennekens et al, 2005). Evaluation of the ten-year risk for CAD in the Clinical Trials of Antipsychotic Treatment Effectiveness (CATIE) Schizophrenia Trial showed significantly higher risk of CAD among male and female schizophrenic patients even when adjusted for body mass index with schizophrenic patients having significantly higher rates of smoking, diabetes, hypertension and low HDL levels (Goff et al, 2005; Hennekens et al, 2005).

Brown et al. evaluated the lifestyle of patients with schizophrenia, and showed that patients ate a diet higher in fat and lower in fiber, smoked excessively and undertook little exercise (Brown et al, 1999). A more recent study comparing the dietary habits of patients with schizophrenia or schizoaffective disorders to age-, gender- and ethnicity-matched population from the National Health And Nutrition Examination Survey (NHANES) found that patients with schizophrenia had significantly greater rates of obesity and consumed significantly fewer calories, carbohydrate, protein and total fat, saturated fat, mono- and polyunsaturated fatty acids, fiber, folate, sodium and alcohol (Henderson et al, 2006). Thus it is likely that obesity in schizophrenia is not solely related to food intake, but perhaps other effects including medication side effects and reduced physical activity. Negative symptoms and lack of insight, likely contribute to an unhealthy lifestyle in schizophrenic patients. Many antipsychotic agents contribute to weight gain and atypical agents also adversely impact lipid profiles. Obesity is also associated with type II diabetes mellitus, which in turn is associated with CAD.

Atypical antipsychotic agents as a group produce greater adverse effects on glucose regulation, independent of their effects on obesity, than traditional agents. Clozapine, olanzapine and quetiapine seem more likely to cause hyperglycemia, weight gain, and hypertriglyceridemia than other atypical antipsychotic drugs. On the other hand, drugs such as ziprasidone, preferred for its low propensity to cause weight gain increase the risk of QT-prolongation with possible increases in the risk of sudden cardiac death, which could worsen CAD related outcomes (Taylor, 2003). In addition to the adverse cardiometabolic risk profile, patients with schizophrenia and coronary artery disease may be less likely to receive evidence-based care of their coronary artery disease, which may also contribute to adverse cardiac outcomes (Druss et al, 2000; 2001).

Since many of the risk factors associated with CAD are modifiable, they are legitimate targets for intervention. Since behavioral modification may be difficult to achieve in patients with schizophrenia and since commonly used antipsychotics may further worsen the cardiometabolic risk profile of these patients, attempts to modify the cardiometabolic risk of these patients by pharmacologic interventions may be of greater importance compared to similar patients without schizophrenia.

Omega-3 Fatty Acids and Coronary Artery Disease Risk

Reductions of high levels of cholesterol and triglycerides involve a number of strategies including dietary modification, lipid-lowering agents and moderate exercise. In schizophrenic patients the effectiveness of diet and exercise are likely to be modest, due largely to the compromised cognitive function and treatment complications. Lipid lowering agents, such as HMG-CoA reductase inhibitors, also known as statins, are indeed effective to reduce low-density lipoprotein (LDL) cholesterol (Chilton, 2003). Statins also have beneficial effects on plasma TG and HDL-cholesterol although they differ in the degree to which they modify the levels of these lipoproteins (Gaw, 2003). Other lipid-lowering drugs, such as niacin and clofibrate, have also been used to reduce plasma TG and LDL-cholesterol, and increase HDL-cholesterol.

However, the use of statins are associated with their own set of side effects including muscle pain, weakness, fatigue, burning and tingling in the hands and feet, memory problems, hepatitis, erectile dysfunction, and rarely significant myositis (Phillips et al, 2002). A recent study by Frolkis et al (2002) also warned that the observed reductions (26%) in LDL cholesterol by statins were significantly less than those projected (34%) by package insert guidelines. In addition, approx. 10% of patients had an increase in LDL cholesterol.

The idea that omega-3 polyunsaturated fatty acids (PUFAs) reduce the risk of coronary artery disease was based on the low rates of heart disease among the Japanese and Greenland Eskimos (Bang et al, 1980; Kagawa et al, 1982). There is now strong evidence that consumption of fish, especially those species with high content of PUFAs, confers protection from ischemic heart disease (Kris-Etherton et al, 2002; Marckmann and Groenbeck, 1999) and that this relationship is particularly strong for coronary heart disease mortality and sudden cardiac death, which has been reported to be on average 52 percent lower in men consuming fish at least once weekly versus men consuming less (Albert et al, 1998).

Although fish have a number of important nutritive qualities, it is likely that their beneficial effects in terms of cardiac outcomes are due to omega-3 fatty acids, which are also found in high concentration in flax seeds, are present in high concentration in fish and flax seeds. The most widely used omega-3 fatty acids for the treatment of hyperlipidemias are the two active molecules in fish oil, eicosapentaenoic acid (EPA) and docosahexaenoic acid (DHA). Increased plasma levels of these fatty acids predicted a considerable reduction in sudden cardiac death (Albert et al, 2002), a results consistent with another report demonstrating that a combined intake of 5.5 g/month of EPA plus DHA (equivalent to one portion of fatty fish per week) was associated with a 50% lower incidence of primary cardiac arrest compared with individuals not consuming fish (Siscovick et al, 2000). This effect appears to be related to enrichment of membrane phospholipids with omega-3 fatty acids and a resulting risk reduction for abnormal electrical conductivity (Siscovick et al, 2000). The ability of omega-3 fatty acids to shift LDL particles from small dense to large buoyant forms (Bell et al, 1996; Suzukawa et al, 1995) provides further evidence of beneficial effect of omega-3 fatty acids on improved cardiac outcomes. Other potential mechanisms by which fish oils may reduce risks of coronary heart disease include effects on thrombosis mediated by thromboxane inhibition or prostacyclin synthesis, fibrinolytic activity, blood viscosity, or even blood pressure as well as anti-inflammatory effects (Kris-Etherton et al, 2002).

Table 1. Selected Trials of Dietary Intervention and Coronary Events Using Fish Oil

Trial	Patients	Dietary Intervention	Duration	Change in Cholesterol, %	Change in CHD*, %
DART	1015 male MI patients	Fish twice per week or fish oil (1.5 g/d)	2 years	Not reported	-16 CHD events -29† total mortality
GISSI-Prevenzione	5666 MI patients, mostly male	Fish oil (EPA+DHA, 1 g/d)	3.5 years	0	-30† CV mortality -45† sudden death
Indian Experiment on Infarct Survival 4	242 MI patients, mostly male	Fish oil (EPA, 1.08 g/d) or mustard oil (ALA, 2.9 g/d)	1 year	0	-30 in fish oil group -19 in mustard oil group

ALA= α-linolenic acid; CHD=coronary heart disease; DART=Diet and Reinfarction Trial; DHA=docosahexaenoic acid; EPA=eicosapentaenoic acid; GISSI=Gruppo Italiano per lo Studio della Sopravvivenza nell'Infarto Miocardico; MI=myocardial infarction

* Change in CHD refers to the percentage difference in coronary event rates in the treatment group compared with the control group.

† $P < 0.05$

The American Heart Association currently recommends consumption of two portions of fish per week (Kraus et al, 2000), particularly those fish rich in omega-3 fatty acids. Clinical trials have supported evidence from epidemiological studies (Marckmann and Groenbeck, 1999) that higher intake of fish may particularly benefit patients with coronary heart disease (Table 1) (Kris-Etherton et al, 2002). In one secondary prevention trial (GISSI-Prevenzione trial, 1999), a supplement containing 0.85 g/d of EPA and DHA resulted in reductions of 45 percent and 30 percent in sudden and cardiovascular death. Total mortality and sudden death were reduced after 3 and 4 months of treatment, respectively, consistent with reduction of arrhythmia by the treatment. Based on the overall evidence to date, the American Heart Association has recommended that supplemental EPA plus DHA at doses of up to 1 g/d may be considered for risk reduction in patients with coronary heart disease, in consultation with their physician (Kris-Etherton et al, 2002).

Fish oil supplements can be used in combination with fibrates, niacin, or statins to treat hypertriglyceridemia. Omega-3 fatty acids have been concentrated into capsules and in doses of 3 to 6 g/d decrease fasting and postprandial triglycerides, by decreasing chylomicron and very low density lipoprotein production. At least 6 g/d is usually required for substantial triglyceride-lowering effect, and many patients require 9-12 g/d. However, the effect of fish oils on total, LDL, and HDL cholesterol is less clear. In general, fish oils are well tolerated and appear to be safe, at least at doses up to 3 g. The large number of capsules required for a therapeutic effect, the associated dyspepsia, limited supply of high-quality omega-3 fatty acid supplements, free of contaminants and fishy aftertaste has limited the clinical use of these agents. Although fish oil administration is associated with a prolongation in the bleeding time, no increase in bleeding has been seen in clinical trials.

Recently, further improvements across the atherogenic components of plasma lipid profile have been achieved by the co-administration of statins with omega-3 fatty acids, niacin or fibrate (Gaw, 2003). Particularly, both statin therapy and diets high in omega-3 fatty acids appear to improve morbidity and mortality by modifying the underlying atherothrombotic disease process (Ballantyne, 1999; Lee and Lip, 2003). Thus, omega-3 fatty acids are just as effective as, or have a benefit superior to, statins in secondary prevention (Bhatnagar and Durrington, 2003). However, combination therapies have been associated with side effects and may be poorly tolerated. For example, combining a statin with either niacin or a fibrate may increase the risk for myopathy (Xydakis and Ballantyne, 2002). Thus, further investigations are needed to evaluate the optimal treatment in reducing the CAD risk.

Therapeutic Implications

Studies suggest that treatment with omega-3 fatty acids improve the psychiatric status of schizophrenic patients (reviewed in chapter 8). Further, omega-3 fatty acid treatment has proven to be beneficial to patients who have higher CAD risk. Since schizophrenic patients are more likely to be at increased risk for developing CAD, due to lifestyle and treatment-related factors, the possibility of utilizing omega-3 fatty acid supplementation to simultaneously address CAD risk and psychiatric symptom severity in patients with schizophrenia is an interesting possibility.

There are, however, ethical issues to be considered. For example, should patients with hypercholesterolemia not immediately be started on proven treatments such as statins? Therefore, it is imperative to develop a body of evidence establishing the efficacy of omega-3 fatty acids in reducing CAD in patients with schizophrenia. Head-to-head comparisons between statins and omega-3 fatty acids will need to be conducted. In the meantime, a reasonable approach might be to utilize combination treatments (statins + omega-3 fatty acids), which seems to be gaining popularity amongst practitioners and patients.

References

Albert CM, Campos H, Stampfer MJ et al: Blood levels of long-chain n-3 fatty acids and the risk of sudden death. *N Engl J Med* 346:1113, 2002.

Albert CM, Hennekens CH, O'Donnell CJ et al: Fish consumption and risk of sudden cardiac death. *JAMA* 279:23, 1998.

Annual smoking-attributable mortality, years of potential life lost, and economic costs-United States, 1995-1999. *MMWR Morb Mortal Wkly Rep* 51:300, 2002.

Ballantyne CM. Reducing atherothrombotic events in high-risk patients: recent data on therapy with statins and fatty acids. *Curr Atheroscler Rep* 1:6-8, 1999.

Bang HO, Dyerberg J, Sinclai HM: The composition of the Eskimo food in northwestern Greenland. *Am J Clin Nutr* 33:2657-2661, 1980.

Bell JD, Barnard ML, Parkes HG, Thomas EL, Brennan CH, Cunnane SC, Dagnelie PC. Effects of n-3 fatty acids on the NMR profile of plasma lipoproteins. *J Lipid Res* **37**:1664-1674, 1996.

Bhatnagar D, Durrington PN. Omega-3 fatty acids: their role in the prevention and treatment of atherosclerosis related risk factors and complications. *Int J Clin Pract* 57:305-314, 2003.

Brown S., Birtwistle J, Roe L, Thompson C. The unhealthy lifestyle of people with schizophrenia. *Psychological Medicine* 29:697-701, 1999.

Brown, S. Excess mortality of schizophrenia. *Bristish J Psychiatry* 171:502-508, 1997.

Chilton RJ. Lipid and nonlipid benefits of statins. *J Am Oseteopath Assoc* 103:12-17, 2003.

Critchley JA, Capewell S: Mortality risk reduction associated with smoking cessation in patients with coronary heart disease: A systematic review. *JAMA* 2003. 290:86.

Dalmau A, Bergman B, Brismar B. Somatic morbidity in schizophrenia-a case control study. *Public Health* 111:393-397, 1997.

Druss BG, Bradford DW, Rosenheck RA, Radford MJ, Krumholz HM. Mental disorders and use of cardiovascular procedures after myocardial infarction. *JAMA*. 2000 Jan 26;283(4):506-11.

Druss BG, Bradford WD, Rosenheck RA, Radford MJ, Krumholz HM. Quality of medical care and excess mortality in older patients with mental disorders. *Arch Gen Psychiatry*. 2001 Jun;58(6):565-72.

Frolkis JP, Pearce GL, Nambi V, Minor S, Sprecher DL. Statins do not meet expectations for lowering low-density lipoprotein cholesterol levels when used in clinical practice. *Am J Med* 113:625-629, 2002.

Gaw A. HDL-C and triglyceride levels: relationship to coronary heart disease and treatment with statins. *Cardiovasc Drugs Ther* 17:53-62, 2003.

Glassman A, Shapiro PA, Ford DE, et al. Cardiovascular health and depression. *Journal of psychiatric practice* 2003; 9:409-21.

Goff DC, Sullivan LM, McEvoy JP, et al. A comparison of ten-year cardiac risk estimates in schizophrenia patients from the CATIE study and matched controls. *Schizophr Res*. 2005 Dec 1;80(1):45-53. Epub 2005 Sep 28.

Grundy SM, Howard B, Smith S Jr, et al: Prevention Conference VI: Diabetes and Cardiovascular Disease: Executive summary: Conference proceeding for the healthcare professionals from a special writing group of the American Heart Association. *Circulation* 2002. 105:2231.

Gruppo Italiano per lo Studio della Sopravvivenza nell'Infarto Miocardico: Dietary supplementation with n-3 polyunsaturated fatty acids and vitamin E after myocardial infarction: results of the GISSI-Prevenzione trial. *Lancet* 354:447, 1999.

Hajjar I, Kotchen TA: Trends in prevelance, awareness, treatment and control of hypertension in the United States, 1988-2000. *JAMA* 2003. 289;2560.

Harris EC, Barraclough B. Excess mortality of mental disorder. *Br J Psychiatry* 173: 11-53, 1998.

Henderson DC, Borba CP, Daley TB, et al. Dietary intake profile of patients with schizophrenia. Ann Clin Psychiatry. 2006 Apr-Jun;18(2):99-105

Hennekens CH, Hennekens AR, Hollar D, Casey DE. Schizophrenia and increased risks of cardiovascular disease. *American heart journal* 2005; 150:1115-21.

Howard BV, Rodriguez BL, Bennett PH, et al: Prevention Conference VI: Diabetes and Cardiovascular Disease: Writing Group I: Epidemiology. *Circulation* 2002. 105:e132.

Kagawa Y, Nishizawa M, Suzuki M, et al: Eicosapolyenoic acids of serum lipids of Japanese islanders with low incidence of cardiovascular disease. *J Nurt Sci Vitaminol* 28:441-453,1982

Krauss RM, Eckel RH, Howard B, et al: AHA Dietary Guidelines-revision 2000: A statement for healthcare professionals from the Nutrition Committee of the American Heart Association. *Circulation* 102:2284, 2000.

Kris-Etherton PM, Harris WS, Appel LJ: Fish consumption, fish oil, omega-3 fatty acids and cardiovascular disease. *Circulation* 106:2747, 2002.

Lee KW, Lip GYH. The role of omega-3 fatty acids in the secondary prevention of cardiovascular disease. *QJ Med.* 96:465-480, 2003.

Lett HS, Blumenthal JA, Babyak MA, et al. Depression as a risk factor for coronary artery disease: evidence, mechanisms, and treatment. *Psychosomatic medicine* 2004; 66:305-15

Marchioli R, Barzi F, Bomba E, et al: Early protection against sudden death by n-3 polyunsaturated fatty acids after myocardial infarction: Time course analysis of the results of the Gruppo Italiano per lo Studio della Sopravvivenza nell'Infarto Miocardico (GISSI)-Prevenzione. *Circulation* 105:1897. 2002.

Marckmann P, Gronbaek M: Fish consumption and coronary heart disease mortality: A systematic review of prospective cohort studies. *Eur J Clin Nutr* 53:585, 1999

Phillips PS, Haas RH, Bannykh S, et al. Statin-associated myopathy with normal creatine kinase levels. *Ann Intern Med* 137:581-585, 2002.

Siscovick DS, Raghunathan T, King I et al: Dietary intake of long-chain n-3 polyunsaturated fatty acids and the risk of primary cardiac arrest. *Am J Clin Nutr* 71:208S, 2000.

Stamler J, Daviglus ML, Garside DB, et al: Relationship of baseline serum cholesterol levels in 3 large cohorts of younger men to long-term coronary, cardiovascular, and all-cause mortality and to longevity. *JAMA* 2000. 284:311.

Suzukawa M, Abbey M, Howe PRC, Nestel PJ. Effects of fish oil fatty acids on low density lipoprotein size, oxidizability, and uptake by macrophages. *J Lipid Res* 36:473-484, 1995.

Taylor D. Ziprasidone in the management of schizophrenia : the QT interval issue in context. *CNS Drugs*. 2003;17(6):423-30.

Xydakis AM, Ballantyne CM. Combination therapy for combined dyslipidemia. *Am J Cardiol* 90:21-29, 2002.

In: Fatty Acids and Oxidative Stress...
Editors: R. Reddy and J.K. Yao, pp. 193-200 ISBN: 978-1-60021-677-0

Chapter XIII

Can Augmentation Enhance Antipsychotic Treatment Response in Schizophrenia? Omega-3 Fatty Acid and Antioxidant Supplementation

Sahebarao Mahadik, Denise Evans, Anilkumar Pillai, Nitin Shendarkar, Meera Narasimhan and Peter Buckley

Introduction

There is ample evidence that oxidative cell injury is an important contributor to the pathophysiology of several brain disorders, chief among these are schizophrenia, Parkinson's disease, Alzheimer's disease, and Down's syndrome (Fiskum, 1996; Mahadik and Mukherjee, 1996; Reddy and Yao, 1996; Simonian and Coyle, 1996), and reviewed in chapter X. Indeed, oxidative injury may account for a progressively deteriorating course and poor prognosis in such illness, a perspective that holds particular intrigue for schizophrenia and begs several fundamental questions: At what stage of the primary pathogenesis is such an effect is exerted? Does peroxidative cell injury contributes to adverse neurodevelopmental events and could that trigger a pathogenic cascade of psychiatric illness such as schizophrenia?

Several biological studies have shown that antipsychotics - both first generation agents (FGAs) and second generation agents (SGAs) have either pro or anti-oxidant properties (Fisher, 1991; Jeding et al., 1995). Moreover, FGAs have been found to increase the oxidative stress-mediated neuronal damage in animals (Mahadik et al., 1988; 2000a; Cadet and Perumal, 1990; Jeste et al., 1992) and it has been hypothesized previously that tardive dyskinesia may be as a result of neuroleptic-induced oxidative injury (Cadet and Lohr, 1987).

On the other hand, there is evidence from preclinical and clinical studies in nonpsychiatric populations (if identified in its early stages) by adjunctive use of dietary antioxidant supplements, e.g., vitamins A, E and C, and β-carotene (Burton and Ingold, 1989; Rice-Evans and Diplock, 1993). Furthermore, some of the cellular oxidative injury, particularly peroxidative loss of membrane essential fatty acids (e.g., ω6 and ω3 EFAs) can be restored quickly by their supplementation (Carlson et al., 1986; Dougherty et al., 1987; Simopoulos, 1991). These findings are also supported by several studies which reported beneficial effects of both antioxidant and EFA supplementation in improving clinical outcome schizophrenia (Mahadik and Gowda, 1996; Mahadik and Scheffer, 1996; Fenton et al, 2000). In view of the debilitating nature of schizophrenia, the accruing concern regarding the adverse effect profile of SGAs which are now the primary treatment for schizophrenia, and allied with provocative initial evidence that fatty acid-antioxidant therapy may reverse weight gain and metabolic disturbances, serious consideration of these agents as an augmentation strategy is both warranted and heuristic.

Prevention and Reversal of Oxidative Neuronal Injury

Supplementation of Antioxidants

The studies in animals and humans have indicated that the dietary intake of antioxidants is effective in the prevention of oxidative cell injury, and that the intake of EPUFAs may correct such injury throughout the body, but especially in the brain. Studies in schizophrenia patients have indicated that antioxidant defense is impaired, and that the content of membrane phospholipid EPUFAs is reduced and lipid peroxidation is increased (Fenton et al, 2000); dietary supplementation with antioxidants has been considered as a mechanistically based strategy to prevent these changes and to improve symptoms (Mahadik and Mukherjee, 1996a; Reddy and Yao, 1996; Mahadik and Gowda, 1996; Mahadik and Evans, 1997).

The choice of the type of antioxidant(s), and its amount and length of treatment, may depend on the type and levels of oxyradicals, the nature (e.g., lipids or proteins or DNA oxidation) and extent of cell injury, and the contribution of pharmacotherapy for the primary disease (Rice-Evans and Diplock, 1993). In schizophrenia, many of these factors have not been investigated. However, since there is increased lipid peroxidation, the use of vitamin E, a hydrophobic antioxidant, should theoretically be preferable. In addition, since potency of vitamin E is maintained by vitamin C, which can also prevent the intracellular peroxidation, parallel use of vitamin C could be of value.

The most common dietary antioxidants are the vitamins A, E and C, and β-carotene, ubiquinones and flavonoids; trace elements such as Zn, Cu and Se, are part of the antioxidant enzymes. Recently it has been suggested that γ-tocopherol, which is naturally present in various foods, is more potent and an effective antioxidant than α-tocopherol (Wolf, 1997). β-carotene seems to have minimum antioxidant benefits. The plasma levels of both vitamin E (Liday et al., 1995; McCreadie et al., 1995) and C (Suboticanec et al., 1990) are found to be lower in schizophrenic patients. Based on these data and theoretical grounds, a combination of a hydrophobic agent such as vitamin E, to protect membranes, and a hydrophilic agent such as vitamin C, in intracellular protection provide complete antioxidant defense.

Several studies have found that the therapeutic doses of antioxidants are generally way below their toxic doses (Meyers et al., 1996). A total of 14 studies have been done using primarily vitamin E (800 to 1600 IU daily) supplement in patients with chronic schizophrenia who have developed tardive dyskinesia (TD) (Mahadik and Gowda, 1996; Reddy and Yao, 1996; Adler et al., 1993; 1999). All but two studies (Shriqui et al., 1992; Adler et al., 1999) have found vitamin E beneficial in controlling some symptoms of TD and symptoms. These studies clearly indicate that vitamin E (a daily dose of 800 to1600 IU) is efficacious, if given to patients under 45 years of age and within 5 years of the expression of TD. This observation has been interpreted as indicating that the use of antioxidant supplements may prevent the deteriorating course at early stages of illness (Peet et al., 1993; Mahadik and Gowda, 1996; Mahadik and Scheffer, 1996b). A study of antioxidant therapy in first episode schizophrenia has yet to be undertaken to confirm or refute this speculation.

Indeed, several issues need to be addressed in evaluating the effect of antioxidant treatment in schizophrenia. Adjunctive antioxidant treatment should be comprised of vitamin E along with 1000 mg vitamin C daily in order to provide antioxidant defense for cell membrane and cytosol. Based on animal studies (Sohal et al., 1994), caloric restriction: daily intake of about 1600 calories for those living a sedentary life has been considered to decrease oxidative stress. A dietary survey at our center revealed that most patients have a daily caloric intake around 2400-2800 per day. In our experience, the most difficult task is to get patients to reduce their caloric intake to a reasonable level. Smoking cessation likely is also the most difficult step of all. The evidence is increasing to suggest that the treatment strategy of psychotic illness should include limiting the oxidative stress, and this should be implemented as early as possible during the course of their illness. This approach is also consistent with more general attention to the physical health of patients with schizophrenia (Marder et al, 2004).

We do not wish to detract from the importance of developing safer and more efficient antipsychotic. Some recent *in vitro* and *in vivo* studies in animals (supra) indicate that treatment with some SGAs may be neuroprotective against oxidative cell injury by inducing antioxidant defense (Li et al., 1999; Parikh et al., 2003; Terry et al., 2003). However, FGAs alone might not be sufficient to contain the deteriorating course of severe mental disorders. If indeed, as the emerging data are indicating, there is ongoing insult from oxidative stress and impaired essential fatty acid metabolism these must be addressed.

Supplementation of EPUFAs

The previous section was targeted at prevention of oxidative damage. In this section, we discuss the promotion of recovery of previously damaged structure. Boosting lower levels of EPUFAs is one approach. EPUFAs are made by plants and must be consumed in diet. EPUFAs are two types, based on the position of double bonds, ω6 and ω3. Predominant neuronal membrane ω6-EPUFA is AA (20:4n-6) and ω3-EPUFA is DHA (22:6n-3). These make almost 25% of the brain weight and their turnover is very rapid. High levels of AA are present in both vegetables and terrestrial animal tissues and DHA is present primarily in fish and vegetables (Simopoulos, 1991). The levels of DHA are steadily declining and levels of AA are increasing in American diet (Mahadik et al., 1999). AA and DHA are also preferentially destroyed by oxyradicals (Mahadik et al., 1999a). Furthermore, these are found lower in schizophrenic patients with schizophrenia (see Section 3.2.1; Fenton et al., 2000). AA can be incorporated directly from the diet or that is made from linoleic acid, its most abundant precursor. Brain DHA either comes preferentially from dietary DHA or made from its immediate precursor, EPA (Anderson et al., 1990; Connor et al., 1993). Therefore, supplementation of ω3-EPUFAs, particularly EPA and DHA has been found to be efficacious in studies discussed below.

Clinical trials of ω6 and ω3 fatty acid supplementation in schizophrenia have been summarized (Mahadik and Evans, 1997; Fenton et al., 2000). All of the supplement studies to date have been conducted as adjunctive treatment with FGAs of the doctor's choice. The DGLA supplement in placebo-controlled trials has been found prophylactic in reducing TD, and improved cognition and symptoms (Vaddadi et al., 1989; Vaddadi, 1996). However more recently, EPA or EPA plus DHA supplements in double blind placebo-controlled trials have been found more effective than ω6-EPUFAs in reducing symptoms - both positive and negative symptoms in patients on FGAs (Mellor et al., 1995; Peet et al., 1996; Peet and Mellor, 1998; Shah et al., 1998). These supplements were found to improve the levels of RBC membrane EPUFAs and psychopathology, particularly negative symptoms (Peet et al., 1996; Peet and Mellor, 1998). This was important since earlier studies by Glen et al (1994) had shown lower levels of DHA in patients with predominant negative symptoms.

Clinical trials are also going on in drug-free patients with recent onset of psychosis (Ho et al., 2000). Two gm compared to 1 or 4 gm of EPA supplement for 12 weeks was better to improve the PNASS in all the patients, and 40% patients (6/14) did not require medication at the end of the treatment. It is also important to point out that, in all the supplementation studies, the levels of RBC membrane EPUFAs (e.g., AA, EPA and DHA) were found to inversely correlate with symptom severity increased EPUFAs with reductions in symptoms. Supplementation of ω3 fatty acids has also been done in double blind, placebo-controlled study in bipolar disorder (Stoll et al., 1999). They reported that the ω3fatty acid patient group had a significantly longer period of remission than the placebo group ($P = 0.002$).

Among the metabolites of the EPUFAs, lower levels of PGE1 and higher levels of PGE2 have been found in patients with schizophrenia (Horrobin et al., 1978; Kaiya et al., 1989). Based on these observations, intravenously administered PGE1 (Kaiya et al., 1984) or boosting its synthesis by Penicillin and EFAs (Vaddadi, 1979) were found to improve symptoms in some patients. However, since the primary problem now seems to be altered levels of membrane EPUFAs and supplementation of these seems to improve the membrane levels as well as symptoms, most of the attention has been on the supplementation of EPUFAs (Fenton et al., 2000).

An epidemiological report has also found an association ($r = 0.97$) of higher intake of fat from fish and vegetables (highly enriched in ω3 EPUFAs and antioxidants) in patients with a better outcome of schizophrenia (Christinsen and Christinsen, 1988).

Preferred Supplementation of a Combination of Antioxidants and Omega-3 EFAs

It is also important to bear in mind that most of augmentation approaches in the treatment of schizophrenia are fraught with either noteworthy side effects of their own and/or have singlestick effects upon the adverse effect profile of SGAs (Buckley and Shendarkar, 2005). Thus, supplementation with antioxidants and omega-3 EFA represent a novel and relatively safe say to boost treatment response. As discussed earlier, it is proposed that ongoing oxidative cell injury be stopped and structural damage reversed in order to preserve and restore function. It is proposed that antioxidants alone may stop ongoing oxidative injury and that EPUFAs restore the cellular structure. Thus the combined use of the antioxidants and EPUFAs may be necessary for optimal treatment of oxidative cell injury. Unfortunately, studies reported so far involve the use of either an antioxidant (Adler et al., 1993; Mahadik and Mukherjee, 1996a; Reddy and Yao, 1996) or EPUFAs with only trace amounts of antioxidants (Fenton et al., 2000; Mahadik and Evans, 1997). Additionally, studies to date show wide variability across patient age and duration of illness, variables that can influence the state of membrane lipid metabolism and antioxidant defense. There is also a pressing need to examine such strategies in patients receiving SGAs - because these agents may be a better substrate for dietary supplementation. Moreover, the proposition that dietary supplementation may counteract weight gain or metabolic disturbances from SGAs is extremely appealing, although, as yet untested. What is required then is a placebo controlled study of dietary supplementation of over 4 months duration in a more homogeneous schizophrenia population being treated with SGAs. Other potential research strategies are outlined in Table 1.

Table 1. Potential Dietary Supplementary Studies in Schizophrenia

- Placebo-controlled, add-on to SGAs
- Placebo-controlled add on to patients with antipsychotic-induced weight gain
- Placebo-controlled add-on to SGAs in first episode schizophrenia patients
- Open-labeled pilot study in high risk/prodromal patient population
- Double-blind comparison of dietary supplementation with another augmentation strategy (eg glycine) in treatment-refractory schizophrenia patients.

Conclusions

While the pathobiology of schizophrenia remains obscure, converging evidence points to generalize the cell membrane pathology - most likely due to reduced levels of phospholipid bound EPUFAs, primarily AA and DHA. There is increasing evidence to indicate that these membrane EPUFA changes may be partly related to their loss by oxyradical-mediated peroxidation and partly by reduced intake as well as defective metabolism. Since AA and DHA are critical in brain development and function (primarily due to their role in membrane signal transduction) these EPUFA alterations may underlie the broad range of neurodevelopmental structural and behavioral abnormalities observed in schizophrenia. The loss of membrane EPUFAs by peroxidation can be prevented by supplementation of antioxidants such as vitamins E and C, and the levels of membrane EPUFAs can be then restored by their supplementation. Initial supplementation studies with vitamin E, and EPA or EPA plus DHA are encouraging and should provide impetus for more complex and definitive studies of dietary supplementation of SGAs. Given the low risk of adverse effects of dietary supplementation in contrast with other augmentation strategies, this is an appealing approach especially in an era of pharmacotherapy of schizophrenia that is characterized by polypharmacy.

References

Adler LA, Peselow E, Rotrosen J, et al. Vitamin E treatment of tardive dyskinesia. *Am J Psychiatry.* 1993;150:1405-1407.

Adler LA, Rotrosen J, Edson R, et al. Vitamin E treatment for tardive dyskinesia. *Arch J Psychiatry.* 1999;56:836-841.

Anderson GJ, Connor WE, Corliss JD. Docosahexaenoic acid is the preferred dietary n-3 fatty acid for the development of the brain and retina. *Pediatric Res.* 1990;27:89-97.

Buckley PF, Shendarkar N. Treatment-refractory schizophrenia. *Current Opinion in Psychiatry.* 2005;(in press).

Burton GW, Ingold KU. Vitamin E as an in vitro and in vivo antioxidant. *Ann N Y Acad Sci.* 1989;570:7-22.

Cadet JL, Lohr JB. Free radicals and the developmental pathophysiology of schizophrenic burnout. *Integr. Psychiatry.* 1987;5:40-48.

Cadet JL, Perumal AS. Chronic treatment with prolixin causes oxidative stress in rat brain. *Biol Psychiatry.* 1990;28:738-740.

Carlson SE, Carver JD, House SG. High fat diets varying in ratio of polyunsaturated to saturated fatty acid and linoleic to linolenic acid: A comparison of rat neural and red cell membrane phospholipids. *J. Nutrition.* 1986;116:718-725.

Christinsen O, Christinsen E. Fat consumption and schizophrenia. *Acta Psychiat. Scand.* 1988;78:587-591.

Connor WE, Lin DS, Neuringer M. Is the docosahexaenoic acid (DHA, 22:6n-3) content of erythrocytes a marker for the DHA content of brain phospholipids? *FASEB J.* 1993;7:A152.

Dougherty A, Kaplan PD, Gollub JP. Development of side branching in dendritic crystal growth. *Phys Rev Lett.* 1987;58:1652-1655.

Fenton WS, Hibbeln J, Knable M. Essential fatty acids, lipid membrane abnormalities, and the diagnosis of schizophrenia. *Biol Psychiatry.* 2000;47: 8-12.

Fisher AB. Possible role of free radical formation in clozapine (clozaril)-induced agranulocytosis. *Mol Pharmacol.* 1991;40:846-853.

Fiskum G. *Neurodegenerative Diseases: Molecular and cellular mechanisms and therapeutic advances.* George Washington University Medical Center, Department of Biochemistry and Molecular Biology, New York: Plenum Press; Annual Spring Symposia 1996.

Glen AIM, Glen EMT, Horrobin DF, Vaddadi KS, Spellman M, Morse-Fisher N, Ellis K, Skinner FK. A red cell membrane abnormality in a subgroup of schizophrenic patients: evidence for two diseases. *Schizophr Res.* 1994;12:53-61.

Ho L, Pasinetti GM, Horrocks LA, Farooqui AA, Horrobin DF, Peet M, Manev H, Uz T. Eicosanoid pathways and psychiatric disorders. *Biol. Psychiatry.* 2000;47:114S.

Horrobin DF, Ally AI, Karmali RA, Karmazyn M, Manku MS, Morgan RO. Prostaglandins and schizophrenia: further discussion of the evidence. *Psychol Med.* 1978;8:43-48.

Jeding I, Evans PJ, Akanmu D, Dexter D, Spencer JD, Aruoma OI, Jenner P, Halliwell B. Characterization of the potential antioxidant and pro-oxidant actions of some neuroleptic drugs. *Biochem. Pharmacol.* 1995;49:359-365.

Jeste DV, Lohr JB and Manley M. Study of neuropathologic changes in the striatum following 4, 8 and 12 months of treatment with fluphenazine in rats. *Psychopharmacology (Berl).* 1992;106:154-160.

Kaiya H, Nishida A, Imai A et al. Accumulation of diacylglycerol in platelet phosphoinositides turnover in schizophrenia: A biological marker of good prognosis? *Biol Psychiatry.* 1989;26:669-676.

Kaiya H, Takeuchi K, Namba M, Imai A, Nakashima S, Nozawa Y. Abnormal phosphatidylinositol-cycle of platelet membrane in schizophrenia--a preliminary study. *Folia Psychiatr Neurol Jpn.* 1984;38:437-444.

Li X-M, Chlan-Fournet J, Juorio AV, Bennett VL, Shrikhande S, Keegan DL, Qi J, and Boulton AA. Differential effects of olanzapine on the gene expression of superoxide dismutase and the low affinity nerve growth factor. *J Neurosci Res.* 1999;56:72-75.

Liday LA, Pippinger CE, Howard AA, and Lieberman JA. Free radical scavenging enzyme activity and related trace metals in clozapine-induced agranulocytosis. A pilot study. *J Clin Psychopharmacol.* 1995;15:353-360.

Mahadik SP, Evans D. Essential fatty acids in the treatment of schizophrenia. *Drugs of Today.* 1997;33:5-17.

Mahadik SP, Gowda S. Antioxidants in the treatment of schizophrenia. *Drugs of Today.* 1996;32:1-13.

Mahadik SP, Laev H, Korenovsky A, Karpiak SE. Haloperidol alters rat CNS cholinergic system: enzymatic and morphological analyses. *Biol Psychiatry.* 1988;24:199-217.

Mahadik SP, Mukherjee S. Free radical pathology and the antioxidant defense in schizophrenia. *Schizophr. Res.* 1996;19:1-18.

Mahadik SP, Scheffer RE. Oxidative injury and potential use of antioxidants in schizophrenia. *Prostaglandins, Leukotrienes and Essential Fatty Acids.* 1996;55:45-54.

Mahadik SP, Sitasawad V, Mulchandani M. Cultural and socioeconomic differences in dietary intake of essential fatty acids and antioxidants-effects on the outcome. In: Horrobin D, Glen AL and Peet M eds. *Phospholipid Spectrum Disorders in Psychiatric.* Marius Press; 1999:167-179.

Mahadik SP, Sitasawad V, Mulchandani M. Membrane peroxidation and the neuropathology of Schizophrenia. In: Horrobin D, Glen AL and Peet M eds. *Phospholipid Spectrum Disorders in Psychiatric.* Marius Press; 1999:99-111.

Marder SR, Essock SM, Miller AL, Buchanan RW, Casey DE, et al. Physical Health Monitoring of Patients with Schizophrenia. *Am J Psychiatry.* 2004;161:1334-1349.

McCreadie RG, MacDonald E, Wiles D, Campbell G, Pateson JR. The Nithsdale Schizophrenia Surveys XIV: Plasma lipid peroxide and serum vitamin E levels in patients with and without tardive dyskinesia and in normal subjects. *Brit. J. Psych.* 1995;170:1-8.

Mellor JE, Laugharne JD, Peet M. Schizophrenic symptoms and dietary intake of n-3 fatty acids. *Schizophr Res.* 1995;18(1):85-86.

Meyers AI, Andres CJ, Resek JE, McLaughlin MA, Woodall CC, Lee PH. A Rapid and Efficient Approach to Chiral, Nonracemic Aza Sugars from Nonsugars. A Formal Synthesis of 1,4-Dideoxy-1,4-imino-D-lyxitol. *J Org Chem.* 1996;61:2586-2587.

Parikh V, Khan MM, Buckley PF, Mahadik SP. Differential effects of antipsychotics on expression of antioxidant enzymes and membrane lipid peroxidation in rat brain. *J Psychiatr Res.* 2003;37:47-51.

Peet M, Laugharne J, Mellor JE, Ramchand CN. Essential fatty acid deficiency in erythrocyte membranes from chronic schizophrenic patients, and the clinical effects of dietary supplementation. *Prostaglandins, Leukotrienes and Essential Fatty Acids.* 1996;55: 119-122.

Peet M, Laugharne J, Rangarajan N, et al. Tardive dyskinesia, lipid peroxidation, and sustained amelioration with vitamin E treatment. *Internatl Clin Psychopharmacology.* 1993;8:151-153.

Peet MJ, Mellor JE. Double blind placebo controlled trial of n-3 polyunsaturated fatty acids as an adjunct to neuroleptics. *Ninth Schizophrenia Winter Workshop.* 1998:p.54.

Reddy R, Yao J. Free radical pathology in schizophrenia: a review. *Prostaglandin Leukotri Essen Fatty Acids.* 1996;55:33-43.

Rice-Evans CA, Diplock AT. Current status of antioxidant therapy. *Free Radic Biol Med.* 1993;15:77-96.

Shriqui CL, Bradwejn J, Annable L, Jones BD. Vitamin E in the treatment of tardive dyskinesia: a double-blind placebo-controlled study. *Am J Psychiatry.* 1992;149:391-393.

Simonian NA and Coyle JT. Oxidative stress in neurodegenerative diseases. *Annul Rev Pharmacol Toxicol.* 1996;36:83-106.

Simopoulos AP. Omega-3 fatty acids in health and disease, and in growth and development. *Am J Clin Nutr.* 1991;54:438-463.

Sohal RS, Agarwal S, Candas M, Forster MJ, Lal H. Effect of age and caloric restriction on DNA oxidative damage in different tissues of C57BL/6 mice. *Mech Ageing Dev.* 1994;76:215-224.

Stoll AL, Severus WE, Freeman MP, Rueter S, Zboyan HA, Diamond E, Cress KK, Marangell LB. Omega 3 fatty acids in bipolar disorder: a preliminary double-blind, placebo-controlled trial. *Arch Gen Psychiatry.* 1999;56:407-412.

Suboticanec K, Folnegovic-Smalc V, Korbar M, Mestrovic B, Buzina R. Vitamin C status in chronic schizophrenia. *Biol Psychiatry.* 1990;28:959-966.

Terry AV Jr, Hill WD, Parikh V, Waller JL, Evans DR, Mahadik SP. Differential effects of haloperidol, risperidone, and clozapine exposure on cholinergic markers and spatial learning performance in rats. *Neuropsychopharmacology.* 2003;28:300-309.

Vaddadi K, Courtney P, Gilleard CS, Manku MS, Horrobin DF. A double-blind trial of essential fatty acid supplementation in patients with tardive dyskinesia. *Psych. Res.* 1989;27:313-323.

Vaddadi K. Dyskinesias and their treatment with essential fatty acids: a review. *Prostaglandins Leukotri Essen Fatty Acids.* 1996;55:89-94.

Vaddadi K. Penicillin and essential fatty acid supplementation in schizophrenia. *Prostaglandins Med.* 1979;2:77-80.

Wolf G. gamma-Tocopherol: an efficient protector of lipids against nitric oxide-initiated peroxidative damage. *Nutr Rev.* 1997;55:376-378.

Section IV: Theoretical and Methodological Frontiers

In: Fatty Acids and Oxidative Stress...
Editors: R. Reddy and J.K. Yao, pp. 203-221
ISBN: 978-1-60021-677-0

Chapter XIV

Apolipoprotein D and Arachidonic Acid Interactions in the Treatment and Pathology of Schizophrenia

Elizabeth A. Thomas

Introduction

Apolipoprotein D (apoD) is a 29 kDa glycoprotein that was first isolated from human plasma high-density lipoproteins (HDLs) in 1973 (McConathy and Alaupovic 1973). Many studies to date have characterized its structure, expression and binding partners and have demonstrated its induction under diverse pathological conditions. Early studies on apoD have linked it to mechanisms of cell growth arrest and certain forms of cancer. While an association between apoD and cancer continues to be supported in the literature, diverse roles for apoD in neurological disorders continue to develop, especially in the context of psychiatric illnesses. The significance of apoD in psychiatric disorders was first illuminated when apoD was identified in a high-throughput screen of genes whose expression levels were altered by treatment with the atypical antipsychotic drug, clozapine. In subsequent studies, it was shown that apoD is important in the pathology of schizophrenia and bipolar disorder I. While accumulated findings have suggested that apoD is a multifunctional protein with tissue-specific ligands and functions (Rassart et al. 2000), its precise biological function remains unclear. Biochemical and functional data support a relevant physiological interaction between apoD and arachidonic acid (AA), which represents the ligand with the highest affinity for apoD identified to date. Several studies have demonstrated AA deficits in the plasma and brains of schizophrenic subjects supporting the hypothesis of an underlying dysfunction in phospholipids/fatty acid metabolism in the pathophysiology of schizophrenia. The importance of apoD/AA interactions in the pharmacotherapy and pathology of psychiatric disorders will be discussed in this chapter.

ApoD Identification, Expression and Function

Identification of ApoD

The human apoD cDNA was cloned and sequenced in 1986 (Drayna et al. 1986). Its encoded protein consists of 189 amino acids containing 5 cysteine residues and a signal peptide of 20 hydrophobic amino acids. After processing, the mature protein of 169 residues shows a high degree of sequence and structural homology to members of the lipocalin superfamily of proteins, a large and diverse group of lipid transporter proteins characterized by their ability to recognize multiple types of hydrophobic molecules. Like other members of this family, apoD adopts an 8-stranded antiparallel beta barrel structure. ApoD can be glycosylated at either or both of two asparagine residues at positions 65 and 98, which results in various apparent molecular weights for this protein (Albers et al. 1981; Drayna et al. 1986).

ApoD has been characterized in several mammalian and non-mammalian species, including monkeys, baboons, rodents, rabbits, deer, goat, chicken and drosophila (Smith et al. 1990; Provost et al. 1991; Seguin et al. 1995; Vieira et al. 1995; Sanchez et al. 2000). The complete sequences for human, rat, mouse and rabbit have been determined and show a high degree of sequence identity (>72% at the protein level). A bacterial lipocalin (Blc), which exhibits 31% identity with human apoD has also been characterized in *Escherichia coli* (Bishop et al. 1995) and a lipocalin from Arabidopsis showing 32% identity has been identified (Frenette Charron et al. 2002).

Sites of Expression for ApoD

ApoD is ubiquitously expressed throughout the body with abundant expression found in the brain, various peripheral tissues and bodily fluids. In the CNS, apoD is expressed in several cell types: oligodendrocytes, astrocytes, neurons, pia cells surrounding the brain and perivascular cells (Boyles et al. 1990; Smith et al. 1990; Navarro et al. 1995; Patel et al. 1995; Thomas et al. 2001). At least one of these cell types, astrocytes, have been shown to secrete apoD (Patel et al. 1995). Peripheral expression of apoD includes adrenal glands, pancreas, kidneys, placenta, spleen, lungs, ovaries, testes, adipose and peripheral nerves (Drayna et al. 1986; Provost et al. 1991; Seguin et al. 1995). Human apoD protein is poorly expressed in the liver and intestine, which are the major sites of synthesis of other apolipoproteins, but is abundantly found in plasma. Almost all of human plasma apoD is found in association with supramolecular complexes, primarily HDLs, some very high density lipoproteins (VHDLs) and low amounts in low density lipoproteins (LDLs) and very low density lipoproteins (VLDLs) (Blanco-Vaca et al. 1992). Like other cysteine-containing apolipoproteins, apoD exists as a disulfide-linked heterodimer with other apolipoproteins, primarily apolipoprotein A-II in lipoprotein fractions (Blanco-Vaca et al. 1992; Yang et al. 1994). ApoD is also expressed in various bodily fluids, including lacrimal and apocrine secretions (Holzfeind and Redl 1994; Escribano et al. 1995). ApoD is identical to the protein, gross cystic disease fluid protein-24 (GCDFP-24) (Balbin et al. 1990), first described in 1973 as a protein present in high concentrations in human mammary cyst fluid. This fluid represents a major source for the purification of human apoD.

Possible Functions for ApoD

The physiological role for apoD is uncertain, however, it is likely to have multiple functions depending on its localization. The presence of apoD in plasma HDL particles and its complex formation with lecithin:cholesterol acyltransferase, the enzyme responsible for the esterification of cholesterol, has suggested a role in human plasma lipid transport system and cholesterol metabolism (Tall AR 1998)(McConathy et al. 1973). In secreted fluids, it may have a paracrine role. Several studies have investigated its role in response to brain damage or pathology in rodents. ApoD expression is dramatically elevated in response to CNS and PNS insult, such as crush injury to the sciatic nerve and neurotoxic brain lesions in the cortex and hippocampus (Boyles et al. 1990; Spreyer et al. 1990; Ong et al. 1997; Montpied et al. 1999; Terrisse et al. 1999). These studies suggest that apoD is a "response" protein that is important for nervous system repair and maintenance. As a lipid-binding protein, apoD may be responsible for shuttling important molecules to sites of injury and regeneration. In addition to its induction in psychiatric disorders, which will be reviewed herein, apoD is elevated in various neurological disorders, including Alzheimer's disease (Terrisse et al. 1998; Kalman et al. 2000; Belloir et al. 2001; Thomas et al. 2001; Thomas et al. 2003), Multiple sclerosis (Reindl et al. 2001), motoneuron disease (Terrisse et al. 1998), cerebrovascular disease (Terrisse et al. 1998), meningoencephalitis (Terrisse et al. 1998) and normal aging (Kalman et al. 2000; Thomas et al. 2001). In these neurological conditions, apoD may be elevated in response to one or multiple pathological stimuli. One such stimulus that has been associated with several neurological disorders is oxidative stress.

ApoD-Arachidonic Acid Interactions

Binding and functional assays support a relevant physiological interaction for apoD and AA. This interaction may be important to the mechanisms of antipsychotic drug mechanisms and pathways associated with the pathophysiology of psychiatric disorders.

Binding Partners for ApoD

ApoD has been reported to interact with a variety of small ligands, depending on the tissue from which it was isolated (Table 1). Separate from its association with HDLs, apoD was independently isolated as an abundant protein from mammary gross-cystic disease fluid and was found to bind with high affinity to progesterone and pregnenolone, but not to other steroid derivatives, such as testosterone (Pearlman et al. 1973; Houtia et al. 1988; Dilley et al. 1990). While initial reports did not show measurable affinity for cholesterol to apoD, more sensitive methods have demonstrated low affinity binding to cholesterol (Patel et al. 1997). This may be important for reverse cholesterol transport to the liver. Further studies have described apoD interactions with bilirubin (Peitsch and Boguski 1990) and an abundant axillary-secreted odorant, E-3-methyl-2-hexenoic acid (E-3M2H) (Zeng et al. 1996). However, in later studies, binding to cholesterol and E-3M2H was not reproduced (Vogt and Skerra 2001). In the hypothalamus of the brain, apoD has been shown to interact with the long form of the leptin receptor (Liu et al. 2001), but the precise affinity was not determined.

In these various studies, different sources of apoD (purified apoD vs. recombinant apoD protein) and diverse methods for studying apoD interactions were used. In addition, the binding of apoD to many ligands is pH-dependent. These reasons may explain some of the discrepancies in the literature and in the studies described below.

Table 1. Reported binding partners for apoD.

Ligand:	Affinity	Reference:
Arachidonic Acid	6 nM	(Morais-Cabral et al. 1995)
	3 μM	(Vogt et al. 2001)
Progesterone	0.6-3 μM	(Vogt et al. 2001)
	1 μM	(Dilley et al. 1990)
	0.4 μM	(Morais-Cabral et al. 1995)
Pregnenolone	0.75 μM	(Dilley et al. 1990)
	not detected	(Vogt et al. 2001)
Cholesterol	Not detected	(Morais-Cabral et al. 1995)
	8 mM	(Patel et al. 1997)
EP092*	~0.1 μM	(Morais-Cabral et al. 1995)
Bilirubin	1:1 molar ratio	(Peitsch et al. 1990)
	not detected	(Vogt et al. 2001)
	40 nM	(Goessling and Zucker 2000)
E-3M2H	not reported	(Zeng et al. 1996)
	not detected	(Vogt et al. 2001)
Leptin Receptor	not reported	(Liu et al. 2001)

Affinities or molar ratios for the given ligand are shown. *, affinity is estimated from Morais-Cabral. EP092, a thromboxane antagonist; E-3M2H, E-3-methyl-2-hexenoic acid.

ApoD Bind Arachidonic Acid with High Affinity

With dissociation constant of 6 nM, arachidonic acid represents the ligand with the highest affinity for apoD (Table 1). Morais-Cabral and collaborators (Morais-Cabral et al. 1995) were the first to report an interaction between apoD and arachidonic acid (AA) using fluorescence spectroscopy. In this study, several molecules were tested for their affinity to apoD, purified from gross-cystic disease fluid. These included EP092 (a thromboxane antagonist (Wilson and Jones 1985), prostaglandins D2, E1 and F2a, arachidonic acid, 12-HETE, 5,15-diHETE. Of these, only EP092 and arachidonic acid demonstrated significant affinity as judged by the quenching of apoD fluorescence. The AA-apoD binding interaction was confirmed by Lipidex-based methods and size-exclusion HPLC. Additional fatty acids, linoleic (C18:2), oleic (C18:1) and palmitic (C16:0) and one phospholipid (dipalmitoyl phosphatidyl choline), were also tested in that study and only arachidonic acid showed any affinity for apoD.

Another study using bacterially produced recombinant apoD protein demonstrated that apoD binds both AA and progesterone with approximately the same affinity, but did not recognize the other presumed ligands, pregnenolone, bilirubin and the odorant component (Vogt et al. 2001). Thus, although previous reports have indicated a more promiscuous binding behavior for apoD, progesterone and arachidonic acid appear to be the only two high-affinity ligands with reproducible binding activity for apoD. Since AA represents the ligand with highest reported affinity, it is tempting to speculate that this interaction signifies the most important functional ligand for apoD, especially in relation to psychiatric disorders.

Arachidonic acid (AA) is of particular relevance as an apoD binding partner since it is associated with various physiological processes which are affected in schizophrenia. AA is one of the most important fatty acids in the brain. AA, together with docosahexaenoic acid (DHA), make up ~90% of the polyunsaturated fatty acid content in the CNS (O'Brien and Sampson 1965). In its esterified form, it is a major component of the lipid bilayer of cellular membranes and is released by a number of phospholipases upon extracellular signals. Once released from the cell membrane, AA can affect a multitude of downstream pathways. AA acts as an important second messenger, can directly modulate neurotransmitter systems and serves as a precursor for the synthesis of eicosanoids and prostaglandins, which are involved in a wide variety of biological phenomena. All of these roles for AA have been associated with schizophrenia pathophysiology (Skosnik and Yao 2003).

ApoD/AA Signaling *In Vitro*

The functional relevance of an apoD/AA interaction has been investigated *in vitro*. Studies from our lab have tested the ability of apoD to modulate AA signaling and mobilizaton in cultured embryonic kidney (HEK) 293T cells (Thomas et al. 2003). Immunofluorescent labeling revealed both cytosolic and membrane-bound expression of apoD protein in apoD-transfected HEK293 cells, suggesting that apoD can bind AA stored in membrane phospholipids or free AA in the cytosol. The effects of phorbal esters, which causes the release of AA into the cytosol, was measured in apoD-transfected cells compared to sham-transfected control cells. In the presence of apoD, a reduction in phorbol 12-myristate 13-acetate-induced AA release was observed (Thomas et al. 2003). Consistent with this finding, the membrane content of AA was higher in apoD-transfected cells, as indicated by the maximal amount of AA incorporated into membrane phospholipids. These results suggested that apoD may act to stabilize membrane-associated AA levels. Further studies showed that the addition of exogenous apoD dose-dependently prevented uptake of [^{3}H]AA into cell membranes, but had no effect on uptake of a different radiolabeled fatty acid, [^{3}H]palmitic acid. These findings demonstrate a functional specificity for AA and suggest that apoD can bind free AA in solution, thereby preventing its normal functioning. Hence, the presence of apoD in the cytosol may effectively divert free AA away from the cycloxygenase pathways, thereby limiting eicosanoid synthesis. Many eicosanoids (prostaglandins, leukotrienes, etc.) are mediators of inflammation, which can have numerous deleterious effects and have been implicated in psychiatric disorders (Fenton et al. 2000). Figure 3A depicts AA signaling in a typical neuron and how apoD, acting as an AA chaperone, perhaps acting to modulate AA release from phospholipids membranes and sequestration of AA in the cytosol.

Fatty Acid Alterations in ApoD Knock-Out Mice

In order to explore potential functions for apoD, mice with a targeted deletion in exon 6 of the apoD gene have been generated (Sanchez D, Ganfornina MD, Bastiani M, unpublished studies). A panel of membrane fatty acids and phospholipids were measured in forebrain samples of wt and apoD knock-out mice using high performance liquid chromatography and capillary gas chromatography. Significant increases in linoleic acid (18:2n-6), the immediate precursor to AA, and docosahexaenoic (22:6n3) (DHA) (Figure 1) were detected in apoD knock-out mice compared to wt littermates (Thomas and Yao 2006). Slightly higher levels of AA were observed, but this difference was not statistically significant.

n-6 PUFAs:		**n-3 PUFAs:**
Linoleic acid (18:2n-6)†*		α-Linolenic acid (18:3n-3)
↓	*? 6 desaturase*	↓
γ-Linolenic acid (18:3n-6)		Octadecatetraenoic acid (18:4n-3)
↓	*? 6 elongase*	↓
Dihomo-γ-linolenic acid (20:3 n-6)*		Eicosatetraenoic acid (20:4n-3)
↓	*? 5 desaturase*	↓
Arachidonic acid (20:4n-6)*		Eicosapentaenoic acid (20:5n-3)
↓	*elongase/? 5 elongase*	↓
Adrenic acid (22:4n-6)*		Docosapentaenoic acid (22:5n-3)
	? 4 desaturase	↓
		Docosahexaenoic acid (22:6n-3) †

† indicates those fatty acids significantly elevated in apoD KO mice compared to wt mice. * indicates those fatty acids significantly decreased in apoD KO mouse treated with clozapine (10 mg/kg, 28 days) compared to saline-treated apoD KO mice.

Figure 1. The common pathways for biosynthesis long-chain fatty acids and modulation in apoD knock-out mice on or off clozapine treatment.

ApoD and Arachidonic Acid Modulation by Antipsychotic Drug Therapy

Studies have demonstrated regulation of both apoD and AA levels by antipsychotic drugs, as well as a direct link between apoD and AA in mechanisms of action for clozapine. These findings highlight the importance of an apoD/AA interaction in antipsychotic drug treatment.

Antipsychotic Drugs and Gene Expression

Typical antipsychotic drugs, including haloperidol and chlorpromazine, are dopamine D_2 receptor antagonists. Clozapine is an atypical antipsychotic drug that represents a significant advancement in the treatment of mental illness. Clozapine displays improved clinical efficacy over typical antipsychotics, such as haloperidol, and has been reported to be effective against negative symptoms associated with schizophrenia (Jann 1991; Ciapparelli et al. 2000). It is also effective in patients who are resistant to treatment with classical drugs (Jann 1991; Ciapparelli et al. 2000). The efficacies of such medications in schizophrenia and other mental disorders has been well established in that antipsychotic drugs reduce symptomatology and prevent relapse in a large percentage of patients. In contrast to typical antipsychotics, clozapine exhibits affinity for neurotransmitter receptors, in addition to the dopamine family, including serotonin, histamine, muscarinic and adrenergic receptors (Jann 1991). However, it is possible that the clinical efficacy of clozapine may not be attributed exclusively to its drug-blocking profile, but that additional factors are involved.

The beneficial effects of antipsychotic drugs are thought to result from changes in gene expression. Imaging studies have shown that dopamine receptors are blocked after only a few hours of drug administration (Sedvall et al. 1986), whereas relief of psychotic symptoms may take several weeks of treatment. This temporal discrepancy between receptor occupancy and clinical benefit suggests that molecular changes are occurring downstream from receptor blockade that might attribute to the delayed effects. The unwanted extrapyramidal motor side effects that often accompany treatment with typical antipsychotics also exhibit a similar delayed onset (Marsden and Jenner 1980). Furthermore, antipsychotic drug treatment has been shown to induce expression of immediate early genes, many of which are transcriptional regulatory factors (for review see (Hughes and Dragunow 1995)). Thus, the latency in antipsychotic effects may result from the progression of a cascade of gene expression changes that cause gradual changes in the expression of critical proteins.

Gene expression changes elicited by drug treatment can be incredibly complex, consisting of primary responses, secondary responses and effects unrelated to drug action. Drugs may be acting via mechanisms that reverse underlying pathological deficits in schizophrenia or may cause beneficial effects via alterations of other pathways. Recent studies have taken advantage of the newer high-throughput methods to study global changes in gene expression after drug administration. These types of studies have the advantage of analyzing the expression of thousands of genes simultaneously. It was this type of methodology that led to the discovery of apoD as an antipsychotic drug-regulated gene.

Regulation of ApoD by Antipsychotic Drugs

Because investigating the effects of antipsychotic drug exposure on gene expression in human brain is not possible, studies have utilized rodent models to assess the CNS effects of antipsychotic drugs. Using the high-throughput method, Total Gene Expression Analysis (TOGA), a PCR-based method that can detect both low abundance and novel sequences (Sutcliffe et al. 2000), it was found that the expression levels of apoD were elevated in mouse brain in response to 2- week treatment with clozapine (Thomas et al. 2001; Thomas et al. 2003). Differential sensitivity to clozapine exists for various mouse strains. For these particular studies, the mouse strain C57black6J was used at a dose of 7.5 mg/kg clozapine injected IP daily for the 2–week time period (Thomas et al. 2001). *In situ* hybridization and immunohistochemical analyses revealed that apoD mRNA and protein were expressed in several different cell types in the brains of clozapine-treated mice, including neurons and two types of glial cells, oligodendrocytes and astrocytes. In the striatum, both neurons and astrocytes displayed increased apoD protein expression, while in white matter regions, the cell types giving rise to the elevated apoD levels were astrocytes and oligodendrocytes (Thomas et al. 2001). Consistent with these studies, others have also reported an increase in apoD protein expression in rat brain after administration of clozapine. One study demonstrated elevated apoD protein levels in cytosolic brain fractions using ELISA after chronic clozapine administration in the drinking water for 45-90 days (Khan et al. 2002). Another study detected apoD protein using Western blot analysis and showed increased levels after 2 weeks clozapine treatment in the cortex, striatum and hippocampus (Grobin et al. 2003).

Table 2. The effects of typical and atypical antipsychotic drugs on apoD expression levels

Atypical Antipsychotics:	Species:	Tissue	Dose mg/kg; duration:	Effect on ApoD Expression:	Reference:
Clozapine	Mouse	Striatum, Cortex	7.5; wks	↑	(Thomas et al. 2001)
Clozapine	Rat	Cortex	20; 45 days	↑	(Khan et al. 2002)
Clozapine	Rat	Striatum, Cortex	7.5; 2 wks	↑	(Grobin et al. 2003)
Clozapine	Human	Plasma	Varied; >1 year	↑	(Mahadik et al. 2002)
Olanzapine	Rat	Cortex	10; 2 wks	↑	(Grobin et al. 2003)
Olanzapine	Rat	Striatum	10; 2 wks	-	(Grobin et al. 2003)
Olanzapine	Rat	Striatum, Cortex Hippocampus	10; 45 days	↑	(Khan et al. 2003)
Risperidone	Rat	Striatum, Cortex Hippocampus	2.5; 45 days	↑	(Khan et al. 2003)

Typical Antipsychotics:	Species:	Tissue	Dose mg/kg; duration:	Effect on ApoD Expression:	Reference:
Haloperidol	Rat	Striatum, Cortex Hippocampus	2.0; 45 days	↓	(Khan et al. 2003)
Haloperidol	Rat	Striatum, Cortex	4.0; 2 wks	-	(Grobin et al. 2003)
Haloperidol	Mouse	Striatum	4.0; 2 wks	-	(Thomas et al. 2001)
Haloperidol	Mouse	Striatum	4.0; 2 wks	↓	Unpublished studies*
Thioridazine	Mouse	Striatum	10; 2 wks	↓	Unpublished studies*

ApoD levels reflect mRNA or protein expression as determined by Northern blot analysis, in situ hybridization, immunohistochemistry, Western blot analysis or ELISA. "-" indicates no change in apoD expression. *, unpublished studies, Thomas, E. A. and Sutcliffe, J. G.

Further work has provided evidence for a differential regulation of apoD expression by atypical versus typical antipsychotics (Table 2). Khan and colleagues have reported that, in addition to clozapine, two other atypical drugs, olanzapine and risperidone, cause elevation of apoD protein levels in the striatum, cortex and hippocampus (Khan et al. 2003). Grobin et al. also demonstrated an increase in apoD in the cortex after olanzapine administration, but not in the striatum (Grobin et al. 2003). In contrast, treatment with haloperidol has been shown to have the opposite effect on apoD expression. ApoD protein levels, measured by both ELISA and immunohistochemistry, were found to be significantly decreased in the hippocampus and motor cortex of rats treated with haloperidol in the drinking water for 90 days, but not 45 days (Khan et al. 2003). Interestingly, treatment with olanzapine and risperidone was found to reverse the decreases in apoD expression elicited by haloperidol treatment (Khan et al. 2003). Other studies have found no effect after haloperidol treatment in rodents (Thomas et al. 2001; Grobin et al. 2003) (Table 2). Overall, these studies suggest that apoD may be a distinguishing factor in the clinical actions of atypical versus typical drugs.

ApoD levels have also been measured in human plasma samples from drug-treated patients using ELISA. Clozapine treatment caused an elevation of apoD levels in the plasma of never-medicated, first-episode schizophrenic patients (Mahadik et al. 2002). In contrast, lowered apoD levels were reported in the serum of chronic schizophrenic subjects treated with a history of typical antipsychotic drug treatment (Thomas et al. 2001). Another study measuring plasma levels of apoD before and after treatment with a mixture of typical and atypical drugs found no difference among samples (Yao et al. 2005). In light of the potential opposing actions of atypical versus typical drugs on apoD levels, the findings from the latter study may have been confounded by drug treatment.

Antipsychotic Drugs and Arachidonic Acid

There is growing evidence that schizophrenia is associated with lipid abnormalities, primarily reductions in essential polyunsaturated fatty acids, such as AA and docosahexaenoic (DHA) (see paragraph on "AA in psychiatric disorders"). Early studies supporting this hypothesis were performed in chronic schizophrenic patients, hence, it was not known whether these membrane abnormalities occur early in illness or were a result of chronic treatment with antipsychotic drugs. In this regard, the ability of antipsychotic drugs to reverse AA and DHA deficits in schizophrenic patients has been investigated. Clozapine administration (8-12 weeks) caused a significant increase in AA and DHA levels in red cell membranes of schizophrenic patients (Horrobin et al. 1997). 6-month treatment with other atypical antipsychotics, olanzapine and risperidone, also reversed deficits in AA and DHA levels detected in schizophrenic patients (Evans et al. 2003). This effect appears to be specific to atypical antipsychotics as studies have shown no difference in the decrease of polyunsaturated fatty acids (particularly 18:2(n-6) and 20:4(n-6)) in both haloperidol-treated and drug-free patients (Yao et al. 1994). There was also no difference in those patients who had withdrawn from haloperidol for more than 5 weeks (Yao et al. 1994). Studies using rodents have not demonstrated similar increases in AA and DHA levels in clozapine-treated mice (Yao et al. 2005), however, these mice, unlike schizophrenic patients, have normal levels of AA and DHA prior to treatment. Hence, the effects of antipsychotic drugs on fatty acid levels may depend on underlying pathology.

An ApoD/Arachidonic Acid Interaction Important to Clozapine Function

While much of the evidence for an apoD link to AA being important to antipsychotic drug function has been circumstantial, a study using apoD knock-out mice has demonstrated a direct role for an apoD/AA interaction in the mechanism of clozapine function. Male and female wt and apoD knock-out mice were treated with clozapine (10 mg/kg) the drinking water for 28 days. A panel of membrane fatty acids and phospholipids were measured in forebrain using high performance liquid chromatography and capillary gas chromatography (Thomas and Yao, 2006). While few significant changes were observed in wt mice in response to clozapine treatment, several differences were detected in the clozapine-treated apoD knock-out mice. Most interestingly, significant differences in the levels of polyunsaturated fatty acids comprising the biosynthetic pathway of n-6 long-chain fatty acids: linoleic acid (18:2n-6), arachidonic acid (20:4n-6), dihomo-γ-linolenic acid (20:3n-6) and adrenic acid (22:4n-6) were detected (Figure 1).

Side Effects of Antipsychotic Drugs

Side effects of antipsychotics may also be mediated via an apoD/AA pathway. Studies have demonstrated neurotoxic effects of typical antipsychotic drugs, such as haloperidol

(Galili et al. 2000; Usuki et al. 2002), while atypical antipsychotics have neuroprotective effects (Bai et al. 2004). Previous studies on rodents have also suggested a neuroprotective role for apoD (Boyles et al. 1990) hence, these studies suggest that increased expression of apoD induced by atypical drugs may represent a novel mechanism of their favorable clinical effects compared to haloperidol and may play a vital role in treatment outcome in schizophrenia. The extrapyramidal side effects associated with typical antipsychotic drug usage may also be explained by apoD/AA mechanisms. A decrease in apoD expression caused by haloperidol would be associated with increased release of AA from the membrane and higher levels of cytosolic AA (see Figure 3). Prostaglandins are known to induce catalepsy in rats (Bhattacharya and Rao 1987; Ono et al. 1992), hence, elevated levels of AA, which serves as a precursor to prostaglandin synthesis, may produced or augment catalepsy. Atypical drugs have a less propensity to cause unwanted motor side effects, which may be explained by an elevation of apoD causing less release of AA from membranes and less available free AA.

ApoD and Arachidonic Acid in Psychiatric Disorders

ApoD is also emerging as important molecule associated with the pathology of psychiatric disorders. Elevated protein and mRNA levels have been reported in post-mortem samples from subjects with schizophrenia and in the plasma of never-medicated schizophrenic patients. A role for AA in the pathology of schizophrenia has long been supported. An interaction between these two molecules is likely to play an essential role in mechanisms of pathology in these disorders.

ApoD Expression in the Post-Mortem Brain from Subjects with Schizophrenia and Bipolar Disorder I

Initial studies measuring apoD levels in brain samples from subjects with schizophrenia focused on the dorsolateral prefrontal cortex, a region widely implicated in the

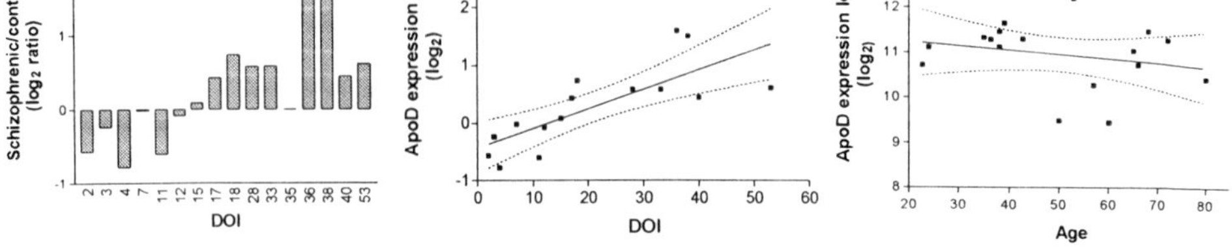

Figure 2. Expression levels of apoD in schizophrenic at different stages of illness and correlation of expression levels with duration of illness. Panel A: expression ratios of apoD mRNA (schizophrenic/control) are decreased early in illness, but increased at intermediate and chronic stages of illness. Panel B: Expression levels of apoD in schizophrenic subjects are correlation with duration of illness (DOI), but no correlations are observed in control subjects with age (Panel C).

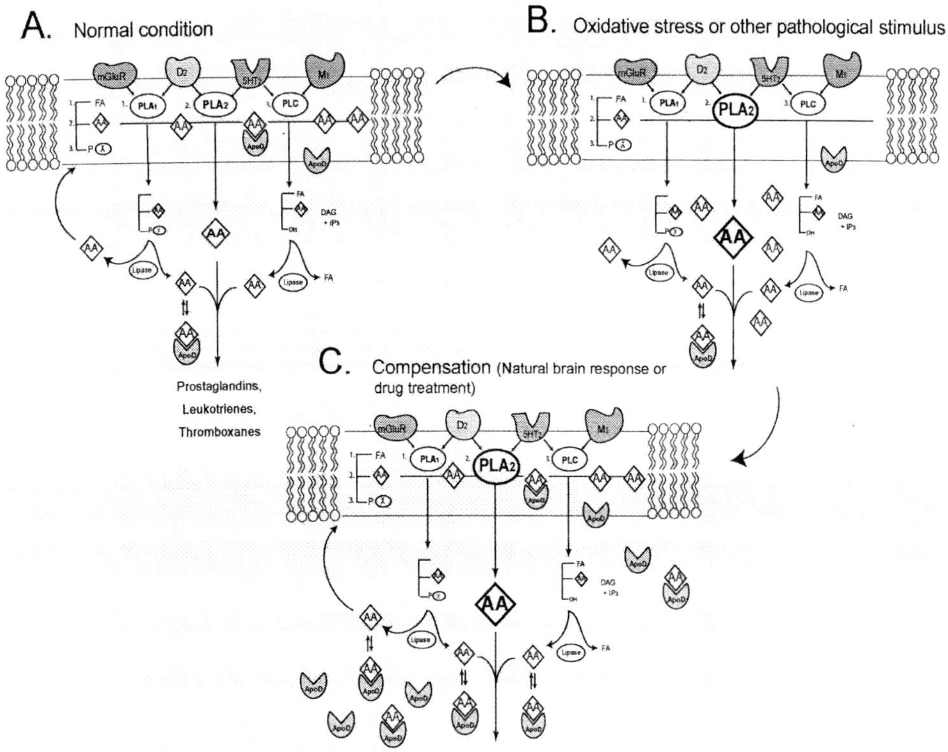

Figure 3. Schematic view of arachidonic acid signaling and potential role for apoD intervention. Upon activation of extracellular receptors, such as dopamine, serotonin, muscarinic or glutamate receptors, phospholipase A_1 (PLA_1), A_2 (PLA_2) and C (PLC) are responsible for the release of phospholipids from the membrane at the given positions, 1, 2, or 3, respectively. AA serves as a precursor for the synthesis of prostaglandins, leukotrienes and other lipid molecules. Under a state of oxidative stress, as in schizophrenia, there is an increase in lipid peroxidation and PLA_2 activity. This results in decreased membrane-bound arachidonic acid (AA) and elevated cytosolic AA levels. A compensatory response resulting from the brain's natural response to pathology or as a result of antipsychotic drug treatment, leads to elevated levels of apoD. Elevated apoD levels can act to stabilize membrane AA or bind free AA, thereby inhibiting lipid peroxidation, modulate re-incorporation of AA into the membrane and/or inhibit synthesis of eicosanoids. FA, fatty acid; P, phosphate; X, choline, ethanolamine, inositol or serine.

pathophysiology of schizophrenia (for review see (Goldman-Rakic and Seleman 1997). Using Western blot analysis, increased levels of apoD protein were detected in the dorsolateral prefrontal cortex of schizophrenic subjects compared to their age- and sex-matched controls (Thomas et al. 2001) Further studies used ELISA methodology to measure apoD levels in tissue homogenates from 12 different brain regions of control subjects and those diagnosed with either schizophrenia or bipolar disorder I. It was found that both disorders are accompanied by apoD increases in several areas, many of which are shared by the disorders. Nevertheless, there are areas of apoD increases that discriminate between schizophrenia and bipolar disorder. Brain regions that exhibited elevated apoD levels in both disorders were the dorsolateral and lateral prefrontal cortex, orbitofrontal cortex and caudate. Additionally, in schizophrenic subjects, elevated apoD proteins levels were observed in the thalamus and amygdala. In subjects with bipolar disorder I, additional regions of increased apoD expression were the parietal and cingulate cortices. No significant differences were detected in the occipital cortex, hippocampus, substantia nigra or cerebellum of either disease group.

Studies using microarray analysis, which allows for global measurements of gene expression differences, support an elevation of apoD mRNA in chronic schizophrenic subjects (Thomas, E.A. and Narayan, S, unpublished studies). Global mRNA profiling using Affymetrix microarrays was performed on lateral prefrontal cortical brain samples obtained post-mortem from 64 individual subjects: 32 schizophrenics divided into three groups, short duration of illness (<5 years), intermediate duration of illness (7-18 years) and long duration of illness (22-53 years), and 32 control subjects age- and sex-matched to the schizophrenic subjects. A pair-wise analysis of the data revealed decreases in expression of apoD in subjects with short duration of illness and increases in subjects at intermediate stages and chronic illness (Figure 2A). Furthermore, the expression levels of apoD in all schizophrenic subjects were correlated with duration of illness (r^2=0.552; p=0.0009) (Figure 2B), while no correlations were observed in controls subjects as a function of age (r^2=0.049; p=0.338) (Figure 2C). These studies support the hypothesis that apoD is gradually responding to an underlying pathophysiological response causing schizophrenia, or alternatively could be elevated as a result of long-term drug treatment. The former hypothesis is preferred since most of the schizophrenic subjects used in the microarray studies had been treated with typical antipsychotic drugs, which do not result in an elevation of apoD expression (see Table 2).

ApoD Elevation: Pathology Driven or Antipsychotic Drug Effect?

One of the universal drawbacks of using post-mortem tissue to study pathological features of psychiatric disease is the potential influence of long-term antipsychotic drug treatment. All of the schizophrenic subjects and most of the bipolar subjects examined for apoD expression had been treated with typical antipsychotic drugs (haloperidol, chlorpromazine, fluphenazine and thioridizine) prior to death, so initially there was a question whether the elevated apoD CNS levels detected might have resulted from long-term drug treatment. However, evidence from rodent studies suggests that typical antipsychotic drugs reduce or have no effect on apoD levels in the brain, and all of the subjects tested had been treated with typical antipsychotics. Another argument against elevated apoD levels being a drug effect is the disease-specific pattern that was observed in schizophrenia versus bipolar disorder I, despite the subjects in both groups being similarly treated with antipsychotic drugs prior to death. Furthermore, no correlations between drug dose and apoD levels were observed in any of the subjects tested (Thomas et al. 2001; Thomas et al. 2002). Since apoD is thought to be a "response" protein, we have hypothesized that elevated apoD levels are a compensatory response initiated by the brain as a result of a pathological stimulus. This is supported by our studies showing higher levels of apoD in chronic schizophrenia.

Plasma Levels of ApoD in Schizophrenic Patients

Also addressing the issue of drug-induced changes in apoD expression are studies that have measured apoD levels in the plasma of never-medicated schizophrenic patients. A most interesting study by Mahadik and colleagues (Mahadik et al. 2002) measured apoD levels in plasma samples from 22 never-medicated first-episode patients with schizophrenia or schizophreniform disorder. Significant increases in plasma apoD levels were detected in these patients compared to 14 normal control subjects, an effect that was further augmented in chronic patients, most of whom were treated with clozapine (Mahadik et al. 2002). However, in another schizophrenic cohort consisting of first-episode, neuroleptic naïve patients, the mean levels of apoD in RBC membranes were not significantly different from that of normal controls or neuroleptic-naive patients with other types psychiatric disorders (Yao et al. 2005). Since it has been suggested that apoD is a "response" protein, elevated levels may not be appear until sometime after the initial onset of psychiatric symptoms. This is supported by the microarray study demonstrating elevated levels of apoD primarily in subjects with chronic schizophrenia (Figure 2). Alternatively, due to the known heterogeneity of schizophrenia, it is possible that apoD levels are elevated only in a subgroup of schizophrenic patients, perhaps those with primary AA deficits and/or those in the study by Mahadik et al., (2002).

Arachidonic Acid in Psychiatric Disorders

It was hypothesized back in the late 70s that schizophrenia is a disorder resulting from abnormal fatty acid metabolism (Feldberg 1976; Horrobin 1977). Amounting evidence now supports this theory, especially the role of AA at the heart of dysfunction (Skosnik et al. 2003). Several studies have demonstrated a marked depletion of AA in membranes of red blood cells, fibroblasts, and brain tissue and an abnormal incorporation/esterification of AA into platelet membranes, of patients with schizophrenia (reviewed by (Skosnik et al. 2003) and elsewhere in this book). Many of these studies were performed on neuroleptic-naïve patients, indicating that these deficits are pathology-driven (Khan et al. 2002; Evans et al. 2003; Reddy et al. 2004; Yao et al. 2005). In addition, an impairment of AA-related signal transduction is observed in schizophrenic patients, as revealed by reduced niacin-initiated skin flushing (Aylward et al. 1998; Shah et al. 2000; Puri et al. 2002). These effects may be related to alterations in the activity of the enzyme phospholipase A_2 (PLA_2), which is responsible for AA incorporation into and release from phospholipid membranes. Elevations in the activity of PLA_2 (the calcium-independent isoform) has been demonstrated in the serum and cortex of schizophrenic patients (Gattaz et al. 1987; Ross et al. 1997; Ross et al. 1999).

Plasma ApoD and AA are Correlated in Subjects with Schizophrenia

The ability of apoD to bind AA may provide a link between the elevated levels of apoD reported in schizophrenia and the deficits in AA signaling and mobilization. By binding AA, apoD may affect facilitate or inhibit the incorporation of AA into membrane phospholipids and may affect substrate availability for PLA_2-induced release of AA and the subsequent eicosanoid signaling cascade. This idea was tested in a study that measured levels of RBC membrane fatty acids and apoD levels in the same first-episode, neuroleptic-naive (FENN) patients with schizophrenia and normal controls. The following fatty acids were tested: 16:0, 18:0,18:1, 24:1, 18:2(n-6), 20:3(n-6), 20:4(n-6), 22:4(n-6), 22:5(n-3) and 22:6(n-3). In the FENN patients with schizophrenia, plasma apoD levels were highly significantly correlated to the levels of membrane polyunsaturated fatty acids, arachidonic acid and docosapentaénoic acid (Yao et al. 2005). No correlations to other fatty acids were observed in schizophrenic patients. No correlations to any fatty acids were observed in control subjects. The lack of correlation in control subjects may indicate that the apoD/AA correlation is pathology driven.

Oxidative Stress as Pathogenic Stimulus in Schizophrenia

In schizophrenia there exists defects in the antioxidant defense system that can lead to increased oxidative damage (Yao et al. 2001). It is well known that oxidative stress can lead to the type of membrane abnormalities that have been observed in schizophrenia (Yao et al. 2001). Thus, free radical-mediated pathological processes may provide an explanatory model for the specific membrane alterations in schizophrenia. Consequently, oxidative stress may also represent the pathological stimulus for apoD upregulation in psychiatric disorders. Recent studies have demonstrated that another member of the lipocalin lipid-transporter family, human tear lipocalin, acts as a scavenger for harmful lipophilic molecules and has been suggested to work as a defense against the deleterious effects of oxidative stress (Lechner et al. 2001) It is possible that apoD may act in a similar manner by binding to and inactivating oxidated derivatives of AA or by preventing AA peroxidation and promoting membrane recovery. Therefore, correlations between plasma apoD and RBC-AA levels observed by Yao and colleagues (Yao et al., 2005) suggests that a pathological stimulus, such as oxidative stress, may lead to the recruitment of apoD, which promotes AA recycling and reincorporation back into the membrane. Figure 3 shows a schematic depiction of the effects of an oxidative stressor, or other pathological stimulus, on membrane PLA_2 and AA levels and the potential response elicited by apoD. This hypothesis is supported by *in vitro* signaling studies on apoD and AA described earlier in this chapter (Thomas et al. 2003). In the absence of pathological stimulus, (i.e. control subjects), levels of plasma apoD and RBC-AA are normal and not correlated with one another.

Summary

ApoD has been characterized as a multiligand, multifunctional protein, with tissue- and context-specific functions. AA has emerged as an important binding partner for apoD and that an apoD/AA interaction is important in the brain. Studies have demonstrated decreases in AA in schizophrenic subjects and that elevated levels of AA are beneficial to psychiatric patients. ApoD may be directly involved in these phenomena. By acting as an AA chaperone, apoD may help restore normal membrane structure and function and prevents an overload of free AA in the cytosol, which can have detrimental effects. Evidence supports a role for an apoD/AA pathway in the pharmacotherapy and pathology of psychiatric disorders. Such a pathway may lead to novel therapeutic approaches.

References

Albers, JJ, MC Cheung, et al. Characterization and immunoassay of apolipoprotein D. *Atherosclerosis* 1981; 39: 395-409.

Aylward, EH, NB Anderson, et al. Frontal lobe volume in patients with Huntington's disease. *Neurology* 1998; 50: 252-258.

Bai, O, H Zhang, et al. Antipsychotic drugs clozapine and olanzapine upregulate bcl-2 mRNA and protein in rat frontal cortex and hippocampus. *Brain Res* 2004; 1010: 81-86.

Balbin, M, JMP Freije, et al. Apolipoprotein D is the major protein component in cyst fluid from women with human breast cystic disease. *Biochem J* 1990; 271: 803-807.

Belloir, B, E Kovari, et al. Altered apolipoprotein D expression in the brain of patients with Alzheimer disease. *J Neurosci Res* 2001; 64: 61-69.

Bhattacharya, SK and PJ Rao. Prostaglandin D2-induced catalepsy in rats: role of 5-hydroxytryptamine. *J Pharm Pharmacol* 1987; 39: 743-745.

Bishop, RE, SS Penfold, et al. Stationary phase expression of a novel Escherichia coli outer membrane lipoprotein and its relationship with mammalian apolipoprotein D. Implications for the origin of lipocalins. *J Biol Chem* 1995; 270: 23097-23103.

Blanco-Vaca, F, DP Via, et al. Characterization of disulfide-linked heterodimers containing apolipoprotein D in human plasma lipoproteins. *J Lipid Res* 1992; 33: 1785-1796.

Boyles, JK, LM Notterpek, et al. Accumulation of apolipoproteins in the regenerating and remyelinating mammalian peripheral nerve. Identification of apolipoprotein D, apolipoprotein A-IV, apolipoprotein E and apolipoprotein A-I. *J Biol Chem* 1990; 265: 17805-17815.

Boyles, JK, LM Notterpek, et al. Identification , characterization, and tissue distribution of apolipoprotein D in the rat. *J Lipid Res* 1990; 31: 2243-2256.

Ciapparelli, A, L Dell'Osso, et al. Clozapine for treatment-refractory schizophrenia, schizoaffective disorder, and psychotic bipolar disorder: a 24-month naturalistic study. *J Clin Psychiatry* 2000; 61: 329-334.

Dilley, WG, DE Haagensen, et al. Immunologic and steroid binding properties of CDFP-24 protein isolated from human breast gross cystic disease fluid. *Breast Cancer Res Treat* 1990; 16: 253-260.

Drayna, D, C Fielding, et al. Cloning and expression of human apolipoprotein D cDNA. *J Biol Chem* 1986; 261.

Escribano, J, J Ortego, et al. Isolation and characterization of cell-specific cDNA clones from a subtractive library of the ocular ciliary body of a single normal human donor: transcription and synthesis of plasma proteins. *J Biochem (Tokyo)* 1995; 118: 921-931.

Evans, DR, VV Parikh, et al. Red blood cell membrane essential fatty acid metabolism in early psychotic patients following antipsychotic drug treatment. *Prostaglandins Leukot Essent Fatty Acids* 2003; 69: 393-399.

Feldberg, W. Possible association of schizophrenia with a disturbance in prostaglandin metabolism: a physiological hypothesis. *Psychol Med* 1976; 6: 359-369.

Fenton, WS, J Hibbeln, et al. Essential fatty acids, lipid membrane abnormalities, and the diagnosis and treatment of schizophrenia. *Biol Psychiatry* 2000; 47: 8-21.

Frenette Charron, JB, G Breton, et al. Molecular and structural analyses of a novel temperature stress-induced lipocalin from wheat and Arabidopsis. *FEBS Lett* 2002; 517: 129-132.

Galili, R, Mosberg, et al. Haloperidol-induced neurotoxicity--possible implications for tardive dyskinesia. *J Neural Transm* 2000; 107: 479-490.

Gattaz, WF, M Kollisch, et al. Increased plasma phospholipase-A2 activity in schizophrenic patients: reduction after neuroleptic therapy. *Biol Psychiatry* 1987; 22: 421-426.

Goessling, W and SD Zucker. Role of apolipoprotein D in the transport of bilirubin in plasma. *American journal of physiology. Gastrointestinal and liver physiology.* 2000; 279: G356-365.

Goldman-Rakic, PS and LD Seleman. Functional and anatomical aspects of prefrontal pathology in schizophrenia. *Schizophrenia Bulletin* 1997; 23: 437-458.

Grobin, A, S Gizerian, et al. (2003). Olanzapine increases apolipoprotein D protein levels in rat frontal cortex. Society for Neuroscience, Washington DC.

Holzfeind, P and B Redl. Structural organization of the gene encoding the human lipocalin tear prealbumin and synthesis of the recombinant protein in Escherichia coli. *Gene* 1994; 139: 177-183.

Horrobin, DF. Schizophrenia as a prostaglandin deficiency disease. *Lancet* 1977; 1: 936-937.

Horrobin, DF, AIM Glen, et al. Clozapine: elevation of membrane unsaturated lipid levels as a new mechanism of action. *Schizophr Res* 1997; 24: 214.

Houtia, NE, JC Maziere, et al. Phenothiazines inhibit cholesteryl ester formation in J 774 monocyte-like cells. *J Clin Chem Clin Biochem* 1988; 26: 673-678.

Hughes, P and M Dragunow. Induction of immediate-early genes and the control of neurotransmitter-regulated gene expression within the nervous system. *Pharmacological Reviews* 1995; 47: 133-178.

Jann, MW. Clozapine. *Pharmacotherapy* 1991; 11: 179-195.

Kalman, J, W McConathy, et al. Apolipoprotein D in the aging brain and in Alzheimer's dementia. *Neurol Res* 2000; 22: 330-336.

Khan, MM, DR Evans, et al. Reduced erythrocyte membrane essential fatty acids and increased lipid peroxides in schizophrenia at the never-medicated first-episode of psychosis and after years of treatment with antipsychotics. *Schizophr Res* 2002; 58: 1-10.

Khan, MM, VV Parikh, et al. Effects of chronic exposure of antipsychotics on apolipoprotein D in rat brain. *Biol Psychiatry* 2002; 51: 168S.

Khan, MM, VV Parikh, et al. Antipsychotic Drugs Differentially Modulate Apolipoprotein D in Rat Brain. *J Neurochem* 2003;86:1089-1100.

Lechner, M, P Wojnar, et al. Human tear lipocalin acts as an oxidative-stress-induced scavenger of potentially harmful lipid peroxidation products in a cell culture system. *Biochem J* 2001; 356: 129-135.

Liu, Z, GQ Chang, et al. Apolipoprotein D interacts with the long-form leptin receptor: a hypothalamic function in the control of energy homeostasis. *Faseb J* 2001; 15: 1329-1331.

Mahadik, SP, MM Khan, et al. Elevated plasma level of apolipoprotein D in schizophrenia and its treatment and outcome. *Schizophr Res* 2002; 58: 55-62.

Marsden, CD and P Jenner. The pathophysiology of extrapyramidal side effects of neuroleptic drugs. *Psychological Medicine* 1980; 10: 55-72.

McConathy, WJ and P Alaupovic. Isolation and partial characterization of apolipoprotein D: a new protein moiety of the human plasma lipoprotein system. *Fedn Eur Biochem Socs Lett* 1973; 37: 178-182.

Montpied, P, F de Bock, et al. Hippocampal alterations of apolipoprotein E and D mRNA levels in vivo and in vitro following kainate excitotoxicity. *Epilepsy Res* 1999; 35: 135-146.

Morais-Cabral, JH, GL Atkins, et al. Arachidonic acid binds to apoliprotein D: implications for the protein's function. *FEBS Lett* 1995; 366: 53-56.

Navarro, A, J Tolivia, et al. Pattern of apoliprotein D immunoreactivity in human brain. *Neurosci Lett* 1995; 254: 17-20.

O'Brien, JS and EL Sampson. Lipid composition of the normal human brain: gray matter, white matter, and myelin. *J Lipid Res* 1965; 6: 537-544.

Ong, WY, Y He, et al. Differential expression of apoliprotein D and apoliprotein E in the kainic acid-lesioned rat hippocampus. *Neurosci* 1997; 79: 359-367.

Ono, N, T Abiru, et al. Influences of cyclooxygenase inhibitors on the cataleptic behavior induced by haloperidol in mice. *Prostaglandins Leukot Essent Fatty Acids* 1992; 46: 59-63.

Patel, R, D Lange, et al. Probing the structure of the ligand binding cavity of lipocalins by fluorescence spectroscopy. *Protein Eng* 1997; 10: 621-625.

Patel, SC, K Asotra, et al. Astrocytes synthesize and secrete the lipophilic ligand carrier apolipoprotein D. *Neuroreport* 1995; 6: 653-657.

Pearlman, WH, JL Gueriguian, et al. A specific progesterone-binding component of human breast cyst fluid. *J Biol Chem* 1973; 248: 5736-5741.

Peitsch, MC and MS Boguski. Is apolipoprotein D a mammalian bilin-binding protein? *New Biol* 1990; 2: 197-206.

Provost, PR, L Villeneuve, et al. Localization of the major sites of rabbit apolipoprotein D gene transcription by in situ hybridization. *J Lipid Res* 1991; 32.

Puri, BK, SR Hirsch, et al. A volumetric biochemical niacin flush-based index that noninvasively detects fatty acid deficiency in schizophrenia. *Prog Neuropsychopharmacol Biol Psychiatry* 2002; 26: 49-52.

Rassart, E, A Bedirian, et al. Apolipoprotein D. *Biochim Biophys Acta* 2000; 1482: 185-198.

Reddy, RD, MS Keshavan, et al. Reduced red blood cell membrane essential polyunsaturated fatty acids in first episode schizophrenia at neuroleptic-naive baseline. *Schizophr Bull* 2004; 30: 901-911.

Reindl, M, G Knipping, et al. Increased intrathecal production of apolipoproten D in multiple sclerosis. *J Neuroimmunol* 2001; 119: 327-332.

Ross, BM, C Hudson, et al. Increased phospholipid breakdown in schizophrenia. Evidence for the involvement of a calcuim-independent phospholipase A2. *Arch Gen Psychiatry* 1997; 54: 487-494.

Ross, BM, S Turenne, et al. Differential alteration of phospholipase A2 activities in brain of patients with schizophrenia. *Brain Research* 1999; 821: 407-413.

Sanchez, D, MD Ganfornina, et al. Characterization of two novel lipocalins expressed in the Drosophila embryonic nervous system. *Int J Dev Biol* 2000; 44: 349-359.

Sedvall, G, L Farde, et al. Imaging of neurotransmitter receptors in the living human brain. *Arch Gen Psychiatry* 1986; 43: 995-1005.

Seguin, D, M Desforges, et al. Molecular characterization and differential mRNA tissue distribution of mouse apolipoprotein D. *Mol Brain Res* 1995; 30: 242-250.

Shah, SH, GK Vankar, et al. Unmedicated schizophrenic patients have a reduced skin flush in response to topical niacin. *Schizophr Res* 2000; 43: 163-164.

Skosnik, PD and JK Yao. From membrane phospholipid defects to altered neurotransmission: is arachidonic acid a nexus in the pathophysiology of schizophrenia? *Prostaglandins Leukot Essent Fatty Acids* 2003; 69: 367-384.

Smith, KM, RM Lawn, et al. Cellular localization of apolipoprotein D and lecithin:cholesterol acyltransferase mRNA in rhesus monkey tissues by in situ hybridization. *J Lipid Res* 1990; 31: 995-1004.

Spreyer, P, H Schaal, et al. Regeneration-associated high level expression of apolipoprotein D mRNA in endoneurial fibroblasts of peripheral nerve. *Embo J* 1990; 9: 2479-2484.

Sutcliffe, JG, PE Foye, et al. TOGA: an automated parsing technology for analyzing expression of nearly all genes. *Proc Natl Acad Sci USA* 2000; 97: 1976-1981.

Terrisse, L, J Poirier, et al. Increased levels of apolipoprotein D in cerebrospinal fluid and hippocampus of Alzheimer's patients. *J Neurochem* 1998; 71: 1643-1650.

Terrisse, L, D Sequin, et al. Modulation of apolipoprotein D and apolipoprotein E expression in rat hippocampus after entorhinal cortex lesion. *Brain Res Mol Brain Res* 1999; 70: 26-35.

Thomas, EA and Yao, JK. Clozapine Specifically Alters the Arachidonic Acid Pathway in Mice Lacking Apolipoprotein D. *Schizophrenia Research* In press.

Thomas, EA, PE Danielson, et al. Clozapine increases apolipoprotein D expression in rodent brain: towards a mechanism for neuroleptic pharmacotherapy. *J Neurochem* 2001; 76: 789-796.

Thomas, EA, B Dean, et al. Increased CNS levels of apolipoprotein D in schizophrenic and bipolar subjects: implications for the pathophysiology of psychiatric disorders. *Proc Natl Acad Sci U S A* 2001; 98: 4066-4071.

Thomas, EA, B Dean, et al. Differences in Neuroanatomical Sites of ApoD Elevation Discriminate Schizophrenia and Bipolar Disorder. *Submitted* 2002.

Thomas, EA, RC George, et al. Antipsychotic drug treatment alters expression of mRNAs encoding lipid metabolism-related proteins. *Mol Psychiatry* 2003; 8: 983-993, 950.

Thomas, EA, RC George, et al. Apolipoprotein D modulates arachidonic acid signaling in cultured cells: Implications for psychiatric disorders. *Prostaglandins, Leukotrienes and Essential Fatty Acids* 2003.

Thomas, EA, S Laws, et al. Apolipoprotein D levels are elevated in prefrontal cortex of subjects with Alzheimer's disease: No relation to apolipoprotein E expression or genotype. *Biological Psychiatry* 2003.

Thomas, EA, LN Sautkulis, et al. Apolipoprotein D mRNA expression is elevated in PDAPP transgenic mice. *J Neurochem* 2001; 79: 1059-1064.

Thomas, EA and JG Sutcliffe (2003). The role of apolipoprotein D in the mechanisms of action of clozapine and in psychiatric disorders. Phospholipid Spectrum Disorders in Psychiatry and Neurology. M. Peet, I. Glen and D. F. Horrobin. Lancashire, Marius Press: 257-269.

Usuki, E, JR Bloomquist, et al. Metabolic studies on haloperidol and its tetrahydropyridinyl dehydration product (HPTP) in C57BL/6 mouse brain preparations. *Neurotox Res* 2002; 4: 51-58.

Vieira, AV, K Lindstedt, et al. Identification of a circulatory and oocytic avian apolipoprotein D. *Mol Reprod Dev* 1995; 42: 443-446.

Vogt, M and A Skerra. Bacterially produced apolipoprotein D binds progesterone and arachidonic acid, but not bilirubin or E-3M2H. *J Mol Recognit* 2001; 14: 79-86.

Wilson, NH and RL Jones. Prostaglandin endoperoxide and thromboxane A2 analogs. *Adv Prostaglandin Thromboxane Leukot Res* 1985; 14: 393-425.

Yang, CY, ZW Gu, et al. Structure of human apolipoprotein D: locations of the intermolecular and intramolecular disulfide links. *Biochemistry* 1994; 33: 12451-12455.

Yao, JK, RD Reddy, et al. Oxidative damage and schizophrenia: an overview of the evidence and its therapeutic implications. *CNS Drugs* 2001; 15: 287-310.

Yao, JK, EA Thomas, et al. Association of plasma apolipoproteins D with RBC membrane arachidonic acid levels in schizophrenia. *Schizophr Res* 2005; 72: 259-266.

Yao, JK, DP van Kammen, et al. Red blood cell membrane dynamics in schizophrenia. II. Fatty acid composition. *Schizophr Res* 1994; 13: 217-226.

Zeng, C, AI Spielman, et al. A human axillary odorant is carried by apolipoprotein D. *Proc Natl Acad Sci U S A* 1996; 93: 6626-6630.

In: Fatty Acids and Oxidative Stress...
Editors: R. Reddy and J.K. Yao, pp. 223-236

ISBN: 978-1-60021-677-0

Chapter XV

Long-Chain Polyunsaturated Fatty Acids and Perinatal Origins of Schizophrenia

Undurti N. Das

Introduction

Despite years of research, aetiopathogenesis of schizophrenia is not clear. Schizophrenia may result from altered neuronal membrane structure and metabolism (Joy et al, 2000), and/or dysregulation of the inflammatory response system (Boin et al, 2001).

Prenatal and Perinatal Influences on Psychopathology

Brain growth and development occurs from 2nd trimester of pregnancy to 5years of age. This suggests that a range of prenatal and perinatal events may have an important role in the pathobiology of various neuropsychiatric conditions including schizophrenia. Investigations directed to identify such prenatal and perinatal factors reported that major depression and schizophrenia were associated with not being breast-fed, maternal emotional problems, and maternal viral infections ((Allen et al, 1998; McCreadie, 1997; Shi et al, 2003). It is believed that breast-feeding potentiates cognitive development (Lucas et al, 1992), although some studies did not support this conclusion (Gale and Martyn, 1996; Malloy and Berendes, 1998). Since schizophrenia is preceded by childhood cognitive impairments (Crow et al, 1995; Jones et al, 1994; McCreadie, 1997), it was opined that breast-feeding might be a factor in the aetiology of schizophrenia (Peet et al, 1997). Human breast milk contains significant amounts of polyunsaturated fatty acids (PUFAs), such as gamma-linolenic acid (GLA), dihomo-gamma-linolenic acid (DGLA), arachidonic acid (AA), eicosapentaenoic acid (EPA), and docosahexaenoic acid (DHA) that form an important component of many cell membranes and also of neuronal membranes. Lack of some of these PUFAs (especially AA, EPA, and DHA) has been shown to have adverse effects on brain development and growth (Das,

2002a;2003a;2003b; Makrides et al, 1995), and so it was thought that decreased availability of these fatty acids as a result of sub-optimal breast-feeding could explain the deviant cognitive and neurological development seen premorbidly in schizophrenia (Done et al, 1991; Dworkin et al, 1993; Jones et al, 1994). Although this is an interesting concept, several other studies did not support the effect of breast-feeding in protecting against the risk of later schizophrenia (Amore et al, 2003;Lesak et al, 2000; Sesaki et al, 2000). Nevertheless, some studies did suggest that breast-feeding might be associated with decreased incidence of schizophrenia in later life (Allen et al, 1998). Some of the reasons for these conflicting reports could be: the duration of breast-feeding and the PUFA content of breast milk. It is likely that longer the duration of breast-feeding the greater will be its impact on the newborn. Thus, subjects who received breast-feeding for more than 6 to 12 months are likely to have obtained significantly larger amounts of PUFAs compared those who were breast-fed for much shorter duration. Similarly, another confounding factor is the PUFA content of breast milk. It is known that the fatty acid composition of the breast milk depends to a large extent on the diet and large variations in the PUFA content of the breast milk have been reported in women of different regions and countries. In view of this, it is not surprising that there were both positive and negative reports with regard to the effect of breast-feeding in protection against the development of schizophrenia in later life.

Early Fetal Environment and Schizophrenia

Associations between suicidal behavior, low maternal age and schizophrenia have been suggested (Oquendo and Barra-Garcia, 2004). Similarly, a causal relation between obstetric complications and schizophrenia has been reported (Jones et al, 1998). This suggests that there is a close relationship between impaired fetal growth and the onset of adult diseases (Das, 2002b). It has been proposed that inadequate fetal nutrition can permanently alter the body's structure, physiology, and metabolism (fetal programming) that predispose them to develop chronic illnesses in adulthood (Bateson et al, 2004). It is known that a given genotype can give rise to different phenotypes, depending on environmental conditions. Many organisms express specific adaptive responses to their environments that can include short-term changes in physiology and behavior. However, responses to the environmental influences, especially when they occur during pregnancy, may be expressed by the offspring during their adult life, and may not be seen/expressed by the mother. One such environmental factor that can have a life long impact on the offspring could be the nutrition during pregnancy. For example, in mice food restriction can slow ageing; enhance cellular maintenance and repair processes, while reducing fertility (Merry, 2004). It is particularly important to note that changes triggered by environmental factors may have long term consequences in the offspring, especially if these events are induced during sensitive, often brief, periods of development. Outside these sensitive periods, the influence of the adverse environmental event (say for instance that of malnutrition or infection) on the offspring may be little or none. It is relevant to note that such responses to the environment, which includes maternal infections, may be expressed in the fetus rather than in the mother. This may apply to the development of schizophrenia in adult life as well. For instance, maternal exposure to glucocorticoids in pregnancy induces hypertension, insulin resistance, obesity and altered muscle mass as well as alterations in the hypothalamic-pituitary-adrenal axis in the adult progeny (Waterland and Garza, 1999). Similarly, it is possible that maternal infection leads to excess production of pro-inflammatory cytokines that are neurotoxic resulting in the development of schizophrenia in adult life. The source of these pro-inflammatory cytokines could be infiltrating macrophages, T cells and the neurons themselves (Das, 2002b; Fricchione et al, 1996). Exposure of fetal neurons to such noxious stimulus may render them more susceptible to further damage even by sub-optimal doses of pro-inflammatory cytokines. Hence, damage to the fetal neurons by various environmental agents such as maternal viral infection may be induced during sensitive, often brief, periods during development of the brain. Outside these sensitive periods, the influence of the neurotoxic agents may be little or no effect. It is this critical window period during which neuronal damage occurs may predispose the fetus to develop schizophrenia in adult life.

Recent studies suggested an association between fetal malnutrition during middle and late gestation and psychiatric disorders (Jones et al, 1998; Neugebauer et al, 1999). Higher frequencies of obstetric, neonatal, and maternal complications have been recorded in adolescents who had psychiatric disorders (Salk et al, 1985). This suggested that infants subjected to complications during pregnancy and at birth might be more vulnerable to environmental events precipitating psychiatric conditions. But other investigators could not confirm some of these observations (Barker et al, 1995; Mittendorfer-Rutz et al, 2004). Although these studies have reported mainly the influence of maternal and perinatal events on the incidence of suicide (Mittendorfer-Rutz et al, 2004), it is possible that they may also influence the development of schizophrenia in adulthood. Mittendorfer-Rutz et al (2004) reported an inverse association between birth weight and risk of suicide that is in accord with earlier findings that depression and schizophrenia are associated with low birth weight (Allebeck et al, 1991; Nilsson et al, 2001). These findings imply the importance of maternal and perinatal factors and their influence on the development of central nervous system and psychiatric conditions such as schizophrenia in adulthood.

In the light of these observations it is interesting to note that fluoxetine, a selective serotonin reuptake inhibitor, has been suspected to raise the risk of suicidal behavior in children. Furthermore, disruption of the serotonin transporter (5-HTT) early in brain development affects the development of brain circuits that deal with stress response (Ansorge et al, 2004). In a longitudinal study of New Zealanders, it was reported that people with a polymorphism that reduced their 5-HTT activities were more likely than others to become depressed in response to stressful experiences (Caspi et al, 2003). Serotonin (5-HT) acts as a trophic factor modulating developmental processes such as neuronal division, differentiation, migration, and synaptogenesis. Inhibition of 5-HTT functions has been found to alter brain development in subtle ways. For instance, genetic inactivation of 5-HTT affects barrel formation in the somatosenory cortex and alters segregation of retinal axons. Mice lacking the 5-HTT gene showed reduced dorsal raphe firing rates and fewer serotonergic neurons (Rumajogee et al, 2004). Similar to 5-HTT $^{-/-}$ mice, rats treated with clomipramine, a tricyclic antidepressant that shows a moderate selectivity for 5-HTT over norepinephrine reuptake site, from P8 to P21 (pregnancy day 8 to 21) were found to exhibit altered emotional behavior as adults and reduced firing rate of raphé neurons (Kinney et al, 1997). These results suggest that alterations in the structure and function of serotonergic nuclei can contribute to the altered behavioral responses observed. It is also known that developmental and/or genetic factors affecting anxiety- or depression-related behaviors can alter hippocampal structure, amygdala function and receptor expression in the prefrontal cortex (Gross et al, 2002; Hariri et al, 2002; Liu et al, 1997; Smythe et al, 1994). All these structures are known to receive significant serotonergic innervation. The findings of Ansorge et al (2004) demonstrated that 5-HTT function modulates the development of brain systems involved in emotional and stress related responses and that low expressing 5-HTT variants may act during fetal brain development to modify brain circuits or gene expression that predisposes carriers of these alleles to emotional disorders and, possibly, schizophrenia. It is likely that low-expressing 5-HTT variants act during development of brain, especially during perinatal period, to modify brain circuits or gene expression that predisposes carriers of these alleles to emotional disorders such as schizophrenia. If this is true, this suggests that nutritional and other factors acting during fetal, perinatal, and infancy periods may influence not only the growth and development of brain but also ability of the neurons to synthesize, secrete, and express receptors for various neurotransmitters in such a way that these early life events would ultimately have an impact in the development of schizophrenia in adult life. Support to this concept comes from the observation that early fetal environment and parental factors have a major impact on suicidal behavior in adolescents and young adults (Allebeck et al, 1991; Ansorge et al, 2004; Barker et al, 1995; Mittendorfer-Rutz et al, 2004; Neugebauer et al, 1998; 1999; Nilsson et al, 2001; Oquendo and Barra-Garcia, 2004; Salk et al, 1985). In this context, it is interesting to note that one of the factors that is known to have an impact on the development of schizophrenia in later life is the presence or absence of maternal infections during pregnancy.

Relationship between Maternal Infections and Schizophrenia in Later Life

Maternal viral infection has been reported to increase the risk for schizophrenia in the offspring. For instance, it was documented that mice born to mothers who had respiratory tract infection at mid-gestation showed features of schizophrenia (Shi et al, 2003). Prenatal immune challenge disrupted sensorimotor gating in adult rats (Borrell et al, 2002) and these animals showed increased serum levels of interleukins-2 (IL-2) and IL-6, suggesting that prenatal immune events play do have a role in the pathogenesis of schizophrenia (Editorial, 2003). There is considerable evidence now to suspect that specific immunological abnormalities do occur in schizophrenia. One of the causes for restricted fetal growth could be maternal infection. The increased plasma concentrations of ILs observed in the maternal plasma secondary to infections are known to cause anorexia. This may cause malnutrition in the mother that ultimately leads to fetal growth restriction. Furthermore, these pro-inflammatory cytokines themselves may have deleterious actions on the growth and development of fetal brain, if the maternal infection occurs during the critical periods of brain growth. It is also possible that maternal malnutrition itself may render them more susceptible to infections that, in turn, interferes with normal feeding behavior and thus, exacerbates malnutrition. This complex web of events would eventually have a larger impact on the fetus than the mother. This once again underscores the importance of maternal and fetal nutrition to prevent adult diseases.

Epidemiological findings revealed that second-trimester maternal infection increased the risk of schizophrenia in the children. Large population studies showed that the incidence of schizophrenia is higher among people who were born in urban settings that are associated with a higher risk of maternal influenza infection during pregnancy. This is supported by the observation that mice born to mothers who were exposed to respiratory infection at mid gestation showed abnormal behavior as adults which resemble those seen in schizophrenics. In these studies, no virus could be detected in the affected offspring, suggesting that maternal immune system itself could lead to the brain changes in their offspring. The fact those similar changes in the offspring were observed following stimulation of maternal immune system using a synthetic double-stranded RNA that evokes an anti-viral immune response lends further support to this view (Editorial, 2003). Mice born to infected mothers showed changes in cortex and hippocampus, loss of Purkinje cells in the cerebellum, similar changes have also been described in schizophrenic patients. Although the concept that human maternal infection is responsible for schizophrenia remains to be proved, if this is right it offers the potential to prevent this condition.

Pro-Inflammatory Cytokines and Schizophrenia

The concept that maternal infection during pregnancy is responsible for the development of schizophrenia in their offspring in adult life is supported by the observation that there are specific abnormalities in the concentrations of pro-inflammatory cytokines in these patients. Serum and cerebrospinal fluid (CSF) IL-2, IL-6, IL-8 and tumor necrosis factor-α (TNF-α) levels were reported to be elevated in patients with schizophrenia (Ebrinc et al, 2002; Kowalski et al, 2001; Maes et al, 2002; McAllister et al, 1995; Zhang et al, 2002). Relapse-prone patients, examined both while medicated and after drug withdrawal, had significantly higher levels of CSF IL-2 than patients who did not relapse and the use of risperidone not only decreased interferon-γ (IFN-γ) production (Cazzullo et al, 2002) but also enhanced IL-10 (a suppressor of Th_1 response). Haloperidol and perazine decreased the release of IL-1β and TNF-α from monocytes of schizophrenic patients (Kowalski et al, 2001). IL-2 treatment-induced behavioral changes can be completely blocked by a selective dopamine D1 receptor antagonist (SCH 23390) or by a relatively high dose of a D2 antagonist (sulpiride) (Zalcman, 2002) indicating that IL-2 induces and/or increases psychiatric abnormalities by causing aberrations in central dopaminergic transmission. Rat cortical cultures exposed to IL-1β, IL-6 and TNF-α showed dose-dependent decreases in the number of neurons immunoreactive for MAP-2 antibody, suggesting decreased neuronal survival (Marx et al, 2001). These pro-inflammatory cytokines interfere with the actions of various neurotransmitters and induce features of schizophrenia (Kin et al, 2000). Based on these data, it is reasonable to believe that injury; infection and inflammation-induced increase in pro-inflammatory cytokines during gestational period have the potential to cause injury to fetal neurons that in turn increase their risk of schizophrenia that explains the association between schizophrenia and maternal infection during pregnancy. This suggests that suppression of the production of pro-inflammatory cytokines could be of significant benefit in schizophrenia. In this context, it is interesting to note that polyunsaturated fatty acids (PUFAs) not only inhibit the production of pro-inflammatory cytokines but also possess neuroprotective properties. Thus, PUFAs may behave as endogenous anti-inflammatory and anti-schizophrenia molecules.

Both EPA and docosahexaenoic acid (DHA, 22:6 ω-3), derived from EPA, and which are EFAs have been shown to give rise to a series of potent anti-inflammatory compounds called as "resolvins". Studies showed that resolvins inhibit leukocyte infiltration and aid in the resolution of inflammation (Serhan et al, 2002).

Polyunsaturated Fatty Acids and Schizophrenia

Human brain is rich in PUFAs and forms an important component of the neuronal cell membranes. ω-3 and ω-6 fatty acids are essential for fetal growth and development (Das, 2002a). Newborn infants and pre-term infants have limited capacity to form EPA, DHA and AA. Dietary AA improves first year growth of pre-term infants (Carlson et al, 1993), whereas EPA and DHA increase birth weight by prolonging gestation and by increasing the fetal growth rate (Baguma-Nibasheka et al, 1999; Olsen et al, 1990). In view of this, low-birth weight infants are likely to have decreased tissue and plasma concentrations of various PUFAs. EPA, DHA, and AA, inhibit TNF-α and IL-6 production; enhance eNO generation ((Das, 2000a), inhibit 3-hydroxy-3-methylglutaryl coenzyme A (HMG-CoA) reductase and angiotensin converting enzyme activities (Das, 2001a; Kumar et al, 1997), and function as endogenous ligands for PPARs, and suppress leptin gene expression (Roseland et al, 2001). Thus PUFAs suppress inflammation, regulate cholesterol metabolism, enhance the production of adiponectin, and decrease insulin resistance. PUFAs augment brain acetylcholine levels and thus, enhance parasympathetic activity and increase heart rate variability (Das, 2001b). This suggests that PUFAs may function as endogenous statins (Das, 2001a), β-blockers, ACE inhibitors, anti-inflammatory, and anti-hypertensive molecules. PUFAs bind to PPARs (similar to thiazolidinediones) and thus, are capable of enhancing adiponectin levels and reduce insulin resistance. In view of these various actions, it is no surprise that PUFAs also have an important role in schizophrenia, especially, since inflammation plays an important role in its pathobiology.

Reduced levels of membrane DHA, EPA, and AA, and increased levels of peroxidation products have been reported in schizophrenics (reviewed in chapters 4, 8 and 10). However, the reductions in levels of both AA and DHA were much smaller in medicated versus never-medicated patients and especially those of AA and DHA levels were much higher in chronic medicated patients than drug-naïve first-episode patients (Khan et al, 2002). It is possible that majority of the antipsychotics modulate metabolism of PUFAs and thus, brings about their beneficial actions in schizophrenia. This is supported by the observation that chronic valproate and lithium treatment decrease AA turnover in brain phospholipids (Chang et al, 1996; 1999; 2001). In this context, it is interesting to note that EPA and DHA have inhibitory effects on protein kinase C, inositol monophosphatase, and inositolpolyphosphatase, thereby leading to a decreased inositol-1,4,5-triphosphate ($InsP_3$) response activity(Chaudhry et al, 1987; Kim et al, 2001; Sperling et al, 1993; Williams et al, 2002), events that are responsible for the their beneficial actions in schizophrenia.

PUFAs have the ability to suppress the production of pro-inflammatory cytokines IL-2, IL-6 and TNF-α both in vitro and in vivo (Das, 1994; Endres et al, 1989; Kumar and Das, 1994; Kumar et al, 1992). Oral supplementation of EPA is useful in schizophrenia (reviewed in chapter 11). Thus, the beneficial effects of EPA in schizophrenia could be attributed to its ability to suppress the production of pro-inflammatory cytokines, which are elevated in this condition. In addition, EPA, DHA and AA have potent neuroprotective and cytoprotective actions and prevent apoptosis of neurons (Kim et al, 2000; 2001; Mahadik et al, 2001; Martin et al, 2002). These results suggest that PUFAs (especially EPA) acts at different levels to bring about its beneficial actions in schizophrenia.

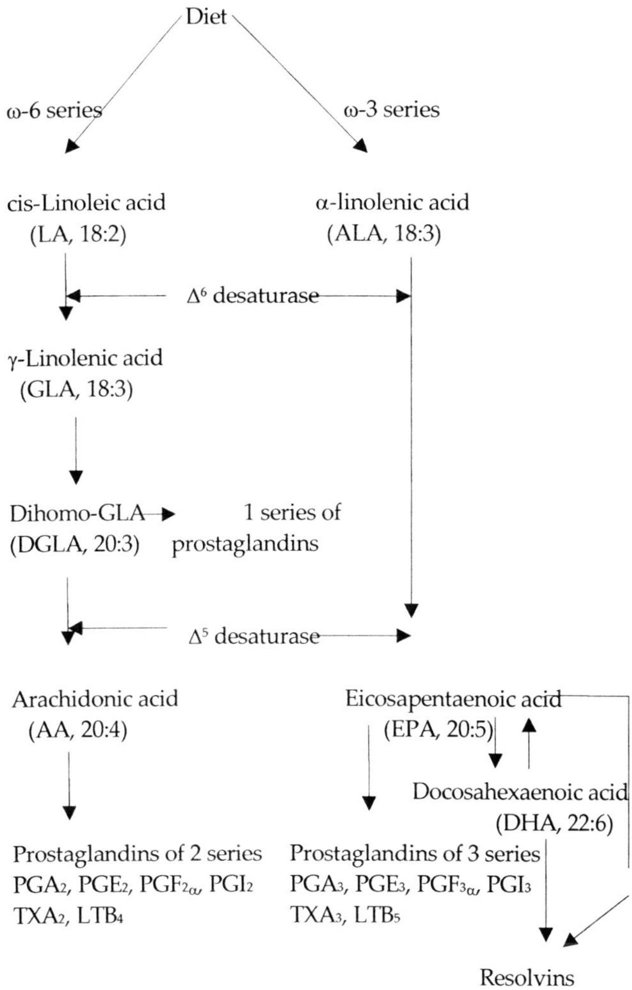

Figure 1. Metabolism of essential fatty acids.

Conclusions

It is possible that injury, infection and inflammation increase the production of pro-inflammatory cytokines during pregnancy both in the mother and the fetus that in turn damage the developing fetal brain by inducing apoptosis of developing neurons, alter the balance between dopaminergic and serotoninergic neurons and predispose them to develop schizophrenia in adult life. Since DHA, EPA and AA have neuroprotective action and suppress the production of pro-inflammatory cytokines, it is possible that sub-clinical deficiency of these fatty acids may lead to enhanced production of pro-inflammatory cytokines due to the absence of their (PUFAs) negative feed back control on the synthesis and action of pro-inflammatory cytokines. Sub-clinical deficiency of PUFAs could also be due to increased production of pro-inflammatory cytokines. It was reported that GLA and EPA have cytoprotective and genoprotective actions (Das et al, 1985; 1989; Lonergan et al, 2002). This suggests that decreased levels of PUFAs may render the fetal brain more susceptible to the cytotoxic actions of cytokines. In view of this, it will be interesting to study whether supplementation of PUFAs to high-risk pregnant and lactating mothers prevents/postpones the development of schizophrenia in their progeny (Figure 2). This is supported by the reported observation that fewer schizophrenic patients than normal were breast-fed (McCreadie, 1997), although this remains controversial. The possible beneficial effect of breast-feeding in the prevention of schizophrenia can be attributed to the presence of significant amounts of PUFAs in human breast milk.

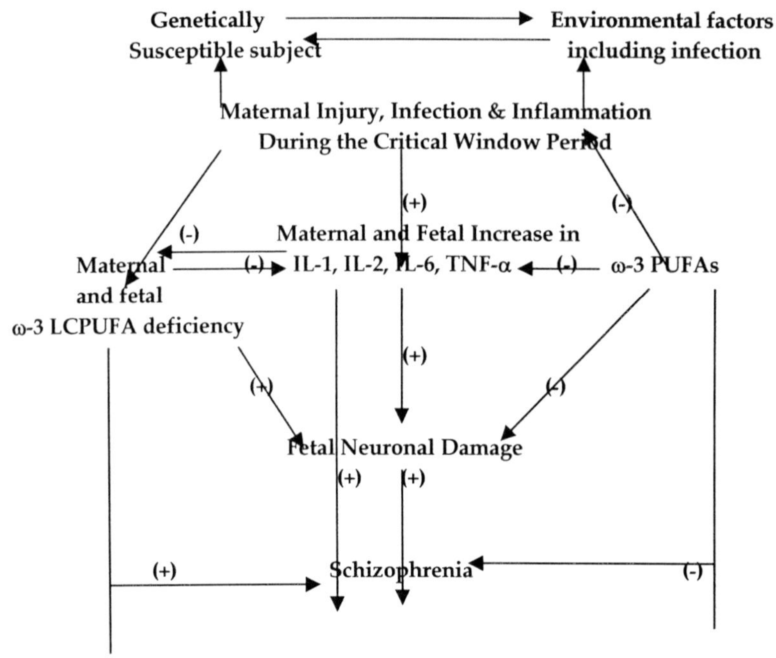

(+) Indicates increase in the synthesis/action or the disease process.

(-) Indicates decrease in the synthesis/action or decrease in the incidence of the disease.

Figure 2. Scheme showing possible interaction(s) between PUFAs, maternal infection, injury, and inflammation, fetal neuronal damage and schizophrenia and bipolar disorders.

References

Allebeck P, Allgulander C, Henninggsohn I, Jakobsson SW. Causes of death in a cohort of 50465 young men-validity of recorded suicide as underlying cause of death. *Scand J Soc Med* 1991; 19: 242-247.

Allen NB, Lewinsphn PM, Seeley JR. Prenatal and perinatal influences on risk for psychopathology in childhood and adolescence. *Dev Psychopathol* 1998; 10: 513-529.

Amore M, Balista C, McCreadie R G, Cimmino C, Pisani F, Bevilacqua G, Ferrari G. Can breast-feeding protect against schizophrenia? Case-control study. *Biol Neonate* 2003; 83: 97-101.

Ansorge MS, Zhou M, Lira A, Hen R, Gingrich JA. Early-life blockade of the 5-HT transporter alters emotional behavior in adult mice. *Science* 2004; 306: 879-881.

Aravindakshan M, Sitasawad S, Debsikdar V, Ghate M, Evans D, Horrobin D F, Bennett C, Ranjekar P K, Mahadik S P. Essential polyunsaturated fatty acid and lipid peroxide levels in never-medicated and medicated schizophrenia patients. *Biol Psychiatry* 2003;53:56-64.

Assies J, Lieverse R, Vreken P, Wanders R J, Dinhemans P M, Linszen D H. Significantly reduced docosahexaenoic and docosapentaenoic acid concentrations in erythrocyte membranes from schizophrenic patients compared with a carefully matched control group. *Biol Psychiatry* 2001; 49: 510-522.

Baguma-Nibasheka M, Brenna J T, Nathaniesz P W. Delay of preterm delivery in sheep by omega-3 long-chain polyunsaturates. *Biol Reprod* 1999; 60: 698-701.

Barker DJP, Osmond C, Rodin J, Fall CHD, Winter PD. Low weight gain in infancy and suicide in adult life. *BMJ* 1995; 311: 1203.

Bateson P, Barker D, Clutton-Brock T, Deb D, D'Udine B, Foley RA, Cluckman P, Godfrey K, Kirkwood T, Lahr MM, McNamara J, Metcalfe NB, Monaghan P, Spencer HG, Sultan SE. Developmental plasticity and human health. *Nature* 2004; 430: 419-421.

Boin F, Zanardini R, Pioli R, Altamura C A, Maes M, Gennarelli M. Association between–G308A tumor necrosis factor alpha gene polymorphism and schizophrenia. *Mol Psychiatry* 2001; 6: 79-82.

Borrell J, Vela J M, Arevalo-Martin A, Molina-Holgado E, Guaza C. Prenatal immune challenge disrupts sensorimotor gating in adult rats. Implications for the etiopathogenesis of schizophrenia. *Neuropsychopharmacology* 2002; 26: 204-211.

Carlson S E, Werkman S H, Peeples J M, Cooke R J, Tolley E A. Arachidonic acid status correlates with first year growth in preterm infants. *Proc Nat Acad Sci USA* 1993; 90: 1073-1077.

Caspi A, Sugden K, Moffitt TE, Taylor A, Craig IW, Harrington H, McClay J, Mill J, Martin J, Braithwaite A, Poulton R. Influence of life stress on depression: moderation by a polymorphism in the 5-HT gene. *Science* 2003; 301: 386-390.

Cazzullo C L, Sacchetti E, Galluzzo A, Adorni A, Pegoraro M, Bosis S, Colombo, F, Trabattoni D, Zagliani A, Clerici M. Cytokine profiles in schizophrenic patients treated with risperidone: a 3-month follow-up study. *Prog. Neuropsychopharmacol Biol Psychiatry* 2002; 26: 33-39.

Chang MC, Bell JM, Purdon AD, Chikhale EG, Grange E. Dynamics of docosahexaenoic acid metabolism in the central nervous system: lack of effect of chronic lithium treatment. *Neurochem Res* 1999; 24: 399-406.

Chang MC, Contreras MA, Rosenberger TA, Rintala JJ, Bell JM, Rapoport SI. Chronic valproate treatment decreases the in vivo turnover of arachidonic acid in brain phospholipids: a possible common effect of mood stabilizers. *J Neurochem* 2001; 77: 796-803.

Chang MC, Grange E, Rabin O, Bell JM, Allen DD, Rapoport SI. Lithium decreases turnover of arachidonate in several brain phospholipids. *Neurosci Lett* 1996; 220: 171-174.

Chaudhry A, Laychock SG, Rubin RP. The effects of fatty acids on phosphoinositide synthesis and myo-inositol accumulation in exocrine pancreas. *J Biol Chem* 1987; 262: 17426-17431.

Crow TJ, Done DJ, Sacker A. Childhood precursors of psychosis as clues to its evolutionary origins. *Eur Arch Psychiatry Clin Neurosci* 1995; 245: 61-69.

Das U N. The lipids that matter from infant nutrition to insulin resistance. *Prostaglandins Leukot Essen Fatty Acids* 2002a; 67: 1-12.

Das UN. *A Perinatal Strategy to Prevent Adult Disease: A role for long-chain polyunsaturated fatty acids.* Kluwer Academic Publishers, Boston, 2002b.

Das U N. Beneficial effect of eicosapentaenoic acid and docosahexaenoic acid in the management of systemic lupus erythematosus and its relationship to the cytokine network. *Prostaglandins Leukot Essen Fatty Acids* 1994; 51: 207-213.

Das U N. Long-chain polyunsaturated fatty acids in the growth and development of the brain and memory. *Nutrition* 2003a, 19: 62-65.

Das U N. Can memory be improved? A discussion on the role of *ras*, GABA, acetylcholine, NO, insulin, TNF-α, and long-chain polyunsaturated fatty acids in memory formation and consolidation. *Brain Development* 2003b; 25: 251-261.

Das U N. Essential fatty acids as possible mediators of the actions of statins. *Prostaglandins Leukot Essen Fatty Acids* 2001; 65: 37-40.

Das U N. The Brain-lipid-heart connection. *Nutrition* 2001; 17: 260-263.

Das UN, Devi GR, Rao KP, Rao MS. Prostaglandins and their precursors can modify genetic damage induced by benzo (a,) pyrene and gamma-radiation. *Prostaglandins* 1985; 29: 911-920.

Das UN, Devi GR, Rao KP, Rao MS. Prostaglandins can modify gamma-radiation and chemical-induced cytotoxicity and genetic damage both in vitro and in vivo. *Prostaglandins* 1989; 38: 689-699.

Done DJ, Johnstone EC, Frith CD, et al. Complications of pregnancy and delivery in relation to psychosis in adult life: data from the British Perinatal Mortality Survey. *BMJ* 1991; 302: 1576-1580.

Dworkin RH, Cornblatt BA, Friedmann R, et al. Childhood precursors of affective vs social deficits in adolescents at risk for schizophrenia. *Schizophrenia Bulletin* 1993; 19: 563-577.

Ebrinc S, Top C, Oncul O, Basoglu C, Cavuslu S, Cetin M. Serum interlukin 1 alpha and interleukins 2 levels in patients with schizophrenia. *J Int Med Res* 2002; 30: 314-317.

Editorial. Of mice and mental illness. *Nature Neuroscience* 2003, 6: 323.

Endres S, Ghorbani R, Kelley V E, et al. The effect of dietary supplementation with n-3 polyunsaturated fatty acids on the synthesis of interleukins-1 and tumor necrosis factor by mononuclear cells. *N Engl J Med* 1989; 320: 265-271.

Fricchione GL, Bilfinger TV, Stefano GB. The macrophage and neuropsychiatric disorders. Neurobiol 1996; 9: 16-29.

Gale CR, Martyn CN. Breast feeding, dummy use, and adult intelligence. *Lancet* 1996; 347: 1057.

Gross C, Zhuang X, Stark K, Ramboz S, Oosting R, Kirby L, Santarelli L, Beck S, Hen R. Serotonin A1 receptor acts during development to establish normal anxiety-like behavior in the adult. *Nature* 2002; 416: 396-400.

Hariri AR, Mattay VS, Tessitore A, Kolachana B, Fera F, Goldman D, Egan MF, Weinberger DR. Serotonin transporter genetic variation and the response of the human amygdala. *Science* 2002; 297: 400-403.

Hibbeln J R, Makino K K, Martin C E, Dickerson F, Boronow J, Fenton W S. Smoking, gender, and dietary influences on erythrocyte essential fatty acid composition among patients with schizophrenia or schizoaffective disorder. *Biol Psychiatry* 2003; 53: 431-441.

Jones P, Rodgers B, Murray R, et al. Child developmental risk factors for schizophrenia in the 1946 birth cohort. *Lancet* 1994; 344: 1398-1402.

Jones PB, Rantakallio P, Hartikainen AI, Isohanni M, Sipila P. Schizophrenia as a long-term outcome of pregnancy, delivery, and perinatal complications: a 28-year follow-up of the 1966 north Finland general population birth cohort. *Am J Psychiatry* 1998; 155: 355-364.

Joy CB, Mumby-Croft R, Joy LA. Polyunsaturated fatty acid (fish oil or evening primrose oil) for schizophrenia. *Cochrane Database Syst. Rev* 2000; 2: CD001257.

Khan M M, Evans D R, Gunna V, Scheffer R E, Parikh V V, Mahadik S P. Reduced erythrocyte membrane essential fatty acids and increased lipid peroxides in schizophrenia at the never-medicated first-episode of psychosis and after years of treatment with antipsychotics. *Schizophr Res* 2002; 58: 1-10.

Kim H-Y, Akbar M, Kim K-Y. Inhibition of neuronal apoptosis by polyunsaturated fatty acids. *J Mol Neurosci* 2001; 16: 223-227.

Kim H-Y, Akbar M, Lau A, Edsall L. Inhibition of neuronal apoptosis by docosahexaenoic acid (22:6n-3). Role of phosphatidylserine in antiapoptotic effect. *J Biol Chem* 2000; 275: 35215-35223.

Kim SHF, Weeber EJ, Sweatt JD, Stoll AL, Marangell LB. Inhibitory effects of omega-3 fatty acids on protein kinase C activity in vitro. *Mol Psychiatry* 2001; 6: 246-248.

Kim Y K, Kim L, Lee M S. Relationships between interleukins, neurotransmitters and psychopathology in drug-free male schizophrenics. *Schizophr Res* 2000; 44: 165-175.

Kinney GG, Vogel GW, Feng P. Decreased dorsal raphe nucleus neuronal activity in adult chloral hydrate anesthetized rats following neonatal clomipramine treatment: implications for endogenous depression. *Brain Res* 1997; 756: 68-75.

Kowalski J, Blada P, Kucia K, Madej A, Herman Z S. Neuroleptics normalize increased release of interleukins-1 beta and tumor necrosis factor-alpha from monocytes in schizophrenia. *Schizophr Res* 2001; 50: 169-175.

Kumar K V, Das U N. Effect of cis-unsaturated fatty acids, prostaglandins, and free radicals on angiotensin-converting enzyme activity in vitro. *Proc Soc Exp Biol Med* 1997; 214: 374-379.

Kumar S G, Das U N, Kumar K V, Madhavi N, Das N P, Tan B K H. Effect of n-6 and n-3 fatty acids on the proliferation and secretion of TNF and IL-2 by human lymphocytes in vitro. *Nutrition Res* 1992; 12: 815-823.

Kumar S G, Das U N. Effect of prostaglandins and their precursors on the proliferation of human lymphocytes and their secretion of tumor necrosis factor and various interleukins. *Prostaglandins Leukot Essen Fatty Acids* 1994; 50: 331-334.

Leask SJ, Done DJ, Crow TJ, Richards M, Jones PB. No association between breast-feeding and adult psychosis in two national birth cohorts. Br J Psychiatry 2000; 177: 218-221.

Liu D, Diorio J, Tannenbaum B, Caldji C, Francis D, Freedman A, Sharma A, Pearson D, Plotsky PM, Meaney MJ. Maternal care, hippocampal glucocorticoid receptors, and hypothalamic-pituitary-adrenal responses to stress. *Science* 1997; 277: 1659-1662.

Lonergan P E, Martin D S D, Horrobin D F, Lynch M A. Neuroprotective effect of eicosapentaenoic acid in hippocampus of rats exposed to γ-radiation. *J Biol Chem* 2002; 277: 20804-20811.

Lonergan PE, Martin DSD, Horrobin DF, Lynch MA. Neuroprotective effect of eicosapentaenoic acid in hippocampus of rats exposed to γ-radiation. *J Biol Chem* 2002; 277: 20804-20811.

Lucas A, Morley R, Cole TJ, et al. Breast milk and subsequent intelligence quotient in children born pre-term. *Lancet* 1992; 239: 261-264.

Maes M, Bocchio Chiavetto L, Bignotti S, Battisa Tura G J, Pioli R, Boin F, Kenis G, Bosmans E, de Jongh R, Altamura C A. Increased serum interleukins-8 and interleukins-10 in schizophrenic patients resistant to treatment with neuroleptics and the stimulatory effects of clozapine on serum leukemia inhibitory factor receptor. *Schizophr Res* 2002; 54: 281-291.

Mahadik S P, Evans D, Lal H. Oxidative stress and role of antioxidant and omega-3 essential fatty acid supplementation in schizophrenia. *Prog Neuropsychopharmacol Biol Psychiatry* 2001; 25: 463-493.

Makrides M, Neumann M, Simmer K, Pater J, Gibson R. Are long-chain polyunsaturated fatty acids essential nutrients in infancy? *Lancet* 1995; 345: 1463-1468.

Malloy MH, Berendes H. Does breast feeding influence intelligence quotients at 9 and 10 years of age. *Early Human Development* 1998; 50: 209-217.

Martin D S D, Lonergan P E, Boland B, Fogarty M P, Brady M, Horrobin D F, Campbell V A, Lynch M A. Apoptotic changes in the aged brain are triggered by interleukin-1β-induced activation of p38 and reversed by treatment with eicosapentaenoic acid. *J Biol Chem* 2002; 277: 34239-34246.

Marx C E, Jarskog L F, Lauder J M, Lieberman J A, Gilmore J H. Cytokine effects on cortical neuron MAP-2 immunoreactivity: implications for schizophrenia. *Biol Psychiatry 2001*; 50: 743-749.

McAllister C G, van Kammen D P, Rehn T J, Miller A L, Gurklis J, Kelley ME, Yao J, Peters J L. Increases in CSF levels of interleukins-2 in schizophrenia: effects of recurrence of psychosis and medication status. *Am J Psychiatry* 1995; 152: 1291-1297.

McCreadie RG. The Nothsdale Schizophrenia Surveys. 16. Breast-feeding and schizophrenia: preliminary results and hypotheses. *Br J Psychiatry* 1997; 170: 334-337.

Merry RJ. Oxidative stress and mitochondrial function with aging-the effects of calorie restriction. *Aging Cell* 2004; 3: 7-12.

Mittendorfer-Rutz E, Rasmussen F, Wasserman D. Restricted fetal growth and adverse maternal psychosocial and socioeconomic conditions as risk factors for suicidal behaviour of offspring: a cohort study. *Lancet* 2004; 364: 1135-1140.

Neugebauer R, Hoek HW, Susser E. Prenatal exposure to wartime famine and development of antisocial personality disorder in early adulthood. *JAMA* 1999; 282: 455-462.

Neugebauer R, Reuss ML. Association of maternal, antenatal and perinatal complications with suicide in adolescence and young adulthood. *Acta Psychiatr Scand* 1998; 97: 412-418.

Nilsson PM, Nyberg P, Ostergren PO. Increased susceptibility to stress at a psychological assessment of stress tolerance is associated with impaired fetal growth. *Int J Epidemiol* 2001; 30: 75-80.

Olsen S F, Olsen J, Frische G. Does fish consumption during pregnancy increase fetal growth? A study of the size of the newborn, placental weight and gestational age in relation to fish consumption during pregnancy. *Int J Epidemiol* 1990;19: 971-977.

Oquendo MA, Baca-Garcia E. Nurture versus nature: evidence of intrauterine effects on suicidal behaviour. *Lancet* 2004; 364: 1102-1104.

Oquendo MA, Baca-Garcia. Nurture versus nature: evidence of intrauterine effects on suicidal behaviour. *Lancet* 2004; 364: 1102-1104.

Peet M, Brind J, Ramchand C N, Shah S, Vankar G K. Two double-blind placebo-controlled pilot studies of eicosapentaenoic acid in the treatment of schizophrenia. *Schizophr Res* 2001; 49: 243-251.

Peet M, Horrobin D F, E-E Multicentre Study Group. A dose-ranging exploratory study of the effects of ethyl-eicosapentaenoate in patients with persistent schizophrenic symptoms. *J Psychiatr Res* 2002; 36: 7-18.

Peet M, Poole J, Laugharne J. Infant feeding and the development of schizophrenia. *Schizophrenia Res* 1997; 24: 255-256.

Puri B K, Richardson A J, Horrobin D F, Easton T, Saeed N, Oatridge A, Hajnal J V, Bydder G M. Eicosapentaenoic acid treatment in schizophrenia associated with symptom remission, normalization of blood fatty acids, reduced neuronal membrane phospholipid turnover and structural brain changes. *Int J Clin Pract* 2000; 54: 57-63.

Reseland J E, Haugen F, Hollung K, Solvoll K, Halvorsen B, Brude I R, Nenseter M S, Christiansen E N, Drevon C A. Reduction of leptin gene expression by dietary polyunsaturated fatty acids. *J Lipid Res* 2001; 42: 743-750.

Rumajogee P, Verge D, Hanoun N, Brisorgueil MJ, Hen R, Lesch KP, Hamon M, Miquel MC. Phenotype in the absence of 5-HT autoreceptors or the 5-HT transporter: involvement of BDNF and camp. *Eur J Neurosci* 2004; 19: 937-944.

Salk L, Lipsitt LP, Sturner WQ, Reilly BM, Levat RH. Relationship of maternal and perinatal conditions to eventual adolescent suicide. *Lancet* 1985; 1: 624-627.

Sasaki T, Okazaki Y, Akaho R, et al. Type of feeding during infancy and later development of schizophrenia. *Schizophr Res* 2000; 42: 79-82.

Serhan CN, Hong S, Gronet K, Colgan SP, Devchand PR, Mirick G, Monssignac R-C. Resolvins: a family of bioactive products of omega-3 fatty acid transformation circuits initiated by aspirin treatment that counter proinflammatory signals. *J Exp Med* 2002; 196: 1025-1037.

Shi L, Fatemi S H, Sidwell R W, Patterson P H. Maternal influenza infection causes marked behavioral and pharmacological changes in the offspring. *J Neurosci* 2003; 23: 297-302.

Shi L, Fatemi S H, Sidwell R W, Patterson P H. Maternal influenza infection causes marked behavioral and pharmacological changes in the offspring. *J Neurosci* 2003; 23: 297-302.

Smythe JW, Rowe WB, Meaney MJ. Neonatal handling alters serotonin (5-HT) turnover and 5-HT2 receptor binding in selected brain regions: relationship to the handling effect on glucocorticoid receptor expression. *Brain Res Dev. Brain Res* 1994; 80: 183-189.

Sperling RI, Benincaso AI, Knoell CT, Larkin JK, Austen KF, Robinson DR. Dietary omega-3 polyunsaturated fatty acids inhibit phosphoinositide formation and chemotaxis in neutrophils. *J Clin Invest* 1993; 91: 651-660.

Waterland RA, Garza C. Potential mechanisms of metabolic imprinting that lead to chronic disease. *Am J Clin Nutr* 1999; 69: 179-197.

Williams RSB, Cheng L, Mudge AW, Harwood AJ. A common mechanism of action for three mood-stabilizing drugs. *Nature* 2002; 417: 292-295.

Zalcman SS. Interleukin-2-induced increases in climbing behavior: inhibition by dopamine D-1 and D-2 receptor antagonists. *Brain Res* 2002; 944: 157-164.

Zhang X Y, Zhou D F, Zhang P Y, Wu G Y, Cao L Y, Shen Y C. Elevated interleukins-2, interleukins-6 and interleukins-8 serum levels in neuroleptic-free schizophrenia: association with psychopathology. *Schizophr Res* 2002; 57: 247-258.

In: Fatty Acids and Oxidative Stress...
Editors: R. Reddy and J.K. Yao, pp. 237-268
ISBN: 978-1-60021-677-0

Chapter XVI

Unraveling Complexity by Integrative Biochemical Investigation

Ashraf H. Elamin, Billy Day and Jeffrey K. Yao

Introduction

Neuropsychiatric disorders are biologically complex disorders with perturbations in a broad array of neurochemical systems, such as in dopaminergic, serotonergic, and glutamatergic pathways, along with changes in energy metabolism, signal transduction, membrane composition and function, and gene expression. Until recently it has been difficult to examine these inter-connected biological systems as a whole, largely because of methodological limitations, leading to the evidence being collected piecemeal.

Recent advances in high-resolution multidimensional separation techniques that allow examination of complex mixtures of proteins and metabolites from multiple metabolic pathways, are offering new views of the "universe" of metabolic systems that are expressed in given tissue, such as plasma or cerebrospinal fluid (CSF), and can aid in identifying biomarkers of disease states. Such an integrative biological investigation offers quantitative assessment of hundreds to thousands of proteins and metabolites, some of which are different between patients and controls. Integrative investigations have generally been conducted in the absence of a theoretical framework, such as the recent study of postmortem brain tissue that identified pathways relating to oxidative stress and energy metabolism being altered in patients with schizophrenia [Prabakaran et al., 2004]. This chapter reviews the latest methods that fruitfully can be applied to broaden the scope of biological investigations in neuropsychiatric disorders.

Proteomics

Proteomics is defined as a scientific approach used to elucidate all protein species within a cell or tissue. Proteomics was first used to introduce the concept of exploring changes in all proteins expressed by a genome; the tools used in proteomic analysis are borrowed from traditional biochemistry. Proteomics now deals with the study of proteins, their structures, localizations, posttranslational modifications, functions and interactions with other proteins. Aberrations in protein structure and expression levels signal in most cases the presence of disease, rendering the proteins excellent targets in disease diagnostics, prognostics and therapeutics [Monti et al., 2005]. The mapping of protein structures to their function holds the key to a better understanding of cellular functions under both normal and disease states, which is critical for modern drug discovery.

These new goals, however, cannot be easily achieved as intrinsic difficulties increase by several orders of magnitude when moving from genome to proteome research. The static nature of the genome, in fact, cannot be compared to the dynamic properties of the proteome; protein expression profiles change several times during the cell cycle and are heavily affected by a number of intracellular and extracellular stimuli (temperature, stress, apoptotic signals, etc.) Moreover, the occurrence of alternative splicing and post-translational modifications led to a complete rethinking of the old paradigm "one gene-one protein," which is now recognized to inaccurately reflect the true nature of the cellular proteome [Anderson and Anderson, 2002].

Like any area of research, careful attention must be given to proper proteomic experimental planning and design so that meaningful results can be obtained. Significant improvements in protein separation methodologies such as two-dimensional electrophoresis (2DE) and chromatography, as well as in mass spectrometry systems, have increased the resolution, sensitivity and accuracy in protein detection and identification. Nevertheless, some major challenges have arisen, such as the detection of the low abundance proteins, and remain [Stasyk and Huber, 2004]. Other new challenges have also arisen, such as the development of the computing resources that can reliably analyze the vast amount of data that the applications of new technologies generate.

Metabolomics

Metabolomics is the study of global metabolite profiles in a system (cell, tissue or organism under a give set of conditions. Low molecular weight metabolic intermediates (*i.e.*, biochemical intermediates and xenobiotic metabolites that are not nucleic acid nor amino acid polymers, typically with molecular weights of less than 2000 Da; hereafter termed "metabolites") are the result of the interaction of the system's genome with its environment and are not merely the end product of gene expression but also form part of the regulatory system in an integrated matter [Saghatellian and Cravatt, 2005] . Metabolomics (or metabonomics) has been labeled one of the new "omics", joining genomics, transcriptomics, and proteomics as a science employed toward understanding of global systems biology.

The metabolome, which refers to full complement of metabolites within a cell, tissue, or organism, encompasses a diverse array of molecular chemotypes, including peptides, carbohydrates, lipids, nucleosides and catabolic products of exogenous compounds. As a rich source of both signaling and structural molecules, the metabolome and, in particular, its composition and regulation are understandably subjects of intense research interest. By integrating these "small molecule" profiles with transcript and protein expression/activity patterns, researchers hope to determine the contribution that specific metabolites and metabolic pathways make to "health" and "disease" [Fiehn, 2002].

Current Experimental Focus in Proteomics

Although the proteomic analysis approach is sometimes used to characterize all protein species within a given cell, many researchers take advantage of proteomic technology to elucidate protein *changes* between healthy and diseased states. The following describes some of the current foci of researchers employing proteomics techniques.

Protein Interaction Mapping

The mapping of protein-protein interactions is arguably the most significant development in proteomics to date. Bioinformatics developments for the analysis and integration of interaction data sets to build networks and pathways have also increased in recent years. Protein interaction reports fall under three categories: low-throughput experiments that often focus on a single protein; medium-throughput experiments that focus on specific pathway components; and high-throughput experiments that focus on large interaction maps and networks (global mapping) [Lambert et al., 2005]. It is also important to note that, to date, a relatively small amount of research has been focused on mapping the different interactions between the basal and activated states of proteins in high-throughput proteome mapping experiments.

Biomarker Discovery

Biomarker discovery studies have become a trend of research in proteomics due to an increasing need for preclinical and clinical biomarker targets. A biomarker may be defined as a characteristic, or a combination of characteristics, that can be objectively measured and evaluated as an indicator of normal biological processes, pathogenic processes, or a pharmacological response to a therapeutic intervention. Lately, biomarker research has evolved at an accelerated pace. Although authors in many reports claim that they have discovered biomarkers, the reality is that they have discovered *candidate* biomarkers. These candidate biomarkers must be taken through a validation process that is linked to potential usage. For example, the validation of a candidate biomarker for preclinical usage is significantly less demanding than its validation as a clinical surrogate end point or as approved diagnostic marker. Many studies have been performed to identify candidate biomarkers in serum, urine, pancreatic juice, interstitial fluid and cell lysates [Lambert et al., 2005].

Other Organisms Proteomics

Because of full genome mapping exercises, there has been a large increase in research on single cell organisms to elucidate the global mapping of their proteomes. Moreover, these organisms divide rapidly and are amenable to considerable controlled genomic manipulation. Bacterial, fungal, yeast proteomics analyses have shown a large increase in the proteins identified and characterized over the past few years due to the improvement in the proteomics analysis techniques that originally made it difficult to perform such experiments in the past.

Posttranslational Modification (PTM)

Posttranslational modification (PTM) is an integral part of gene regulation. PTMs of proteins, which cannot detected through RNA analyses, may occur at different stages of, *e.g.*, tumor development and be indicative of early or late events of transformation. During PTM, a protein is modified by phosphorylation, acetylation, glycosylation, ubiquitination, farnesylation, sumoylation, methylation, sialylation, and/or a variety of other means. PTMs are important because some modifications can influence subcellular localization and cell signal transduction pathways [Lambert et al., 2005]. All alter the chemical composition of the protein in question, and therefore its identity.

One important PTM is glycosylation. Earlier studies were focused on global glycosylation to detect different glycoforms. However, the focus has recently shifted toward site-specific glycosylation and determination of the nature of the sugar moiety present. Another heavily researched PTM is phosphorylation. There has been significant progress in the biochemical and instrumental analysis of reversible phosphorylation in the past two years. New techniques have been developed to isolate, purify, and enrich phosphopeptides, and large scale phosphoproteomic analyses have been conducted [Lambert et al., 2005]. Significantly, some of these analyses have monitored the phosphorylation dynamics of a system as a stimulus is added.

There has been significant development of bioinformatic tools for proteomics. Major developments have been driven by an increasing interest in the validation of peptide identifications, especially in the context of high-throughput experiments, the increasing use of statistical methods to analyze results and an increasing use of statistical methods to analyze results across multiple samples due to an increased emphasis on differential proteomics. Furthermore, the need to link mass spectrometric data to a multiplicity of genomic sequence entries in databases and the development of common standards for the dissemination of proteomic results have increased the global interest in proteomic bioinformatics.

Clinical Applications in Metabolomics

Metabolomics has been an application-driven science. A natural extension of early single metabolite analyses was the application of metabolomics to identify disease biomarkers. Specific examples to indicate the range of approaches include the successful diagnosis of coronary heart disease, the identification of Lesch-Nyhan syndrome (that results in a serious mental disorder), the correct discrimination of women with epithelial ovarian cancer compared to control subjects and the investigation of host-pathogen interactions in plants. Analysis of blood spots has enabled the detection of inborn errors of metabolism, and the techniques have also been applied to shellfish to identify novel biomarkers for withering syndrome (a disease of abalone) [Saghatellian and Cravatt, 2005; Rochfort, 2005].

Metabolomics techniques have also been applied in natural products research that includes drug discovery and functional food analysis. One of the often-perceived hurdles involved with natural product drug discovery is the possibility of redundancy -- that is, working on multiple samples with the same active agent. The use of a metabolomic approach, with appropriate statistical analysis, has the potential to allow rapid analysis of these complex data.

Analytical Approaches in Proteomics and Metabolomics

1- and 2- Dimensional Electrophoresis

Gel electrophoresis, particularly two-dimensional separations (2-DE), is the classical and still a principal tool for protein profiling of biological samples. This methodology provides high resolution in protein separation and the capability of resolving thousands of proteins in one experiment.

Figure 1. An illustrative 2-DE silver-stained image.

In 2-DE, proteins are first separated on the basis of differences in their net charge through a technique known as isoelectric focusing (IEF), and then through differences in their molecular radii (which are closely aligned to their molecular masses) through polyacrylamide gel electrophoresis (Fig. 1). Use of immobilized pH gradient (IPG) strips is the preferred method of IEF due to the method's reproducibility. 2-DE is also ideal for subsequent protein identification techniques as the method also allows for application of relatively large amounts of proteins in a single sample, plus a convenient and safe means for storing isolated, albeit denatured, proteins.

Gels can be stained for protein detection with several reagents. These typically include, in increasing order of sensitivity, Coomassie blue, silver stain or fluorescent dyes. Images of the gels are acquired by visible lamp, photodiode, laser scanners or fluorescent imagers as applicable, and are subsequently analyzed by specific densitometry software that can be used to discover differences within not only a single gel but also a single spot (if two or more, e.g., fluorescent dyes are used simultaneously). These software programs provide spot detection and ratio analysis for the determination of quantitative protein changes between different samples.

The major advantage of 2-DE is that it enables the simultaneous separation and visualization of thousands of unknown proteins. It should be noted that, due to PTMs and other mechanisms, the proteins on 2-DE gels are often represented by more than one spot, so the number of expressed products is much higher that the number of the corresponding encoding genes.

2-DE has four major disadvantages and limitations. The first is that only proteins that are soluble under the conditions of IEF can be analyzed -- this excludes, e.g., most membrane-integral proteins. The second is that the detection of low and high molecular mass as well as of basic and hydrophobic proteins is inefficient -- 2-DE works best with proteins with isoelectric points (pIs) between 4-10 and molecular weights between 15 kDa-200 kDa. The third is that 2-DE analysis resolves only the major components of a protein mixture. The fourth is that 2-D gel running as well as analysis remains labor intensive and also requires excellent technical skill [Vlahou and Fountoulakis, 2005].

Difference Gel Electrophoresis (DIGE)

Differential gel electrophoresis (DIGE) is the technique that involves labeling of samples with fluorescent dyes, followed by their mixing then separation in the same gel. Further analysis includes visualization of the spots and signal quantification using fluorescent imagers.

The DIGE system is a specialized two-dimensional electrophoresis technique that uses electrophilic (typically as *N*-hydroxysuccinimide esters; maleimide forms are also available) of two or three different fluorescent dyes (currently Cy-3, Cy-5, and Cy2) to covalently label the protein samples being compared. The fluorescent dyes that work best have been designed so that they do not alter the charge of proteins or impart significant electrophoretic mobility differences to identical proteins. The gel is scanned at the wavelengths corresponding to the dyes used. Thus, in DIGE, every protein in one sample superimposes with its differentially labeled counterpart in the second sample. By overlaying the two images, it is easy to visualize those proteins that differ between the two samples [Gorg et al., 2004]. Compared to traditional 2-D gel techniques for protein profiling, DIGE provides a significantly more efficient, sensitive and reliable way to detect proteins whose levels are altered between control and treated samples.

Affinity Tags

The identification of interacting proteins in stable complexes *in vivo* essentially relies on affinity-based procedures (Fig. 2). The basic idea is to express the protein of interest with a suitable tag to be used as a bait to fish its specific partners out from a cellular extract. Isolation of the entire multi-protein complex can then be accomplished by taking advantage of the availability of several anti-tag systems, immobilized on, *e.g.*, agarose-supports.

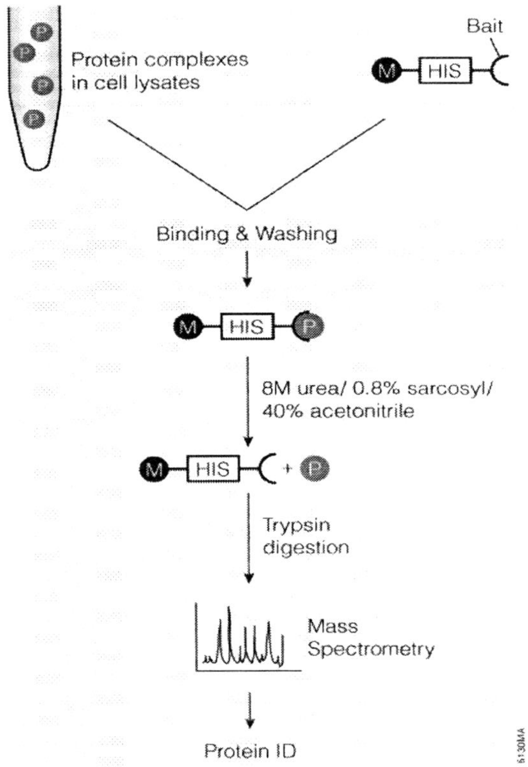

Figure 2. Affinity tag systemic workflow explaining the different steps for the affinity tag experiments analysis.

With some commercially available protein expression systems, the protein bait can be produced as a hybrid protein fused to glutathione *S*-transferase (GST-fused protein), to small peptide epitopes (FLAG, HA, or c-myc), containing a poly-His tail for metal-affinity (e.g., Ni^{2+}) isolation, or covalently modified with biotin. All these affinity tag systems provided a general applicability with a large number of proteins and have a minimal effect on the tertiary structure and biological activity of the bait. They can also serve to avoid complexes' instability. The immobilized protein forms stable, non-covalent interactions with specific partners occurring in the cellular extract, whereas the unbound proteins will be eluted during the washing. The protein components specifically recognized by the bait and retained on the immobilized anti-tag support can then be eluted and fractionated by SDS-PAGE [Monti et al., 2005; Cho et al., 2004].

Affinity tags have several disadvantages and pitfalls. One disadvantage is that when protein extracts are prepared by cellular lysis, the architecture of the subcellular compartments is disrupted and proteins that are normally segregated in different organelles can come into contact, generating nonphysiological interactions. The major limitation of the use of affinity tags is that all the interactions among the bait and its protein partners are essentially *in vitro* interactions taking place outside the cell, on the derivatized support [Cho et al., 2004]. These pitfalls and disadvantages might cause false positives, an outcome that should be considered carefully when using affinity tags.

Immunoprecipitation

To overcome most of the disadvantages of affinity-based approaches, alternative strategies, essentially relying on immunoprecipitation techniques, have been introduced. The gene coding for the bait tagged with an epitope against which good antibodies exist (FLAG, HA, c-myc, etc.) is transfected into the appropriate cell line and expressed in the cognate host. Protein complexes are allowed to form *in vivo* within the cell and the cell extracts are immunoprecipitated with anti-tag monoclonal antibodies using suitable experimental conditions to avoid dissociation of the complexes (Fig. 3). The immunoprecipitated material containing the protein bait and its interacting partners can then be fractionated by 1- or 2-DE and the individual protein components identified by different mass spectrometric methodologies [Monti et al., 2005].

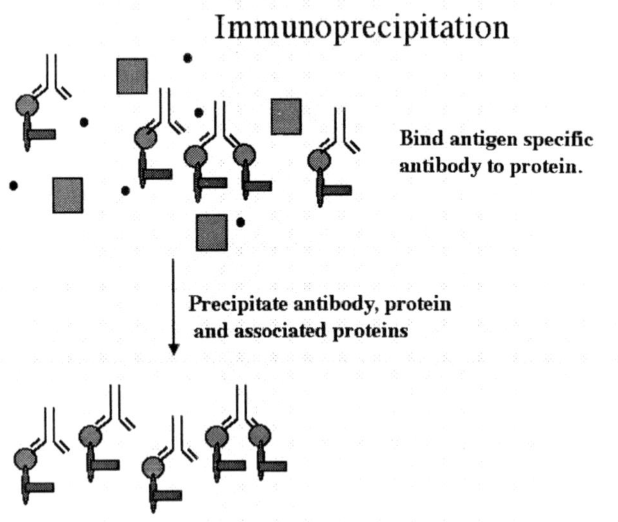

Figure 3. The immunoprecitation procedure.

One of the drawbacks of immunoprecipitation is that the antibodies used are not always suitable for immunoprecipitation protocols, which require a more efficient (and quantitative) recognition of the proper antigen when compared to Western blot or ELISA applications. Cross-recognition of specific antigens or specific binding of proteins to the antibodies, to the peptide tags, or to the insoluble support can lead to false positives. According to recent literature, the use of antibodies specifically directed against the protein bait is not encouraged because the antibody might compete with the interacting proteins for binding of to the bait epitope, thus leading to destabilization of protein-protein interactions and dissociation of the protein complexes. The presence of the tag can alter protein conformation enough to alter or impair complex formation. The last drawback is that the overexpression of the tagged proteins in the host cells may definitely alter the stoichiometric ratio between the bait and the natural partners, often leading to the formation of non-specific and/or nonnatural protein interactions with host proteins [Kobor et al., 1999].

Protein and Peptide Fractionation using Liquid Chromatography

Liquid chromatography (LC) techniques, as well as electrophoretic separation in liquid-phase techniques, are well known methods of protein separation and have been extensively described in literature. Each type of LC has already undergone numerous developments and improvements. LC separation techniques take advantage of differences in the peptide/protein physiochemical properties, such as size (size exclusion chromatography), charge (ion-exchange chromatography) and hydrophobicity (reverse phase chromatography). The electrophoretic techniques include the use of capillary isoelectrofocusing (IEF) and capillary zone and free-flow electrophoresis (FFE)].

The experimental protocol involves dilution of the protein mixture with a buffer that is compatible with the type of stationary phase in use and then elution of the bound material based on some type of competitive attraction [Neverova and Van Eyk, 2004]. For example, in cation-exchange chromatography, elution is differentially induced by use of a gradient of salt. Various types of detectors can be used on the effluent from the columns so as to provide qualitative and occasionally quantitative information on the separated molecules. Unless, however, coupled to mass spectrometers, chromatography almost always indicates only the presence of a protein, without providing its identity.

LC separation is faster than 2-DE and can be applied efficiently for the study of hydrophobic proteins and peptides, as well as for PTMs. In addition, multi-dimensional chromatographic methods (for example ion-exchange and hydrophobic columns used in tandem) have been developed and successfully coupled to mass spectrometric systems. This methodology, when applied to proteolytic digests of proteins in a complex mixture, is known as MUD-PIT (multidimensional protein identification technology) and allows the rapid study and identification of proteins within such mixtures. When combined to metabolic, enzymatic or chemical protein-labeling methods, this methodology allows the detection of quantitative differences between proteins of different classes of samples. Semi-automated systems have been developed that perform all operations from sample application to fraction collection and on-line connection to mass spectrometers for protein identification. The major drawback of the use of MUD-PIT is that it is in terms of the LC plumbing, the LC and instrument systems associated with it are expensive and the computing resources for the identification of proteins from such complex mixtures are still under development.

Chromatography is the most versatile technique available for protein and peptide separation and can be used for single analyte purification or profiling and analysis of complex protein and peptide mixtures. Here are some applications of chromatography as a part of proteomic analysis.

Liquid Chromatography for Small Molecules

The task may be the purification of peptides, hormones, metabolites, or small proteins as pharmaceuticals. Efficient purifications of picogram to multi-gram amounts of peptides are achieved through 1-D and 2-D LC. Multidimensional LC is an essential analytical and preparative chromatographic tool in the pharmaceutical industry. Along with high resolution and sensitivity, multidimensional LC systems offer adequate high throughput and reproducibility [Neverova and Van Eyk, 2004].

LC in Clinical Proteomics

Using proteomic approaches and technologies to search for biomarkers of a disease for diagnosis, prognosis and risk stratification, as well as for monitoring the effectiveness of treatment is termed clinical proteomics. The search for small molecules, peptides or proteins secreted or escaping into body fluids (particularly plasma or serum) represents the next level of separation complexity over purification of a single analyte. In this situation, the protein(s) of interest must be isolated from the many highly abundant proteins in plasma (e.g., albumin), leading to the complexity of dealing with a huge dynamic range of concentrations. The best approach is through the use of affinity chromatography, either for removal of highly abundant proteins or for enrichment of the analyte(s) of interest, due to its extreme selectively based on the molecular recognition between ligand and target.

Affinity chromatography is a powerful tool for the study of protein-protein interactions. Most common affinity ligands are peptides, oligonucleotides (RNA or DNA fragments) and antibodies that can be generated by high-throughput combinatorial biochemistry [Neverova and Van Eyk, 2004].. Furthermore, there are numerous examples of affinity-based separation of proteins, glycoprotein, peptides containing cysteine, tryptophan or metheonine, histidine-tagged proteins and peptides, and phosphopeptides purified using metal-containing resins.

Liquid Chromatography Approaches that Expand the View of the Proteome

Analysis of the entire proteome of a cell, organ, organism or a body fluid is difficult if not impossible, particularly in higher order eukaryotes such as humans. The most common approach is to divide the sample into parts or subproteomes, thereby simplifying protein separation and identification. Even so, subproteomes, such as subcellular fractions (e.g., mitochondrial, nuclear or cytoplasmic/soluble cellular pools) can have thousands of proteins. In this case, Mud PIT LC is a good approach. The multidimensional separation achieved by biphasic column chromatography in combination with mass spectrometry used in Mud-PIT approach has led to the identification of 1484 proteins in a soluble part of a *S. cerevisiae* cell lysate, including low-abundance proteins, proteins with extreme pI values, and proteins with transmembrane domains [Neverova and Van Eyk, 2004].

In a typical Mud-PIT experiment, the entire protein sample is subjected to proteolytic digestion prior to the separation. Trypsin is the common choice as the proteolytic enzyme because it cleaves specifically at lysine or arginine, two highly abundant amino acids found throughout the sequences of most proteins. The digestion produces an even more complex mixture than the original mixture. As a result, thousands of tryptic peptides must be detected and identified in each fraction submitted for MS analysis, including the unique subset of peptides separated by biphasic chromatography [Neverova and Van Eyk, 2004]. The prime benefit of Mud-PIT is that it facilities high-throughput protein identification of an extremely complex mixture by combining the resolving power of high-performance chromatography with the analytical capacity of tandem mass spectrometry and bioinformatics.

It is important to notice that the above mentioned global approaches, often termed "shotgun proteomics," have only been applied to simple organisms or cellular systems. For experiments designed to identify as many components as possible, too high a level of complexity will result in a less than complete survey of peptides in the mixture. The alternative approach is to devote time to the separation step prior to MS analysis through extensive fractionation in order to detect all components present (that are abundant enough to be detected) and fully utilize the dynamic range of the system. The lack of complete and accurate genome sequences for higher species will become more critical as the analysis of low abundance proteins increases and unique proteins are more frequently encountered [Neverova and Van Eyk, 2004]. Therefore, techniques for enrichment of specific subpopulations of the proteins and peptides will be highly sought..

Labeling Techniques for Relative and Absolute Protein Quantitation Using iTRAQ

Often, the goal in proteomics research is to monitor changes in protein composition under different physiologically relevant conditions, a type of study referred to as differential profiling. While monitoring these qualitative changes is valuable, there is a need to develop quantitative tools that can provide more insight into these experiments. There is a requirement for development of a technique that can address some of the remaining issues in quantitative proteomics such as: 1) improve overall protein and proteome coverage while retaining important PTM information; 2) simultaneously compare multiple samples, for example, normal versus disease versus drug treated samples or time course studies, in the same experiment; 3) quantify, in absolute terms, specific proteins of interest, such as biomarkers or to screen drug targets, and 4) increase statistical relevance needed for quantitative experiments by expanding multiplexing, up to four, to include duplicates or triplicates in the design.

Applied Biosystems's iTRAQ reagents are a multiplexed set of isobaric electrophilic reagents that are amine specific and yield labeled peptides identical in mass and hence also identical in single MS mode, but which produce strong, diagnostic, low mass MS/MS signature fragment ions that allow for quantitation of up to four different samples simultaneously (Fig. 4). In addition to multiplexing, information such as PTMs is not lost using this new chemistry. Since all the peptides are tagged, proteome coverage is expanded and analysis of multiple peptides per protein improves the confidence in those identified. The iTRAQ technique gives the ability of perform absolute quantitation across numerous sample states, for the synchronous uniform comparison of normal, diseased and/or drug-treated states [iTRAQ Kit, Applied Biosystems].

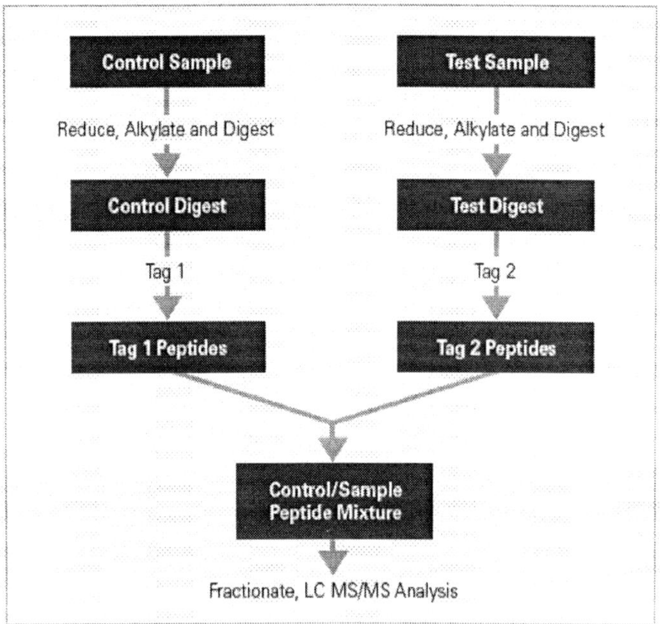

Figure 4. Flow chart showing the iTRAQ workflow.

The technology has been incorporated into a simple workflow involving parallel protein extraction and routine enzymatic digestion common to any LC/MS experimental procedure. Subsequent labeling of peptides occurs in a manner that allows rapid transfer to MS/MS analysis and easy data interpretation without burdensome sample handling requirements. The use of these labels therefore permits simultaneous measurement of relative and/or absolute protein abundance of multiple samples in a single LC MS/MS run.

Identification and Characterization Techniques using Mass Spectrometry

Mass spectrometers (MS) are analytical instruments used for measuring the mass to charge ratio of ions formed from a sample. Structural information can be generated using certain types of mass spectrometers, usually tandem mass spectrometers (MS/MS), and this is achieved by fragmenting the parent ion of an analyte and analyzing the products generated. This procedure is useful for the structural elucidation of organic compounds and peptides, for oligonucleotide sequencing, and for monitoring the existence of previously characterized compounds in complex mixtures with high specificity, by defining the molecular weight and a diagnostic fragment of the molecule simultaneously. For example, MS/MS instruments are particularly useful for the detection (and, if appropriate internal standardization is employed, quantitation) of specific drug metabolites in biological matrixes [Gorg et al., 2004; Aebersold and Mann, 2003; Mann et al., 2001; Shevchenko et al., 1996].

Mass spectrometry is the current method of choice for identification of proteins, since this method offers high analytical sensitivity and the capacity for high throughput. Mass spectrometry has been very effectively coupled to both 2-DE and LC methodologies so as to achieve protein identification. In brief, analysis of protein/peptide molecule (analyte) by mass spectrometry occurs in three major steps: (1) ionization of the analyte and generation of gas-phase ions; (2) separation of the ions according to their mass to charge ratio; and (3) detection of these generated ions. Mass spectrometers can be divided into three fundamental parts, namely the ionization source, the analyzer and the detector [Gorg et al., 2004; Aebersold and Mann, 2003; Mann et al., 2001; Shevchenko et al., 1996].

The sample under investigation has to be introduced into the ionization source of the instrument. Once inside the ionization source the sample molecules are ionized, because ions are easier to manipulate than neutral species. The ions are extracted into the analyzer region of the mass spectrometer where they are separated to their mass (*m*) to charge (*z*) ratios (*m/z*). The separated ions are detected and this signal sent to the data system where the *m/z* ratios are stored together with their relative abundance for presentation in the format of a *m/z* spectrum. The analyzer and the detector of the mass spectrometer, and often the ionization source too, are maintained under high vacuum to give the ions a reasonable chance of traveling from one end of the instrument to the other without any hindrance from air molecules. The entire operation of the mass spectrometer, and often the sample introduction process also, is under complete data system control on modern mass spectrometers.

The ionization methods used for the majority of biochemical and protein/peptide analyses are electrospray ionization (ESI) and matrix assisted laser desorption ionization (MALDI). Both of these methods will be described later. The main function of the mass analyzer is to separate, or resolve, the ions formed in the ionization source of the mass spectrometer according to their *m/z* ratios. There are a number of mass analyzers currently available, the better known of which include quadrupoles, time of flight (TOF), magnetic sectors, both three-dimensional and linear quadrupole ion traps, and the ion cyclotron resonance (ICR) cells used in Fourier transform (FT) mass spectrometers. Each of these shows distinct advantages and weakness with regards to the sensitivity, mass accuracy and mass resolving power they provide. .

Tandem (MS/MS) mass spectrometers are instruments that either have more than one analyzer separated by a collision/fragmentation cell, or a means to eject parent ions not of interest and form then analyze fragment ions from only the precursor ion of interest. MS/MS systems can be used in structural and sequencing studies. Two, three and four analyzers have all been incorporated into the now several commercially available MS/MS instruments, and the analyzers do not necessarily have to be of the same type, in which case the instrument is termed a "hybrid" MS/MS system. More popular MS/MS systems include those of the 3D ion trap, triple quadrupole, TOF/TOF, magnetic sector(s)-quadrupole, quadrupole-TOF geometries, and more recently the linear ion trap-ICR and quadrupole-quadrupole-ICR systems. In this way, combination of the strengths of each analyzer is achieved greatly improving the outcome of the proteomics research.

Electrospray Ionization (ESI)

Electrospray ionization (ESI) is one the atmospheric pressure ponization (API) techniques, and is well suited to the analysis of polar molecules ranging from less than 100 Da to more than 1,000,000 Da in molecular weight.

During ESI, the sample is dissolved in a polar, volatile solvent (e.g., typical reverse phase HPLC solvents) and pumped through a narrow, stainless capillary (75-150 μm i.d.) at a flow rate of between 1 nL/min to 1 mL/min. A high voltage is applied to a metal needle attached to the tip of the capillary, resulting in a strong electric field that forces the dispersion and emergence of the sample as an aerosol of highly charged droplets, a process that is aided by a co-axially introduced nebulizing gas (usually nitrogen, but compressed air also works) flowing around the outside of the capillary. Eventually, the neutral solvent molecules disperse to leave the naked charged sample ions, some of which will pass into the mass analyzer, which is held under high vacuum, due to the potential difference between it and ion. The advantage of nL/min flow rates in ESI (often termed nanospray) is that the flow rate is very low and so not only is far less sample required but also the small volume of sample can last for considerable time, thus enabling multiple experiments to be performed. With biological samples that often contain many protonatable sites, multiple charging of the analyte occurs. This raises the z term in m/z. Nano-LC-ESI systems have recently been developed that enable sample introduction at very low flow rates (<20 nL/min). In this way, decreased sample quantity requirements and increased sensitivity are achieved. Proteins and peptides are usually analyzed under positive ionization conditions, with a trace of volatile organic acid (e.g., formic acid) often added to aid protonation of the sample molecules. A common application of this technique is for a protein digest to be analyzed to generate a list of m/z values, which can be converted to molecular weights for the components present, and then each component to be analyzed further by MS/MS to allow amino acid sequencing [Gorg et al., 2004; Aebersold and Mann, 2003; Mann et al., 2001; Shevchenko et al., 1996].

Matrix Assisted Laser Desorption Ionization (MALDI)

MALDI-MS is used successfully in biochemical areas for the analysis of proteins, peptides, glycoproteins, oligosaccharides, and oligonucleotides. The mass accuracy depends on the type and performance of the analyzer of the MS, but modern instruments should be capable of measuring peptides to within 0.01 % of the molecular weight of the sample.

MALDI is based on bombardment of the sample molecules with laser light to bring about sample ionization. The sample is premixed with a matrix that, in the solid state, strongly absorbs photons at the wavelength of the laser light. For the most consistent and reliable results, a low concentration of sample to matrix works best. The matrix is usually a small, organic molecule. The common matrixes that are used depend upon the physical properties of the analytes; for routine analysis of peptides in MALDI, α-cyano-4-hydroxy-cinnamic acid is most commonly used, and for proteins sinapinic acid acid is most common. Although the exact mechanism of MALDI is incompletely elucidated, the probable steps are laser-induced excitation of the matrix from the ground to an excited singlet state and ionization of the matrix molecule, protonation of the embedded analyte, and sputtering, via thermal dissipation of the absorbed energy, of the matrix and protonated analyte into the gas phase. In this way, energy transfer is efficient and also the analyte molecules are spared excessive direct energy that may otherwise cause decomposition. Most commercially available MALDI systems use a pulsed nitrogen laser of wavelength 337 nm and some a Q-switched Nd/YAG laser of wavelength 355 nm. After ions are sputtered, they are collimated then accelerated. The time-of-flight analyzer separates the ions based on their m/z ratios by measuring the time it takes for ions to travel through a field free region known as the flight, or drift, tube. The heavier ions move slower than the lighter ones. MALDI is a "soft" ionization method that predominately generates singly charged quasimolecular $[M+H]^+$ ions, regardless of the molecular weight. Hence, the spectra are relatively easy to interpret (Fig. 5) but fragmentation of the sample ions does not usually occur and little to know structure information is generated.

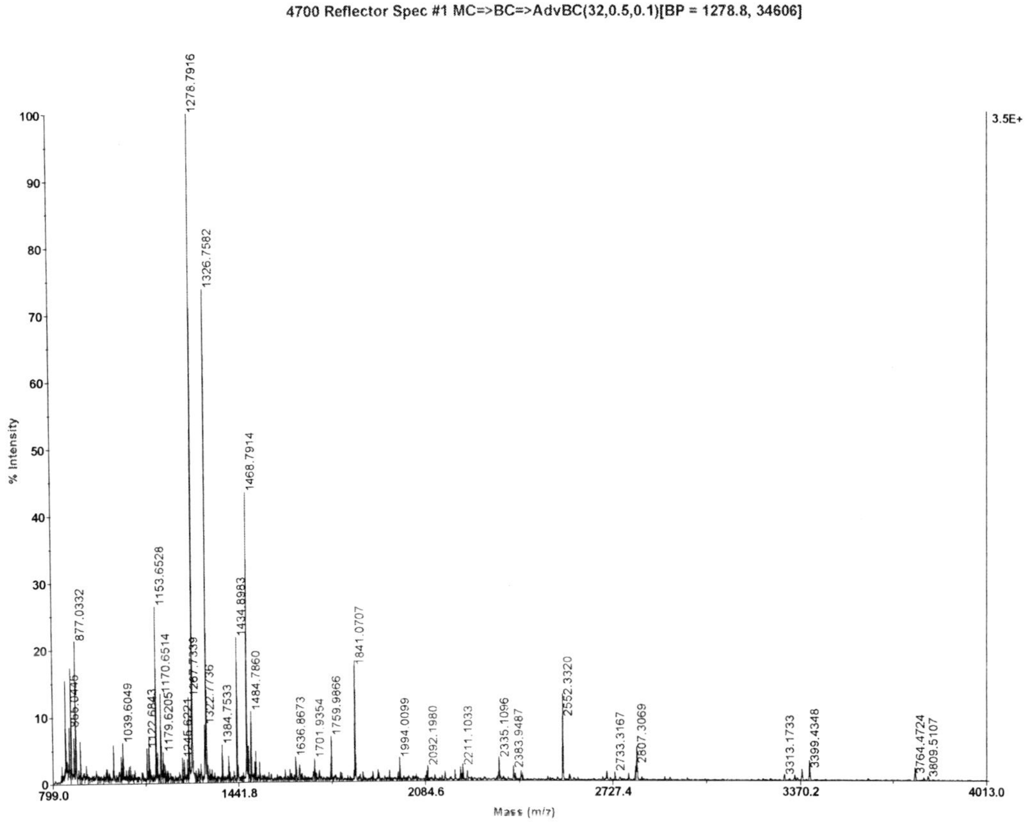

Figure 5. An illustrative MALDI-TOF-MS spectrum from trypsin digested proteins.

Protein and peptides are usually analyzed in the positive ion detection mode, where the protonated molecular ions $[M+H]^+$ usually predominate, although they can be accompanied by salt ion adducts (e.g., $[M+Na]^+$) and a trace of the doubly charged quasimolecular ion $[M+2H]^{2+}$ at approximately half the *m/z* value, and/or a trace of a dimeric species $[2M+H]^+$ at approximately twice the *m/z* value. Together, peptide mass information derived from the MALDI-TOF-MS analysis, in conjunction with peptide sequence information obtained from MS/MS analysis, provides a strong basis for protein identification [Aebersold and Mann, 2003].

Peptide Fingerprinting by MS Analysis.

MALDI and ESI-based MS technologies are being widely employed as protein identification tools. Most widely utilized is the "bottom-up" approach where proteins are digested with endoproteinases and their peptide maps (that is, a list of peptide masses) subsequently obtained and characterized by MS. These maps are then searched against theoretical maps of proteins included in databases (PROWL, EMBL, etc.).

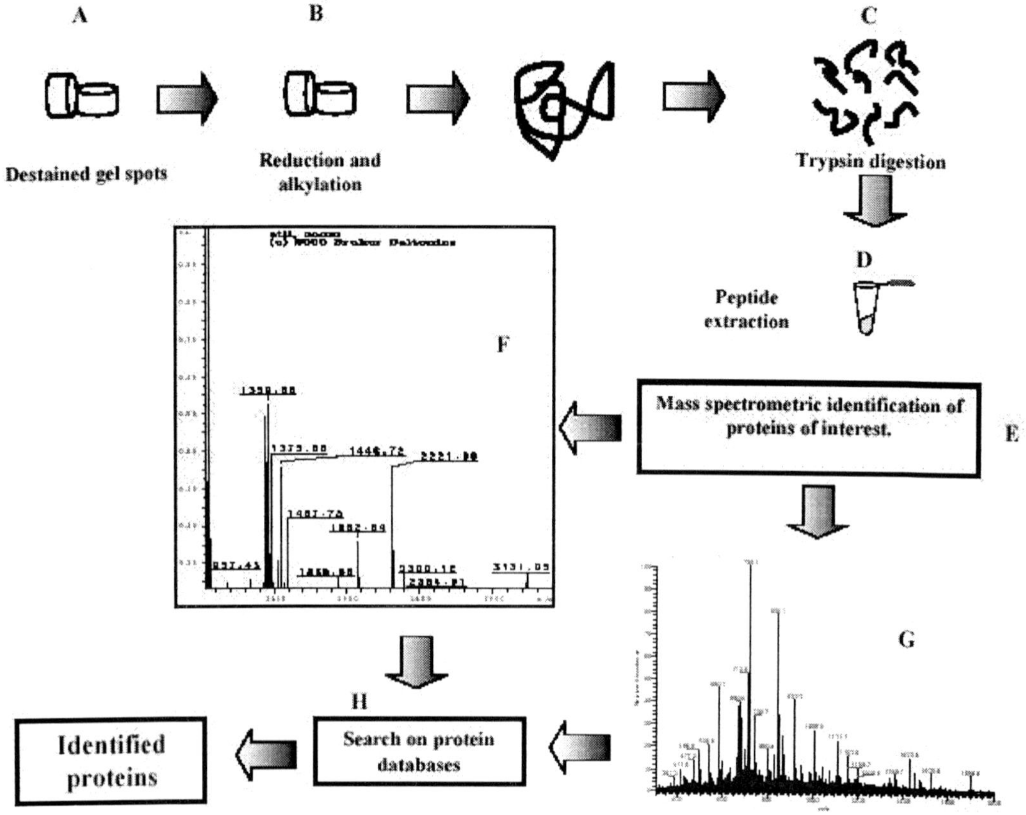

Figure 6. Summary of the procedures for protein identification workflow, step by step.

Figure 6 is an example of a protein identification experiment. In the case of 2D-E gels, for the generation of peptides, proteins spots are excised from the gels and the gel plugs are digested using in-gel trypsin digestion. The extracted peptides from the gel plugs can then be directly analyzed by MS or first separated by LC. The experimental peptide masses are then correlated to the peptide fingerprints (i.e., the list of peptides generated by the theoretical digest) of known proteins through search engines/algorithms such as Mascot, MS Fit, Sequest, etc. Matches are calculated on a probability basis and the degree of similarity of the experimental to theoretical is denoted by various mathematical parameters such as score and delta correlation, depending on the algorithm in use. Typically, a minimum of four peptide molecular weights is recommended to reduce the possibility of the false-positive matches [Gorg et al., 2004; Aebersold and Mann, 2003; Mann et al., 2001; Shevchenko et al., 1996; Resing and Ahn, 2004; Maddalo et al., 2005].

In contrast to the above "bottom-up" (i.e., use of peptide MS data to identify the protein), an alternative "top-down" approach for protein identification has been recently developed which includes the direct MS and MS/MS analysis of intact proteins. In this case, protein fragmentation takes place electronically during the MS analysis (for example using Fourier transform-MS in conjunction with electron capture dissociation). This methodology provides the advantage of studying the protein along with its PTMs as a whole; however, its application is still limited, due to primarily the lack of optimized relevant experimental methodologies, and the lack of specific software for the analysis of MS data from intact proteins, with the exception of the Prosight software developed by Taylor *et al.*

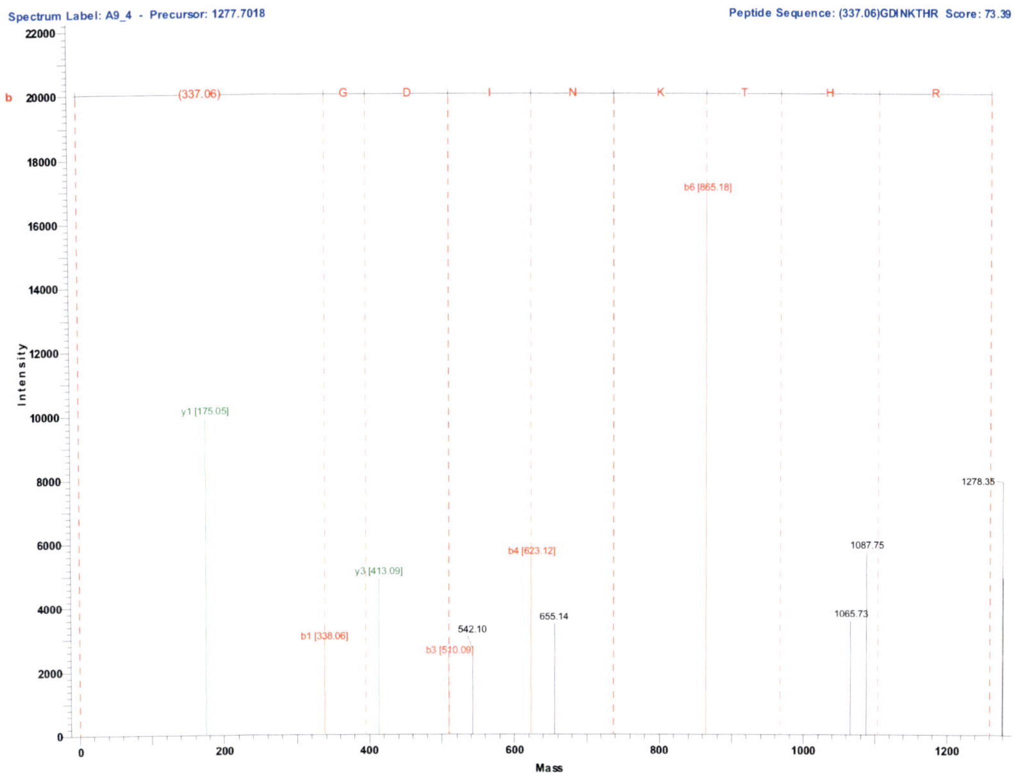

Figure 7. An illustrative MALDI-TOF/TOF-MS/MS spectrum from a peptide of a trypsin digested protein showing the y and b fragment ions.

Peptide sequencing by tandem MS provides an alternative to the traditional Edman degradation procedure. MS/MS is based on the detection and *m/z* determination of the fragment ions that are formed either spontaneously (post-source decay or PSD) or by induced fragmentation (collision-induced dissociation or CID) from an intact ion. Peptide ions, following their ionization and introduction into the gas phase, may be fragmented to produce ions that can be separated in a second mass analyzer to give the *m/z* ratios of these fragments. Information obtained by MS/MS can be utilized for the identification of proteins from protein databases, databases of expressed sequence tags or translated DNA databases (Figure 7). Besides allowing the more reliable identification of protein spots or LC fractions, the MS/MS approach enables the identification of PTMs such as phosphorylation and glycosylation. With the current availability of identification algorithms that analyze the MS/MS results, there is a widespread application of this technology [Maddalo et al., 2005], although the instruments used for it are expensive.

An important consideration when identifying proteins from "shotgun proteomics" datasets is distinguishing protein isoforms. Ambiguity ensues when a given peptide sequence is found in more that one database entry, a common occurrence. Several programs such as Protein Prophet and Isoform Resolver resolve many of these ambiguities, although mass neutral mutations may be overlooked/undetected.

Protein Microarrays

Advances in genomics and proteomics have created a demand for miniaturized, robust platforms for high-throughput (HT) study of proteins. Microarrays, generated by spotting biomolecules on a solid surface at high spatial density, offer these features by allowing investigators to query thousands of targets simultaneously. Given the central role that proteins play in biology and physiology, better methods are needed to study protein abundance and activity in HT mode. Several properties of proteins make building protein microarrays more challenging than building their DNA counterparts. First, unlike the simple hybridization chemistry of nucleic acids, proteins demonstrate a staggering variety of chemistries, affinities

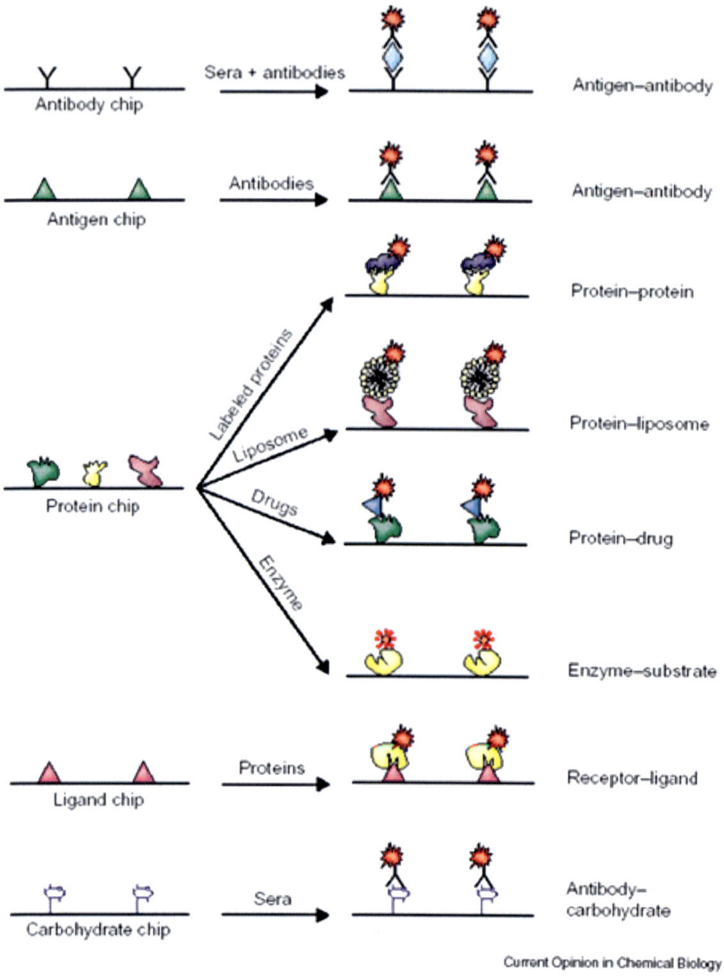

Figure 8. Figure showing the different types of microarrays and their applications.

and specificities. Second, expression and purification of proteins is a tedious task and does not guarantee the functional integrity of the protein. Lastly, many proteins are unstable, which raises concerns about microarray shelf life [LeBaer and Ramachandran, 2005].

Two general strategies have been pursued (Figure 8). The first, abundance-based microarrays, seeks to measure the abundance of specific biomolecules using analyte-specific reagents (ASRs), such as antibodies. Currently there are two types of abundance-based microarrays: capture microarrays and reverse-phase protein (RPP) blots. Capture microarrays are generated by spotting specific capture molecules (for example antibodies, aptamers) on the array surface to trap and assay their targets from complex mixtures. Typically, the capture microarray is probed with a complex sample, and then relative amounts of targeted analytes can be determined by comparison to a reference sample. Capture microarrays had been used by Sreekumar *et al.* to monitor changes in protein abundance in colon carcinoma cells following exposure to ionizing radiation. Miller *et al.* used capture microarrays to identify potential prostate cancer biomarkers in prostate cancer serum.

Although sometimes referred to as an array, the RPP blot is not a true array, in that it is not an arrangement of known elements with defined content. Instead, the RPP blot are produced by spotting the unknown experimental samples themselves (or a series of experimental samples) that are then probed with ASRs in a fashion that is directly analogous to nucleic acid dot blots. The detection of signal on RPP blots relies entirely on the specificity of the ASR. This is particularly important because effective concentration of analytes within the spot is high, and will tend to facilitate spurious interactions. Nishizuka *et al.* used RPP blots to differentiate cancers based on protein abundance. RPP blots will probably find their greatest application in screening experiments across many experimental samples for a handful of well-defined proteins, which will then require a more detailed and quantitative analysis using other methods.

There are some limitations for capture microarrays that need to be addressed before they become common and generally useful. First, high quality ASRs are not available for most targets. Second, most existing antibodies do not function well in the microarray format. It is estimated that only 30% of commercially available antibodies can be used qualitatively and even fewer (20%) can be used quantitatively. Lastly, even antibodies advertised as monospecific demonstrate significant cross-reactivity on the microarray format.

The second type of microarrays is the function-based microarrays. Functional-based microarrays can be used to study the biochemical properties and activities of the target proteins printed on the array. Function-based microarrays can be used to examine protein interactions with other proteins, nucleic acids, lipids, small molecules and other biomolecules. In addition, function-based microarrays can be used to examine enzyme activity and substrate specificity. These microarrays are produced by printing the proteins of interest on the array using methods designed to maintain integrity and activity of the protein, allowing hundreds to thousands of target proteins to be simultaneously screened for function.

The list of potential applications of function-based microarrays is large. Proteins expressed by pathogenic organisms can be screened with serum from convalescent patients to identify immunodominant antigens, leading to good vaccine candidates. A candidate drug could be used to probe a broad range of enzymes to look for unintended binding of targets that might suggest possible toxicities [LeBaer and Ramachandran, 2005] sembly of multiprotein complexes, can shed light on biochemical. Protein interactions networks, including the as pathways and networks. A microarray of a particular class of enzymes could be screened with a candidate inhibitor to examine binding selectivity.

There are some limitations and challenges in using functional-based microarrays. First, the highly lability of proteins raises concerns about their stability and integrity on the microarray surface. Second, it is time consuming and costly to produce proteins of good purity and yield, and many proteins can not be purified at all. Finally, the methods used to attach proteins to the array surface may affect the behavior of the proteins [LeBaer and Ramachandran, 2005].

To further advance the area of protein microarrays there needs to be continued efforts in the area of protein production, surface chemistries, detection schemes and alternative platforms. These efforts must emphasize the need to develop and use high-quality reagents, and robust platforms to minimize the occurrence of false positives and negatives. Although these *in vitro* approaches are a step removed from cells and organisms, they are highly complementary to *in vivo* approaches.

Metabolomics Techniques

Nuclear Magnetic Resonance

Nuclear magnetic resonance (NMR) is a useful tool in metabolomics as it offers a rapid and non-invasive method to comparatively characterize metabolite expression patterns *in vivo* or *in vitro*. NMR analysis has been shown to be sufficiently robust to be of use in clinical studies, and variations in age, gender, and diet have been investigated to further examine the capability of the technique to detect variation due to additional factors [Rochfort, 2005]. These results suggest that these NMR-based techniques can be applied to systems that differ due to genetics and environment. NMR-based metabolomics provides a means to categorize organ-specific toxicity, monitor the onset and progression of toxicological effects, and identify biomarkers of toxicity.

While 1D ^{1}H NMR is the most used method in NMR metabolomics, there have been reports of 2D techniques being used. Urine profiles from cattle treated with anabolic steroids have been analyzed by HMBC techniques with linear discriminant analysis. Despite its relative insensitivity, ^{13}C NMR has also been applied to metabolomics in the study of brain tissue. It is also possible to examine the metabolome of intact tissue or cells through the use of magnetic angle spinning techniques. However, NMR exhibits limited sensitivity and resolution and is therefore capable of detecting only the abundant metabolites in complex samples [Rochfort, 2005].

The next challenge for metabolomics by NMR is quantitation. While proton NMR can be quantitative, the resolution is not as good in the less quantitative carbon dimension, creating problems for separating molecules. More research on quantitation by 2D NMR needs to be done, and it remains to be seen if 2D NMR will offer the level of quantitation required for rapid metabolomic analysis.

Mass Spectrometry (MS)

To increase the depth of metabolite profiling, investigators have turned to MS methods. MS analyses can be performed either directly or in combination with GC or LC separation steps. "Direct inject" MS experiments have the advantage of being fast (minutes per sample), but, like NMR, suffer from limited resolving power. By contrast, gas chromatography (GC)- and LC-MS experiments are considerably slower (1-2 h/sample), but permit more than 1000 metabolites to be analyzed per run and thus sacrifice speed in exchange for substantial increases in sensitivity and "metabolome coverage" [Rochfort, 2005].

GC, especially when interfaced with mass spectrometry (GC-MS), is one of the most widely used and powerful methods. It offers very high chromatographic resolution, but requires the analytes be thermally volatile and thermally stable. This requires that the typically polar (and not very volatile) metabolome be subjected to chemical derivatization (i.e., methoxyamination and silylation). Some modern instruments allow '2D' chromatography, using a short polar column after the main analytical column, which increases the resolution still further. Large and polar metabolites cannot be analyzed by GC. Compared to GC, HPLC has lower chromatographic resolution, but it does have the advantage that a much wider range of analytes can potentially be measured.

GC-MS applications have benefited from large data collections such as the NIST database. For LC-MS, these databases are still in their infancy, and the commonly used ESI techniques are much more dependent on multiple variables than the equivalent techniques employed for GC-MS. For both, MS applications standards will be needed to quantitate specific metabolites [Rochfort, 2005]. There are a number of "lessons learned" from proteomics that can be applied to metabolomics including reproducibility, issues which are being considered. The adoption of standards and reporting of all the experimental conditions will assist in this area.

High-Pressure Liquid Chromatography with Coulometric Multi-Electrode Array System (HPLC-CMEAS).

A major drawback of NMR is that certain sample preparation steps are needed because the individual resonances are sensitive to the chemical and physical environment of a molecule. In addition, it is difficult to measure high abundance and very low abundance metabolites simultaneously. On the other hand, MS is highly beneficial for wide screening of non-destructive metabolites in coupling with HPLC or GC. However, lack of accurate quantification is a major liability for metabolomics assessments.

On the other hand, studies of a series of related metabolites are often complicated by cumbersome and different analytical methods, which require separate and multi-step extraction and chemical reaction procedures. Thus, measurements of multiple parameters are always restricted by the limited sample-size. HPLC-CMEAS provides high specificity and sensitivity for determination of multiple redox-active low-molecular-weight compounds based on differences in oxidation-reduction properties of analytes [Milbury, 1997; Hensley et al., 1998; Kristal et al., 1998 & 1999]. The basic concepts of HPLC-CMEAS technology have been extensively elucidated [Matson et al., 1987; Beal et al., 1990]. In brief, the CMEAS equips four cell packs in series, each consisting of four coulometric electrodes, and set to increasing specified potentials between 0 and 1000 mV, which allows resolution of co-eluting compounds whose oxidation or reduction potentials differ by as little as 60 mV. Multisensor, coulometric detectors can oxidize and measure nearly 100% of the oxidizable compounds in a sample. The coulometric array detectors utilize a combination of retention time and ratio of response across adjacent detectors as a molecular fingerprint to enable identification of a specific peak in HPLC chromatograms [Fig. 9]. More than one thousand electrochemically oxidizable compounds such as antioxidants, oxidative stress markers, and monoamine metabolites, can be analyzed simultaneously in biological samples [Kristal et al., 1998 & 2002].

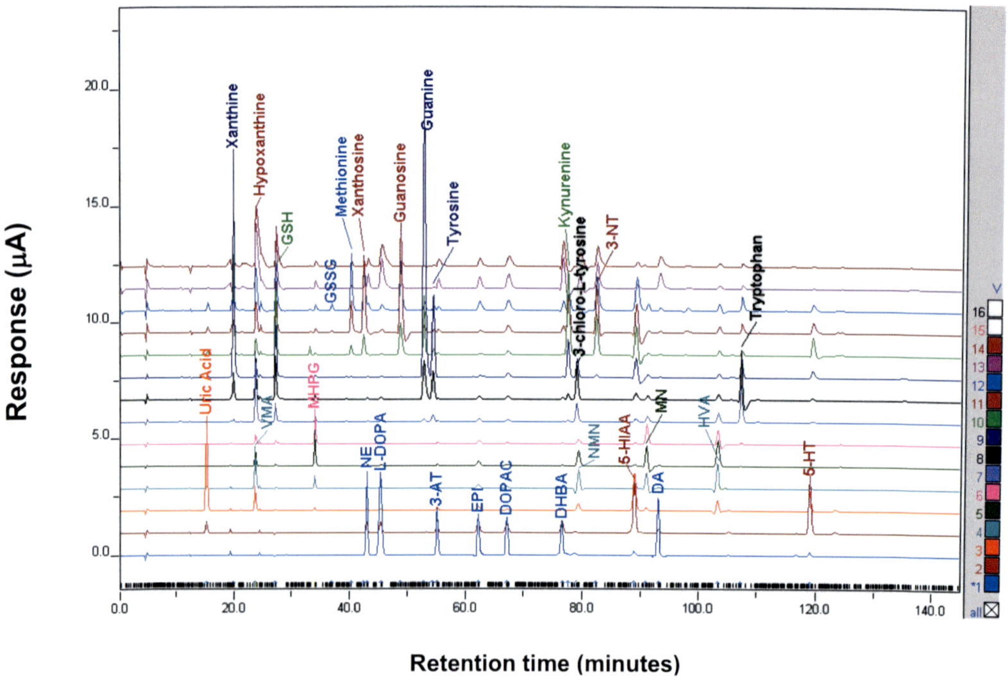

Figure 9. Separation of a standard mixture containing a variety of low-molecular weight, redox-active compounds by HPLC-CMEAS (from Yao and Cheng, 2004).

The data collected from HPLC-CMEAS allow multiple rather than single metabolites to be used in markers for a group, which will greatly improve the predictive diagnostics for phenotypes that directly involve in the oxidative stress. More significantly, these comprehensive analyses that generate metabolic profiles represent not only biomarkers for disease but also metabolic maps that can be used to identify specific genes responsible for disease [Watkins, 2000]. Such metabolic maps or metabolomics provides different perspective to biomedical research in further understanding the effects of therapeutic, nutritional, toxicological and environmental interventions [Watkins and German, 2002].

Applications of Proteomics and Metabolomics Techniques

Proteomics Analysis of Human Cerebrospinal Fluid

Recently, Yuan and Desiderio [2005] have reviewed the variety of analytical methods that have been used for proteomics analysis of the cerebrospinal fluid (CSF), including sample preparation, 2D liquid and gel electrophoresis, mass spectrometry, bioinformatics, and non-gel methods. The CSF is secreted from several different central nervous system (CNS) structures, in particular, from the ventricular chorioid plexus. It circulates within the ventricles of the brain, and surrounds the brain and spinal column. Proteomics analysis offered a comprehensive view to analyze, at a system level, all of the proteins that might result from, or contribute to, each different CNS disorder. The CSF proteome could provide novel biomarkers for the early-stage diagnosis or the staging of a neuronal disease, offer potential insights into the biochemical characterization of affected neuronal populations, and clarify the basic molecular basis of CNS pathologies.

The first step in the sample preparation was to remove salts for the CSF, which contains a high salt concentration (> 150 mmol/L) and a low protein concentration (200-700 μg protein /ml). Salt was removed using several methods including ultrafiltration, dialysis, protein precipitation, and gel filtration spin columns. The highest recovery was obtained with the spin column salt-removal method. The next step was the prefractionation of CSF. Some low-abundance CSF proteins are disease-related, but they are difficult to characterize due to three major factors. First, the CSF contains a very low protein concentration. Secondly, it has high amounts of albumin and immunoglobins -- both constitute ~ 65% of the total protein content in the CSF. Thirdly, there is a wide dynamic range of protein concentrations. Raymachers *et al.* [2000] used affinity chromatography to remove albumin and immunogloblins. They used immobilized Cibacron blue for albumin and immobilized protein G for the immunogloblins. The experimental results showed that most of the albumin and immunogloblins were removed, but there was the appearance of new spots and the disappearance of some proteins. Yuan and Desiderio [in press] developed a prefractionation method that was based on the different hydrophobic properties of the CSF proteins. A CSF sample was prefactionated into three fractions. Some of the high-abundance CSF proteins were removed from two of those three fractions; some low-abundance CSF proteins were preferentially enriched in those two fractions.

Many proteomics analyses of CSF have focused on protein mapping. Proteins are separated with 2D-E, each selected protein is digested with trypsin, the tryptic peptides were analyzed with MS, and either the peptide mass fingerprint (PMF) or the amino acid sequence data were input to a database to search for the matching protein. Coomassie Blue staining and sliver staining were used to visualize proteins on the gels, but more proteins were observed with silver staining because of its relatively high sensitivity [Yuan and Desiderio, 2005].

Davidsson *et al.* [2002] reported using 2-D liquid-phase electrophoresis (2D LPE) for the pre-separation of CSF proteins. This method is similar to 2D-E; the only difference is the liquid IEF, where proteins were collected into 20 cells, and the proteins in each cell were separated with SDS-PAGE. The advantage of using 2D LPE is increased sample-loading for the 2D gel and less discrimination of membrane proteins, which have lower transfer efficiency from the first to second dimension in 2DGE. The pitfall of this method is the decreased level of LPE resolution for protein separation.

MALDI-TOF-MS was used for peptide mass fingerprinting (PMF), and MALDI-post-source decay (PSD) to obtain the amino acid sequence data from a selected (within ± a few Da) precursor ion. LC ESI-MS/MS is the most used method for protein characterization and identifications from CNF fluid.

An important aim of proteomics research is to understand, at the protein level, the basic molecular mechanisms that operate in the cell, CNS or CSF by an analysis of the dynamic proteome that result from the static genome. The comprehensive knowledge of each protein's function in a proteome will help to clarify those cellular, CNS or CSF processes. As PTM processes play critical roles in protein's function, Yuan *et al.* tested three different combination methods-1D-E plus Western blotting, immunoprecipitation plus 2D-E, and 2D-E plus Western blotting to detect phosphotyrosine proteins in human lumbar CSF. Kallikrein-6 precursor, complement C4 γ-chain, gelsolin and ceruloplasmin precursor were identified-on a 2D Western blot as phosphotyrosine proteins.

Glycoproteins play important roles in cell-cell interactions, tissue morphogenesis, immune reactions, and pathologies such as cancer and inflammation. CSF contains many glycoproteins, and each glycoprotein has a variety of different glycosylated isoforms. Hakansson *et al.* [2003] reported the analysis of CSF glycoproteins that were visualized with a commercial glycoprotein detection kit. Selected glycoproteins were digested, each glycopeptide was analyzed with ESI-FT-ICR-MS, and specific sites of N-glycosylation were assigned by analysis of fragment ions produced by infrared multi-photon dissociation.

Mitochondrial Proteomics in Free Radical Research

Mitochondria are recognized as having fundamental roles in cellular processes including energy metabolism via the oxidative phosphorylation system, the Kreb's cycle, and β-oxidation of free fatty acids. Recently, though, interest in mitochondrial physiology has been renewed due to the growing evidence implicating mitochondrial involvement in cellular signaling pathways through modulation of intracellular calcium stores, production of reactive species, the apoptosis process and the interaction of nitric oxide (˙NO) on mitochondrial functions such as respiration and biogenesis. As a source for the formation and target of modifications mediated by reactive oxygen and nitrogen species (ROS/RNS), the mitochondrion is recognized as a site critical in cellular responses to oxidative and nitrosative changes. It should also be mentioned that the mitochondrial proteome is not only determined simply by protein levels but also by PTMs to proteins, which might be significant, particularly in cells undergoing an oxidative or nitrosative stress [Bailey et al., 2005].

While conventional 2D-E is well-suited to identify changes in the levels of more hydrophilic proteins of the mitochondrion, for example, matrix proteins, the analysis of membrane proteins is affected by the fact that many of these proteins precipitate at the basic end of the IEF gels and are thus poorly resolved on conventional 2DGE. To solve and overcome this problem, Capaldi *et al.* [this reference is missing] used a sucrose gradient fractionation technique. Mitochondrial extracts were loaded onto a 10-35% step fraction sucrose gradient and centrifuged overnight, and fractions were collected from the bottom of the tubes after the run. By use of this technique in combination with MS and bioinformatics analyses, more than 600 proteins were identified as being part of the human heart mitochondrial proteome [Bailey et al., 2005].

Mootha *et al.* [2003] used an integrated proteomic and genomic profiling technique to elucidate the mitochondrial proteome of mouse brain, liver, kidney and heart. In this study, mitochondrial proteins from each tissue were solubilized and size separated using gel filtration chromatography into approximately 15-20 fractions, which were then digested and analyzed by LC-MS/MS. A final set of 399 proteins were identified of which 236 proteins were previously identified as mitochondrial and 163 proteins not previously annotated to the mitochondrion. Moreover, DNA microarray analysis demonstrated that mRNA expression levels were for the most part correlated to the detection of these proteins and allowed the authors to generate modules of "neighborhoods" of mitochondrial genes, the expression which is tightly co-regulated, and thus they share similar functional and regulatory mechanisms. Mitochondrial neighborhoods enriched in genes of oxidative phosphorylation, amino acid metabolism, steriodogenesis, heme biosynthesis, transcriptional regulation, and mtDNA repair were identified. This study is highly significant because it is the first to combine state-of-the-art proteomics, genomics and bioinformatics technologies to elucidate the functional and regulatory organization of mitochondrial genes.

Cysteine Modifications

Exposure of mitochondrial proteins to ROS, RNS and electrophilic lipids can lead to the PTM of amino acid residues, which has the potential to alter protein structure and/or induce a permanent loss in function. Recent work also suggests that oxidants can induce reversible modifications, specifically at cysteine residues, which may function to 1) protect cysteines from irreversible modifications or "overoxidation", and 2) modulate protein function in response to stress. Reversible modifications to thiols that are thought to play key roles in modulating a protein's function include the formation of nitrosothiols (P-SNO), sulfenic acids (P-SOH), and protein mixed disulfides with low molecular weight thiols such as glutathione (P-SSG). Each of these thiol modifications has the potential to elicit a biological response that might disrupt or diversify mitochondrial function. Indeed, emerging evidence emphasizes a key link between the induction of the mitochondrial permeability transition, alterations in energy metabolism, and protein thiol oxidation in pathways of cell death induced by an oxidative or nitrosative stress [Bailey et al., 2005].

One approach used to detect *S*-nitrosated proteins, is the use of the three-step "biotin-switch" method developed by Jaffrey and Snyder [2001]. This technique converts nitrosated cysteines to biotinylated cysteines that can be detected and identified following purification with streptavidin-agrose. While the identification of these *S*-nitrosated proteins in mitochondria is important, the functional significance of these modifications remains to be determined, as exposure to high concentrations of NO^+ donors failed to inhibit the activity of these enzymes.

Tyrosine Nitration

The detrimental effects of $^{\bullet}NO$ on mitochondrial function stem largely from the diffusion of $^{\bullet}NO$ into mitochondria and the formation of more reactive secondary intermediates like

$ONOO^-$ and nitrogen dioxide ($^{\bullet}NO_2$), which can nitrate tyrosine residues within mitochondrial proteins. Numerous papers have been published using proteomic techniques to identify nitrated mitochondrial proteins under both normal and pathophysiological conditions and from endogenous and exogenous sources of RNS. Typically, these studies have detected nitrated proteins by using antibodies directed against the 3-nitrotyrosine moiety. While many of these studies have observed an increase in nitrated proteins following disease or nitrosative insult, few studies have been successfully in identifying the nitrated proteins or, more importantly, the specific modified tyrosine residue within the protein.

Murray *et al.* [2003] performed a study to identify the specific tyrosine residues within mitochondrial proteins that are nitrated on exposure to RNS. Using both immunologic and MS approaches, he demonstrated that *in vivo* exposure of intact mitochondria to $ONOO^-$ resulted in the selective nitration of several complex I subunits, which was associated with a loss in enzyme activity. MS/MS analysis demonstrated that subunits NDUFB4 and NDUFA6 contained the highest degree of nitration, with Tyr^{122} in NDUFA6 being the most reactive and Tyr^{46}, Tyr^{50}, and Tyr^{51} being the most heavily nitrated in the NDUFB4 subunit. In addition to these findings, subunits of ATP synthase were found to be nitrated, along with the oxidation of Trp^{61} and Met^{68} residues in the 17.2 kDa subunit of complex I. These findings supported those reported earlier for the unique susceptibility of Trp and Met residues to oxidation in several mitochondrial proteins identified in healthy human heart tissue.

Advancements in proteomics technologies have now made it possible to critically examine the molecular targets within the mitochondrion that contribute to cellular dysfunction. Cutting-edge research in the field has now begun to identify specific molecular modifications to selective amino acids within mitochondrial proteins that occur from exposure to ROS and RNS, under both normal and pathophysiological conditions. Thus, the challenge for future studies will be to incorporate this knowledge into a framework whereby these complex functional and regulatory alterations in the mitochondrial proteome can be used to increase our understanding and treatment of human disease.

Application of HPLC-CMEAS to Investigate Psychiatric Disorders

The use of HPLC-CMEAS for the analysis of metabolites has the potential to yield valuable insights into the likely complex pathophysiological mechanisms that affect multiple metabolic pathways and thereby, offers multiple windows of therapeutic opportunities. The HPLC-CMEAS has been used previously to study oxidative stress markers in neurodegenerative disorders such as Alzheimer's disease, amyotrophic lateral sclerosis (ALS), Huntington's disease, and Parkinson's disease [Albers and Beal, 2000]. For example, differential analysis of normal subjects and ALS patients revealed numerous differences in metabolic profiles that could be easily measured and monitored by CMEAS [Kaddurah-Daouk et al., 2004]. In patients with Huntington's disease, HPLC-CMEAS showed significant reduction of kynurenic acid in CSF compared to controls and other neurologic diseases [Beal et al., 1990]. In addition, metabolome analysis of samples from autistic and control subjects could be separated by cluster analysis using 65 plasma biomarkers [Kaddurah-Daouk et al., 2004].

Complex disorders such as schizophrenia, bipolar disorder, and major depression are associated with multiple etiologies and pathogenetic mechanisms. To identify candidate pathological process(es) that account for the complex assemblage of clinical and biological features in schizophrenia, as an example, it is advantageous to evaluate simultaneously the network of interacting biochemical pathways that produce and remove oxygen and nitrogen free radicals in mammalian cells. A novel rapid multidimensional analytical approach was used to simultaneously examine metabolites derived from different biochemical pathways. These metabolites are a group of low-molecular weight compounds, such as, antioxidants, amino acids, cholamines, vitamins, lipids and nucleotides, which reflect the metabolic processes, including anabolism and catabolism as well as other related cellular processes (for example, signal transduction, regulation, detoxification, etc.). Such an analytic approach has the potential to yield valuable insights into the likely complex pathophysiological mechanisms that affect multiple metabolic pathways and therefore, offer multiple openings of therapeutic opportunities [Yao and Reddy, 2005].

One of applications of HPLC-CMEAS, that it was utilized in the study of glutathione redox state in postmortem brain from patients with schizophrenia for the evaluation of brain metabolites (Yao et al., 2006). A metabolic profile of control versus diseased patients identified multiple metabolites showed that there was presence of antioxidants (such as ascorbic acid, uric acid, glutathione), purine catabolites (such as xanthine, guanosine, hypoxantine), tyrosine metabolites (such as tyrosine, homovanillic) and tyrosine metabolites (such as tryptophan, 5-hydoxy-indoleacetic acid). Several other minor metabolites were identified reflecting oxidative damage (such as glutathione disulfide, 3-nitrotyrosine), tyrosine pathway (dopamine, epinephrine, norepinephrine), tryptophan pathway (kynurenic acid, serotonin) [Yao and Reddy, 2005].

HPLC-CMEAS was used also in the analysis of serum metabolome. Vigneau-Callahan *et al.* [2002] had conducted a series of investigations concerning the questions of maintaining stable patterns of response in rat serum samples in a dietary restriction study. They were able to identify approximately 250 compounds in serum as potential components of serum metabolome. This study and subsequent studies laid the foundation for quantitative analysis of serum metabolites that can provide important information for identifying biomarkers chromatographically, and eventually for pattern-recognition-based approaches in establishing metabolome-based categorical separations.

HPLC-CMEAS was also used in therapeutic monitoring to examine the effects to current and novel treatments for psychiatric disorders. This will help in the investigations of predicators of therapeutic outcome. Changes were detected using HPLC-CMEAS in tryptophan pathways from plasma metabolites in first-episode neruoleptic-naïve patient individual with schizophrenia following treatments with risperidone (second-generation antipsychotic drug) up to 2 years.

Mitochondrial Dysfunction in Schizophrenia: Evidence for Compromised Brain Metabolism and Oxidative Stress

Prabakaran *et al.* [2004] were proposing that oxidative stress and the ensuing cellular adaptations were linked to the schizophrenia disease process and were hoping that this new disease concept my advance the approach to treatment, diagnosis and disease prevention of schizophrenia and related syndromes. The authors employed an integrative approach on a large sample set to define disease-specific abnormalities within schizophrenia post-mortem brain tissue by analyses at the level of mRNA (transcriptome), protein (proteome) and low-molecular weight intermediates (metabolome). Using these approaches, they were able to identify several significantly altered metabolic pathways in schizophrenia brain tissue. In the view of their findings, they proposed that mitochondrial dysfunction and oxidative stress were strongly involved in, and may underlie, the pathophysiology of schizophrenia. These abnormalities might pertain to abnormal oxygen or glucose supply and/or growth factor signaling resulting in global metabolic disturbances and precipitate the clinical syndrome of schizophrenia]. For the proteomic approach, the authors used DIGE, where samples and control were labeled with Cy3 or Cy5. The mixture was first separated by isoelectric focusing followed by separation on SDS-PAGE gels. Protein samples that showed significant changes ($P \leq 0.05$) in schizophrenia with respect to controls were picked from a colloidal Coomassie-stained gel, digested with trypsin and the resulting peptides were analyzed using LC-QTOF-MS/MS. The data obtained were used to identify proteins from primary sequence database using Mascot search engine. The results showed that 215 protein spots were found to be significantly altered in schizophrenia as compared to control samples. Out of the 215 spots, 170 were identified, with 114 corresponding to single known proteins. Of these, some proteins were identified and were associated with mitochondrial function (such as pyruvate kinase, heat shock 70 kDa protein, glyceraldehyde-3-phosphate dehydrogenase), oxidative stress (such as tubulin, enolase, NADH dehyrogenase Fe-S protein), and peroxisomal function (such as malate dehydrogenase, peroxiredoxin). Other prominently altered classes of proteins were cytoskeletal proteins and proteins associated with protein trafficking/turnover .For the metabolomic approach, high-resolution magic angle spinning (HRMAS) ^1H NMR spectroscopy was performed on white and gray matter brain tissue using a 700 MHz instrument. Following Fourier transformation, spectra were integrated across 0.04 ppm spectral regions between 0.4 and 9.4 ppm using a routine developed in Matlab. The output vector representing each spectrum was normalized across the integral regions, excluding the water resonance, and this formed the input to the pattern recognition routine. The authors found that 10 out of approximately 60 identifiable metaboliteswere significantly altered ($P \leq 0.05$) between schizophrenia and control tissues in white matter. In the gray matter, the authors did not observe significant data using univariate statistical analysis. Examples of the metabolites identified in the white matter were adenosine, phosphocholine, acetate and taurine .

Conclusions

Understanding protein function and unraveling cellular mechanisms at the molecular level constitute a major need in modern biology. With the availability of full genome sequences, these goals can be achieved by determining which macromolecules interact with a given protein in a specific manner. Experimental procedures are available now to define the transient macromolecular assembly of proteins needed to fulfill specific and fundamental functions within the cell, or to design the routes along which signaling mechanisms can take place. As the field of proteomics matures, its application to biology is expanding greatly. The information gleamed from proteomic studies will have direct impact on the quality of human life. For example, plant proteomics is accelerating due to the increased availability of genomic sequences for plant species, particularly important crops. Identification of allergens (most of them are proteins, glycoproteins or peptides) will also benefit from technical development of better LC approaches integrated into the arsenal of proteomic tools. Due to the importance of proteins responsible for cellular aging, dementia, and geriatric diseases another area proteomics will be appreciated is gerontology. Certainly, chromatography will have a central role alongside other separation techniques in the application of proteomics in the future.

Effective separation of peptides and proteins, and their quantification and identification have become the central topic in proteomics. Liquid chromatography in particular has undergone extensive improvements and adaptations to become a multidimensional technique of proteins and peptides separation. Advances in multidimensional chromatography along with mass spectrometry will allow high-throughput analysis of minuscule samples. Although the multidimensional separation approaches improve the number of observed peptides, they add a series level of complexity to the overall procedure. This can have negative effect on the reproducibility for well-designed biomarker studies that require multiple biological and analytical repeats.

Yet, improvements and development of bioinformatics packages are needed for the conduction of all steps of proteomic studies. In the case of data acquired by MS, automatic mass assignment, noise subtraction (i.e., removal of background), and mass correction based on standard peptides and mass clustering are very crucial factors for the successful application of mass spectrometry-based approaches for protein profiling. Furthermore, the vast amount of data acquired by the contemporary mass spectrometric methods require databases and mining tools that will allow comparisons of different samples, generation of reference profiles as well as both statistical and graphical information about the sample under examination.

Previously, oxidative modification of proteins was though to represent a detrimental process in which the modified proteins are irreversibly inactivated, leading to cellular dysfunction. While this is still the case in many situations, it is now clear that posttranslational modifications of mitochondrial proteins can be specific and reversible and, thus, may play a key role in normal mitochondrial physiology. Thus, the challenge for the future studies will be to incorporate this knowledge into a framework whereby these complex functional and regulatory alterations in the mitochondrial proteome can be used to increase our understanding and treatment of human diseases.

There are a number of challenges for the developing science of metabolomics. For the science to develop in the same way as genomics and proteomics, standard ontology must be adopted. Adoption of standards will make it easier to integrate the large amount of data that will be generated in metabolomic experiments and will enhance reproducibility and credibility of metabolomics data.

Much of initial work on metabolomics has been of a qualitative nature, often the results of statistical model analysis rather than assessment of individual metabolite flux. This level of qualification and quantification is of increasing importance, as the significance of more biosynthetic pathways is elucidated. GC-MS applications have benefited from large databases. For LC-MS these databases are still in their infancy, and the commonly used ESI techniques are much more dependent on multiple variables than the equivalent EI techniques employed by GC-MS. For both, MS applications standards, particularly stable isotope internal standardization, will be needed to quantitate specific metabolites. For human or animal studies this may be achieved by the synthesis (or microbiological production) of specific isotopically labeled metabolites.

Identification of specific metabolites in complex mixtures by NMR is also problematic. However, it has been demonstrated that many common metabolites can be identified by the application of 2D NMR experiments. The next issue to be addressed in metabolomics by NMR is quantitation. While proton NMR can be quantitative, the resolution is not as good in the less quantitative carbon dimension, creating problems for separating compounds. More research on quantitation by 2D NMR needs to done, and it remains to be seen if 2D NMR will offer the level of quantitation required for rapid metabolomic analysis.

There is no doubt that more research is needed to disentangle the genetic and epigenetic components associated with schizophrenia to distinguish primary from secondary disease phenomena. There is quite a few findings that are published that should enhance the understanding of the disorder and should encourage new directions in schizophrenia research, which in turn should impact on treatment approach, diagnosis and disease prevention of schizophrenia and related syndromes.

References

Aebersold R., and Mann M. (2003) Mass-spectrometry-based proteomics. *Nature*. 422: 198-207.

Albers D.S. and Beal MF. (2000) Mitochondrial dysfunction and oxidative stress in aging and neurodegenerative disease. *J. Neural Transm. Suppl.* 59:133-154.

Anderson N.L., and Anderson N.G. (2002) The human plasma proteome: history, character, and diagnostic prospects. *Mol Cell Proteomics*. 1: 845-67.

Bailey S., Landar A., and Usmar V. (2005) Mitochondrial proteomics in free radical research. *Free Radical Biol Med.* 38: 178-88.

Beal M., Matson W. Bird E., *et al.* (1990) Kynurenine pathway measurements in Huntington's disease striatum: evidence for reduced formation of kynurenic acid. *J Neurochem*. 55: 1327-1339.

Cho S., Park S., Lee D., *et al.* (2004) Protein-protein interaction networks: from interactions to networks. *J Biochem Mol Biol*. 37: 45-52.

Davidsson P., Folkesson S., Christiansson M., Lindber, M., Dellheden, B., Blennow, K., and Westman-Brinkmalm, A. (2002) Identification of proteins in human cerebrospinal fluid using liquid-phase isoelectric focusing as a prefractionation step followed by two-dimensional gel electrophoresis and matrix-assisted laser desorption/ionisation mass spectrometry. *Rapid Commun Mass Spectrometry*. 16: 2083.

Dean B., Keriakous D., Thomas E., and Scarr E. (2005) Understanding the Pathology of Schizophrenia: The Impact of High-Throughput Screening of the Genome and Proteome in Postmortem CNS. *Current Psychiatry Review*. 1: 1-9.

Fiehn O. (2002) Metabolomics-the link between geneotypes and phenotypes. *Plant Mol Biol.* 5: 155-171.

Gorg A., Weiss W., and Dunn M. (2004) Current two-dimensional electrophoresis technology for proteomics. *Proteomics*. 12: 3665-3685.

Hakansson K., Emmett M., Marshall A., Davidsson, P., and Nilsson, C.L.. (2003) Structural analysis of 2D-gel-separated glycoproteins from human cerebrospinal fluid by tandem high-resolution mass spectrometry. *J Proteome Res.* 2: 581.

Hensley K., Maidt M.L., Yu Z., Sang H., Markesbery W.R., and Floyd R.A., (1998) *Electrochemical Analysis of Protein Nitrotyrosine and Dityrosine in the Alzheimer Brain* Indicates Region-Specific Accumulation. *J Neurosci* 18: 8126-8132.

iTRAQ Kit. Applied Biosystems *website: www.appliedbiosystems.com.*

Kaddurah-Daouk, R, Beecher C, Kristal BS, Matson WR, Bogdanov M, and Asa DJ. (2004) Advances in Bioanalytical techniques for metabolomics and metabolic profiling studies in drug discovery. *Pharma Genomics* 4: 46-52.

Kobor M., Lester W., Jansma D., *et al.* (1999) An unusual eukaryotic protein phosphatase required for transcription by RNA polymerase II and CTD dephosphorylation in *S. cerevisiae*. *Mol Cell.* 4: 55-62.

Kristal, B.S., Vigneau-Callahan, K. E., and Matson, W.R. (1998) Simultaneous Analysis of the Majority of Low-Molecular-Weight, Redox-Active Compounds from Mitochondria. *Anal Biochem* 263: 18-25.

Kristal, B.S., Vigneau-Callahan, K. E., and Moskowitz, A.J. (1999) Purine Catabolism: Links to Mitochondrial Respiration and Antioxidant Defenses? *Arch Biochem Biophys* 370: 22-33.

Kristal, B.S., Vigneau-Callahan, K., and Matson, W.R. (2002) Simultaneous analysis of multiple redox-active metabolites from biological matrices. In: D. Armstrong (Ed) *Methods in Molecular Biology*, Vol. 186: Oxidative Stress Biomarkers and Antioxidant Protocols, Humana Press Inc.: Totowa, NJ, pp. 185-194.

Lambert J., Ethier M., Smith J., and Figeys D. (2005) Proteomics: from Gel Based to Gel Free. *Anal Chem.* 77: 3771-3788.

LeBaer J., and Ramachandran N. (2005) Protein microarrays as tools for functional proteomics. *Current Opinions in Chemical Biology.* **9:** 14-19.

Maddalo G., Petrucci F., Iezzi M., *et al.* (2005) Analytical assessment of MALDI-TOF Imaging Mass Spectrometry on thin histological samples. An insight in proteome investigation. *Clinica Chimica Acta.* 357: 210-218.

Mann M., Hendrickson R., and Pandey A. (2001) Analysis of proteins and proteomes by mass spectrometry. Annu Rev Biochem. 70: 437-473.

Matson, W.R., Gamache, P.G., Beal, M.F., and Bird, E.D. (1987) EC array sensor concepts and data. *Life Sci.* 41, 905-908.

Michaud G.A., and Snyder M. (2002) Proteomic approaches for the global analysis of proteins. *Biotechniques.* 33: 1308-1316.

Milbury, P.E. (1997) CEAS Generation of large multi-parameter metabolic databases for determining categorical process involvement of biological molecules. In: I.N. Acworth, et al. (Eds.) *Progress in HPLC-HPCE,* Vol 6, VSP International Science Publication: Utrecht, Holland, pp 127-144.

Monti, M., Orru S., Pagnozzi D., and Pucci P. (2005) Functional Proteomics. *Clinica Chimica Acta.* 357: 140-150.

Mootha, V., Bundenborg J., Olsen J., *et al.* (2003) Integrated analysis of protein composition, tissue diversity, and gene regulation in mouse mitochondria. *Cell.* 115: 629-640.

Murray J., Taylor S., Zhang B., Gosh S.S., and Capaldi R.A. (2003) Oxidative damage to mitochondrial complex I due to peroxynitrite: identification of reactive tryrosines by mass spectrometry. *J Biol Chem*. 278: 37223-37230.

Neverova I., and Van Eyk J. (2004) Role of chromatographic techniques in proteomic analysis. *J Chromat*. 815: 51-63.

Prabakaran S., Swatton J., Ryan M., *et al.* (2004) Mitochondrial dysfunction in schizophrenia: evidence for compromised brain metabolism and oxidative stress. *Mol Psychiatry*. 9: 687-697.

Resing K. and Ahn N. (2004) Proteomics strategies for protein identification. *FEBS Letters*. 579: 885-889.

Rochfort S. (2005) Metabolomics Reviewed: A New "Omics" Platform Technology for Systems Biology and Implications for Natural Products Research. *J Nat Prod*. 68: 1813-1820.

Saghatellian A., and Cravatt B. (2005) Global strategies to integrate the proteome and metabolome. *Current Opinion in Chemical Biology*. 9: 62-68.

Shevchenko A., Wilm M., and Mann M. (1996) Mass spectrometry sequencing of proteins silver-stained polyacrylamide gels. *Anal Chem*. 68: 850-858.

Stasyk T., and Huber L.A. (2004) Zooming in: fractionation strategies in proteomics. *Proteomics*. 12: 3704-3716.

Vigneau-Callahan KE, Shestopalov AI, Milbury PE, Shi, H., Matson WR, and Kristal BS. (2002) Characterization of diet-dependent metabolic serotypes: Analytical and biological variability issues in rats. *J Nutr*. 131: 924S-932S.

Vlahou A., and Fountoulakis M. (2005) Proteomic approaches in the search for disease biomarkers. *J Chromatgr*. 814: 11-19.

Watkins, S.M. (2000) Comprehensive lipid analysis: a powerful metanomic tool for predictive and diagnostic medicine. *Isr Med Assoc J* 2: 722-724.

Watkins, S.M., and German, J.B. (2002) Toward the implementation of metabolomic assessments of human health and nutrition. *Current Opinion in Biotechnology* 13, 512-516.

Yao, J.K., and Cheng, P. (2004) Determination of multiple redox-active compounds by high-performance liquid chromatography with coulometric multi-electrode array system. *J. Chromatogr B* 810, 93-100.

Yao J., and Reddy R. (2005) Metabolic Investigation in Psychiatric Disorders. *Mol Neurobiology*. 31: 193-203.

Yao, J.K., Leonard, S., and Reddy, R.D. (2006) Altered glutathione redox state in schizophrenia. *Disease Markers* 22:83-93.

Yuan X., and Desiderio D. (2005) Proteomics analysis of human cerebrospinal fluid. *J Chromat*. 815: 179-189.

Yuan X., and Desiderio D. (2005) Proteomics analysis of prefractionated human lumbar cerebrospinal fluid. *Proteomics*. 5: 541-550.

Zhang J., Goodlett D., Peskind E., *et al.* (2005) Quantitative proteomic analysis of age-related changes in human cerebrospinal fluid. *Neurobiology of Aging*. 26: 207-227.

Index

#

A

B

C

D

E

F

G

H

I

K

L

M

N

Q

R

T